P9-BYD-860

PREVENTION'S

HEALING

WITH

MOTION

PREVENTION'S
HEALING
WITH
MOTION

*An All-New Approach
to Health and Healing Based
on Simple Mind and Body Exercises*

By the Editors of

Foreword by Dan Hamner, M.D., visiting professor of rehabilitation medicine
at the New York Hospital–Cornell Medical Center

Rodale Press, Inc.
Emmaus, Pennsylvania

Notice

This book is intended as a reference volume only, not as a medical manual. The information given here is designed to help you make informed decisions about your health. It is not intended as a substitute for any treatment that may have been prescribed by your doctor. If you suspect that you have a medical problem, we urge you to seek competent medical help. As with all exercise programs, you should seek your doctor's approval before you begin.

Copyright © 1999 by Rodale Press, Inc.

Illustrations copyright © 1999 by Karen Kuchar

All rights reserved. No part of this publication may be reproduced or transmitted in any form or by any means, electronic or mechanical, including photocopying, recording, or any other information storage and retrieval system, without the written permission of the publisher.

Prevention Health Books is a trademark and *Prevention* is a registered trademark of Rodale Press, Inc.

Printed in the United States of America on acid-free ∞, recycled paper ♻

Library of Congress Cataloging-in-Publication Data

Prevention's healing with motion : an all-new approach to health and healing based on simple mind and body exercises / by the editors of Prevention Health Books.
     p.   cm.
   Includes index.
   ISBN 0–87596–533–4 hardcover
   1. Exercise—Health aspects—Popular works.   2. Physical fitness—Popular works.   3. Health—Popular works.   4. Exercise—Physiological aspects—Popular works.   5. Healing—Popular works.   6. Mind and body—Popular works.
I. Prevention Health Books.   II. Title : Healing with motion
RA781.P8856   1999
615.8'2—dc21   98–44631

Distributed to the book trade by St. Martin's Press

2 4 6 8 10 9 7 5 3 1 hardcover

OUR PURPOSE

"We inspire and enable people to improve their lives and the world around them."

Healing with Motion Staff

Senior Managing Editor: Edward Claflin
Managing Editors: Hugh O'Neill, Jeff Bredenberg
Senior Editor: Matthew Hoffman
Staff Writers: Joely Johnson, Nanci Kulig, Barbara Loecher, Arden Moore, Donna Raskin
Associate Research Manager: Anita C. Small
Lead Researcher: Lori Davis
Editorial Researchers: Jennifer Abel, Jennifer A. Barefoot, Elizabeth A. Brown, Raymond DiCecco, Christine Dreisbach, Jennifer Fiske, Grete Haentjens, Sherry Weiss Kiser, Terry Sutton Kravitz, Sandra Salera Lloyd, Mary S. Mesaros, Paris Mihely-Muchanic, Anne L. Morris, Staci Ann Sander, Amy Seirer, Teresa A. Yeykal, Shea Zukowski
Senior Copy Editor: Jane Sherman
Art Director: Darlene Schneck
Cover and Book Designer: Richard Kershner
Layout Designer: Dale Mack
Illustrator: Karen Kuchar
Manufacturing Coordinators: Brenda Miller, Jodi Schaffer, Patrick Smith
Office Manager: Roberta Mulliner
Office Staff: Julie Kehs, Mary Lou Stephen

Rodale Health and Fitness Books

Vice President and Editorial Director: Debora T. Yost
Executive Editor: Neil Wertheimer
Design and Production Director: Michael Ward
Marketing Manager: Denyse Corelli
Research Manager: Ann Gossy Yermish
Copy Manager: Lisa D. Andruscavage
Production Manager: Robert V. Anderson Jr.
Assistant Studio Manager: Thomas P. Aczel
Manufacturing Managers: Eileen F. Bauder, Mark Krahforst

Prevention Health Books Board of Advisors

Joanne Curran-Celentano, R.D., Ph.D.
Associate professor of nutritional sciences at the University of New Hampshire in Durham

Robert DiBianco, M.D.
Director of cardiology research at the risk factor and heart failure clinics at Washington Adventist Hospital in Takoma Park, Maryland, and associate clinical professor of medicine at Georgetown University School of Medicine in Washington, D.C.

Roger L. Gebhard, M.D.
Practicing gastroenterologist at the VA Medical Center in Minneapolis and professor of medicine in the division of gastroenterology at the University of Minnesota

Gary N. Gross, M.D.
Clinical professor of medicine in the division of allergy and immunology at the University of Texas Southwestern Medical School in Dallas

James G. Garrick, M.D.
Director of the center for sports medicine at Saint Francis Memorial Hospital in San Francisco

Scott Haldeman, M.D., D.C., Ph.D.
Associate clinical professor in the department of neurology at the University of California, Irvine, and adjunct professor at the Los Angeles Chiropractic College

Arthur I. Jacknowitz, Pharm.D.
Professor of clinical pharmacy and chair of the department of clinical pharmacy at West Virginia University in Morgantown

William J. Keller, Ph.D.
Professor and chair of the department of pharmaceutical sciences in the School of Pharmacy at Samford University in Birmingham, Alabama

Robert D. Kerns, Ph.D.
Chief of the psychology service and director of the comprehensive pain management center at the VA Connecticut Health Care System in West Haven and associate professor of psychiatry, neurology, and psychology at Yale University

Charles P. Kimmelman, M.D.
Professor of otolaryngology at Cornell University Medical College and attending physician at Manhattan Eye, Ear and Throat Hospital, both in New York City

Jeffrey R. Lisse, M.D.
Professor of medicine and director of the division of rheumatology at the University of Texas Medical Branch at Galveston

Susan Olson, Ph.D.

Clinical psychologist, transition therapist, and weight-management consultant in private practice in Seattle

David P. Rose, M.D., Ph.D., D.Sc.

Chief of the division of nutrition and endocrinology at Naylor Dana Institute, part of the American Health Foundation, in Valhalla, New York, and an expert on nutrition and cancer for the National Cancer Institute and the American Cancer Society

Maria A. Fiatarone Singh, M.D.

Scientist I in the nutrition, exercise physiology, and sarcopenia laboratory at the Jean Mayer USDA Human Nutrition Research Center on Aging at Tufts University in Boston, associate professor at Tufts University School of Nutrition Science and Policy in Medford, Massachusetts, and professor of exercise and sports science and medicine at the School of Health Sciences at the University of Sydney in Australia

Yvonne S. Thornton, M.D.

Associate clinical professor of obstetrics and gynecology at Columbia University College of Physicians and Surgeons in New York City and director of the perinatal diagnostic testing center at Morristown Memorial Hospital in New Jersey

Lila Amdurska Wallis, M.D., M.A.C.P.

Clinical professor of medicine at Cornell University Medical College in New York City, past president of the American Medical Women's Association, founding president of the National Council on Women's Health, director of continuing medical education programs for physicians, and Master and Laureate of the American College of Physicians

Andrew T. Weil, M.D.

Director of the program in integrative medicine and clinical professor of internal medicine at the University of Arizona College of Medicine in Tucson

E. Douglas Whitehead, M.D.

Associate clinical professor of urology at Albert Einstein College of Medicine in the Bronx, associate attending physician in urology at Beth Israel Medical Center and cofounder and director of the Association for Male Sexual Dysfunction, both in New York City

Richard J. Wood, Ph.D.

Laboratory chief of the mineral bioavailability laboratory at the Jean Mayer USDA Human Nutrition Research Center on Aging at Tufts University in Boston and associate professor at the Tufts University School of Nutrition Science and Policy in Medford, Massachusetts

Susan Zelitch Yanovski, M.D.

Director of the obesity and eating disorders program in the division of digestive diseases and nutrition at the National Institute of Diabetes and Digestive and Kidney Diseases in Bethesda, Maryland

CONTENTS

FOREWORD . . . . **xvii**
Improving the Power of Body and Mind

INTRODUCTION . . . . **xix**
The Magic of Movement

ADULT ATTENTION DEFICIT DISORDER **1**
Restoring Order and Calm

AEROBIC EXERCISE **4**
Heavy Breathing Helps Hearts

AFTERNOON SLUMP **7**
Getting over the Hump

ALEXANDER TECHNIQUE **11**
Posture Skills to Take You Onward and Upward

ANGER **13**
Exercise Restraint

ANKLE PAIN **17**
Soothing a Sprain

ANXIETY **21**
Cut Your Worries Down to Size

ARTHRITIS **24**
Treating It in Water and on Land

BACK PAIN **29**
Put Your Woes behind You

BELCHING **36**
Gulp Less Now, Burp Less Later

Contents

BICYCLING **39**
The Adventurous Exercise

BIOFEEDBACK **43**
Your Best Body Language

BIRTHDAY BLUES **47**
Don't Let Those Milestones Weigh You Down

BOREDOM **49**
Boost Your Ego and Beat the Blahs

BREAST SAGGING **52**
Gaining Support

BREATHING DISORDERS **55**
Get Your Fair Share of Air

BREATHING EXERCISES **61**
The Easiest Way to Ease Stress

BURSITIS AND TENDINITIS **63**
Making Smooth Moves

CABIN FEVER **68**
Working Out When You're Stuck Inside

CAFFEINE WITHDRAWAL **70**
How to Survive Week One

CARPAL TUNNEL SYNDROME **72**
Stop Squeezing the Nerve

CELLULITE **75**
Getting Rid of French Fat

CLUMSINESS **79**
Moving toward a State of Grace

COLDS AND FLU **83**
Effortless Remedies

COMMUTER'S BACK **86**
How to Veer from Pain

CONSTIPATION **89**
Graham Crackers, Cornflakes, and Easing the Squeeze

CREATIVITY BLOCK 93
Exercises in Innovation

DANCING 95
Find Happiness with Some Fancy Footwork

DEPRESSION 98
Step Outside of Your Blues

DIABETES 101
Controlling a Too-Sweet Bloodstream

DRIVER FATIGUE 105
Get in Shape and Stay Safe

ELBOW PAIN 107
Relieving Aches Off the Court

EMPTY NEST SYNDROME 110
Cognitive Calisthenics and Annual Competitions

ERECTILE DYSFUNCTION 112
Try an Uplifting Approach

EYESTRAIN 116
Protecting Your Peepers

FARSIGHTEDNESS 119
Celebrate Birthdays with Clear Vision

FATIGUE 123
Get Moving with Sir Isaac

FIBROMYALGIA 128
The Great Pretender

FINGER STIFFNESS 132
Count on These Moves to Limber Up Your Digits

FLATULENCE 135
Fighting Poison Gas

FOOT PAIN 138
How to Be a Sole Survivor

FORGETFULNESS 142
Dance to Enhance Your Memory

Contents

GLAUCOMA 145
Steps to Slow the Blurring

HANGOVER 149
How to Survive the Moaning After

HAY FEVER 152
Mind over Pollen

HEADACHES 156
Relief Can Be Free and Freeing

HEARTBURN 165
Firefighting 101

HEART DISEASE 169
Giving Your Pump a Perfect Day

HICCUPS 177
Block Those Nerves and Get Quiet

HIGH BLOOD PRESSURE 179
Low-Intensity, Low-Stress Exercise to the Rescue

HIGH CHOLESTEROL 184
Accentuating the Positive

HIP PAIN 188
Get Yourself an Elvis Pelvis

IMAGERY 192
Finding the Healer Within

INCONTINENCE 196
Shoring Up Your Leaky Plumbing

INHIBITED SEXUAL DESIRE 199
Tune Up Your Body and Mind

INNER EAR PROBLEMS 202
Bring the World Back into Balance

INSOMNIA 204
Some Well-Timed Motion Can Help You Sleep

INTERMITTENT CLAUDICATION 208
Turning Off the Pain

JET LAG. **210**
Of Body Clocks and Hard Knocks

JOB BURNOUT. **212**
Re-Igniting Enthusiasm

JOB LOSS. **215**
Bouncing Back Physically and Emotionally

KNEE PAIN. **219**
When Your Hinges Hurt, Take a Stand

LARYNGITIS. **223**
Pamper Your Voice

LAUGHTER. **227**
The Hee-Haws of Good Health

LAZY EYE. **229**
One Side Sees All

LOVE HANDLES. **233**
Come On, Baby, Let's Do the Twist

LOW BLOOD PRESSURE. **236**
Help with a Dizzying Deficiency

LOW SELF-ESTEEM. **239**
Nourishing Your Athletic Ego

MACULAR DEGENERATION. **242**
Aging Eyes See Hope

MASSAGE THERAPY. **246**
Rubbing for Royals and Regular Folks

MEDITATION. **249**
Building Mental Muscles

MENOPAUSE. **252**
Exercising through Changes

MENSTRUAL DISCOMFORT. **256**
Getting Free from "Female Complaints"

MIDLIFE CRISIS. **261**
Searching for Satisfaction

Contents

MOTION SICKNESS **265**
Soothing Moves

MUSCLE CRAMPS **267**
Relief for the Thighs (and Backs and Shoulders) That Bind

MUSCLE WEAKNESS **271**
Getting More Flux in Your Flex

MUSIC **279**
The Healer That You Hear

NEARSIGHTEDNESS **281**
The Far-Reaching Effects of Eye Exercise

NECK PAIN **286**
Nix That Crick

NERVOUS HABITS **289**
A Two-Step Exercise in Quitting

OSTEOPOROSIS **292**
Bone Up on Weight-Bearing Exercise

OVERWEIGHT **296**
Keep Yourself Thin

PANIC ATTACKS **306**
Breathing in Relief

PILATES **310**
Strengthen Yourself to the Core

POSTURE PROBLEMS **313**
How to Get Out of a Slump

PREMATURE EJACULATION **317**
When Slower Is Better

PROGRESSIVE MUSCLE RELAXATION **319**
Relief for Tension, Pain, and Even Insomnia

RAYNAUD'S DISEASE **322**
Delivering Warm Thoughts

REAR-END SAGGING **325**
Make Your Own Happy Ending

REFLEXOLOGY **328**
Fancy Footwork

RESISTANCE TRAINING **331**
More Important the Older You Get

RETIREMENT BLUES. **338**
Make Getting in Shape Your Job Description

RUNNING. **340**
Taking It All in Stride

SENSORY LOSS **345**
Exercises That Sharpen Your Sensitivity

SEX **350**
Shape Up in the Sack

SHINSPLINTS. **352**
Soothing Your Lower Legs

SHOULDER PAIN **355**
Relief Is within Reach

SHYNESS. **358**
Four Steps to Social Fitness

SINUSITIS **362**
Kick That Block!

SMOKING. **365**
Take a Run at Kicking the Habit

SNORING **369**
Active Days Make for Silent Nights

STAGE FRIGHT **372**
Turning the Tension Around

STRESS **374**
Finding Serenity in Chaos

STRETCHING. **380**
Cats Know Where It's At

STUTTERING. **383**
Vocal Exercises to the Rescue

Contents

TAI CHI. 386
Slow-Motion Meditation

TEMPOROMANDIBULAR DISORDER (TMD). 390
Giving Your Jaw a Little TLC

TOOTH GRINDING. 394
Calm Down to Unclench

TYPE-A BEHAVIOR. 397
Ease Anger and Stay Well

UNDERARM SAGGING. 400
Bye-Bye to Batwings

VARICOSE VEINS. 403
Insure Your Legs with an Exercise Policy

WALKING. 405
The No-Excuses Exercise

WATER RETENTION. 411
Sinking the Bloat

WINTER BLUES. 414
Turn Your Brights On

WRINKLES. 417
Smoothing Out the Decades

WRIST PAIN. 422
Wave Adios to Aches

YOGA. 426
A Time-Out to Tune In

CREDITS. 431

INDEX. 433

PREVENTION's
healthy ideas™
For the best interactive guide to healthy active
living, visit our Web site at **www.healthyideas.com**

FOREWORD

Improving the Power of Body and Mind

The future promises medical magic, but we don't have to wait for promises. When it comes to healing, we have the remarkable ability to produce miracles ourselves, with the power of movement.

Movement is life. Cells embrace a constant field of motion with their dynamic nuclei and ever-mobile protoplasm. And there's constant unseen biochemical motion throughout our bodies as we release hormones and neurotransmitters that carry all sorts of messages that dictate how we feel and act.

In my medical and healing practice, movement is the tool that I use for many purposes—prevention, recovery, rehabilitation, and the restoration of overall good health. Some people with severe pain in their shoulders or wrists—candidates for rotator cuff or carpal tunnel surgery—have learned moves from professional boxers that have helped them recover. For others, increased movement improved their digestion, boosted their mental activity, and re-energized their muscles. And some people in their seventies or eighties have benefited from range-of-motion activities that actually restored some freedom of movement that they thought they had lost forever.

Working with these people, seeing the changes in their health, their lives, and their outlook, I have come to have great respect for movement as a healing tool, just as a radiologist respects his x-ray machine, an internist respects his medicines, and a surgeon respects his scalpel.

Of course, there are many differences between the complex equipment and medicines used by these physicians and the self-healing, dynamic, living body that is my primary tool for healing. Unlike the x-ray machine, the pharmaceuticals, and the scalpel, the body constantly reveals its hidden power to repair and renew. As I work with patients—nonathletes as well as athletes—I am constantly reminded of how much we have to learn about the healing processes in our bodies.

In the years that I have specialized in the practice of physical medicine (physiatry), I have wondered whether everything we know about the healing power of movement could be assembled in one book. If so, that book would certainly be a resource that everyone would want to use—not just specialists, not just doctors, but all of us who carry around the potential for healing with movement. Imagine having a compendium of hundreds of remedies and preventive measures that you can implement yourself, with no costly medicines and no elaborate workout equipment—just the body using its inherent ability to self-repair and self-renew.

Healing with Motion is exactly the kind of book that I imagined. Written and edited by the staff of *Prevention* Health Books, it offers readers not only my experience but also the experiences, observations, and research of hundreds of experts in a wide range of fields.

This book represents a profound change in attitude toward exercise. As you'll see, exercise does not have to be synonymous with working out, pumping up, or burning fat. In *Healing with Motion*, it is defined as any movement that is vital to healing. And it goes beyond physical movement to include the mind's action as well.

Both research and intuition tell us that there is a close interrelationship between these two aspects of us. Mental activity affects our physical selves, and both influence the power of healing. The experience of meditating while walking, for instance, is quite different from the experience of race-walking breathlessly, in a panic, to get to an appointment on time. True, both activities involve walking, but there the resemblance ends—and so do the health results. A meditative walk produces a biochemical reaction that can help reduce stress, lower blood pressure, and improve heart health. A panic-filled dash to a high-stress appointment does just the opposite.

That's why *Healing with Motion* has so much about mental as well as physical movement. Yes, the mind *does* exercise—in harmony or in conflict with the body. And when we're using our bodies for healing, mental exercise is just as important as the physical kind.

So here it is—thanks to the expertise and vision of *Prevention* Health Books editors and writers—exactly the book that I imagined being a resource for all of us. Read on, and you'll discover many surprises about the proven power of the body and mind to heal numerous health conditions. And you'll have expert, up-to-date guidance to show you exactly how to bring about that healing.

Dan Hamner, M.D.
Physiatrist, sports medicine expert,
author, and visiting professor of
rehabilitation medicine at the
New York Hospital–Cornell Medical
Center in New York City

INTRODUCTION

The Magic of Movement

We have an instinct for movement.

Just ask any woman who's ever had a baby. Along about the fifth month, a mother can start to feel Junior kicking and jabbing. By nine months, it's an in-utero slugfest. He's moving and shaking, ready to get going.

Movement is in our bones and blood—and in our hearts.

The most gifted members of our species can do extravagant things—a triple axel on ice skates or those things that Michael Jordan somehow does. But even just plain folks have a superb range of movement skills. We can squeeze, pull, push, flex, lift, press, rub, wiggle, shrug, sashay, and shimmy. We're versatile, finely engineered organisms, complete with wrists that flick, toes that curl, spines that twist, and heads that swivel.

But as handy as our hardware is, it's not the star of the movement story—our software is. It's our brains that make movement a special gift. Only a human can do the tango, stitch a quilt, hit a cross-court winner, or plunk out a piano rag. Our intelligence lets us turn our skills to a purpose. Only a human can play golf, master yoga, or even try tai chi. This ability to use movement—deliberately and by design—offers us a remarkable healing opportunity.

We can turn movement into medicine—and use ourselves to heal ourselves.

The Most Natural Cure of All

We're looking for alternatives to conventional health care. Disillusioned with technomedicine and wary of drug treatment and surgical interventions, we're open to learning more about herbs, tonics, and secrets from ancient disciplines, including Chinese medicine and the Ayurvedic tradition. We're exploring it all—meditation, massage, breathing exercises, aromatherapy, music therapy, remedies from indigenous cultures around the world, and a dazzling array of bridges between body and mind.

And yet, in our search for and exploration of new cures, it's easy to forget Dorothy's downhome lesson from *The Wizard of Oz*. Sometimes, the answer you're looking for is right there in your own backyard. Too many of us have overlooked the self-care technique that is the most natural, the most convenient, the safest, the cheapest, and the most empowering of them all.

Getting regular exercise is probably the single best thing that we can do for our bodies. "If we could put all of its health benefits into pill form, it would be the most potent remedy known to man," says Ken Forsythe, M.D., clinical instructor of sports medicine at the University of California, Davis, San Joaquin Hospital. Here's a small sample of the proven health benefits of physical activity.

- ►Protection from heart disease, cancer, diabetes, and stroke
- ►Enhanced immune function and defense against infections
- ►Bone building and posture enhancement
- ►Increased muscle strength and overall flexibility
- ►Stress reduction
- ►Freedom from many kinds of joint and back pain
- ►Slowing of the aging process

Both practitioners of alternative medicine and conventional doctors are in remarkable agreement about the many benefits of healing with motion. Exercise has the power to sharpen our senses, protect our memories, and even inspire self-esteem. Depending on the type of movements we do and how often we do them, regular physical activity can help prevent wrinkles, constipation, headaches, insomnia, jet lag, bloating, posture problems, menstrual pain (for women), erectile dysfunction (for men), and dozens of other common ailments and health problems.

Countless scientific studies have endorsed the conclusion that people who are active lead longer, healthier lives. Preventive, balm, and cure, exercise can help stop disease, improve our sense of well-being, and heal us faster. Through dynamic living, we can access the body's extraordinary self-healing power—at no cost and risk-free.

A New Notion of Exercise: Low-Sweat, Convenient, and Fun

Despite all the evangelism in praise of exercise, most of us aren't doing much of it. The best estimate is that only about 37 percent of all Americans are getting enough exercise to reap the health harvest.

Why? Explanations abound. Surely, one big reason is that busy lives don't give us much time. But another reason is that we have an intimidating notion of exercise. We think that we have to run miles every day or work ourselves to exhaustion before we get the benefits. To many people, the word *exercise* evokes images of sculpted Olympic athletes going for their personal best or hyper television commercials with hired athletic heroes celebrating the exaltation of exertion. We see exercise as a burden, just another obligation. After all, it's called a *work*out, right?

It's time for a new view, one that sets exercise free from the treadmills, the weight machines, and all that perspiration. Yes, in an ideal world, we'd all have the time and discipline to work our muscles and raise our heart rates every day. We would all be slim, well-toned athletes, just a few days away from our next 10-K. But even though we aren't, we still have some great moves within us. Even though we may have a few extra pounds, we can reap the benefits of exercise, according to Dan Hamner, M.D., a physiatrist and sports medicine specialist in New York City. The trick is to see past the huffing and puffing paradigm to all of the gentle avenues for exercise. That's the real meaning of healing with motion—seizing whatever opportunities are within reach, no matter how modest, says Dr.

Hamner. Exercise can be low-sweat, convenient, and even fun.

Aerobic work doesn't have to be a three-mile run; it can be a 20-minute walk with your dog. Strength training doesn't have to be a struggle with those weight machines; it can be a half-hour in your garden. Staying flexible doesn't have to mean pretzeling your body into impossible poses. You can do easy, quick stretches in your kitchen that will keep you nice and limber. You can even sneak curative exercises into your daily routines. Taking a bath can offer the perfect opportunity to strengthen your hands and ward off carpal tunnel syndrome. Driving can be an obligation to a modern lifestyle, or you can use travel time to stop back pain in its tracks. Waxing your car will protect not only its body but your body as well.

Any of hundreds of low-impact activities can keep you well and make you better. Here's a list of healing moves: bird-watching, gardening, golf, softball, sitting in a rocking chair, pulling your tongue, paddling a canoe, making love, walking in the woods, dancing, playing with a Slinky, squeezing your toe, self-massage, meditation, breathing deeply, and even just having a few good laughs. All of these easy moves can enhance the healing process. You don't have to be the king of the calorie burners to do yourself a lot of good. A gentle swim can ease aches. Just tensing and relaxing your muscles can help you manage pain. Listening to Mozart can keep your memory keen.

A Field Guide to Feeling Better: Understanding the Ripples

Every move we make is like the pebble in the pond; it reverberates throughout our bodies.

Every stretch, every small exertion, every breathing exercise has a ripple effect. "We're an exquisite ecology," says Dr. Hamner. "Our nervous systems, our immune systems, and our musculoskeletal systems are all intertwined." Consider everything that happens with even the most modest of moves—making a fist.

As you bend your fingers and clench them into a fist, your brain sends an impulse down to your hand. As you close your hand, you use 14 joints and 17 muscles. You activate flexor tendons in each finger. You stimulate production of a lubricant called synovial fluid. You send vibrations through your arm muscles and electrical signals back up to your brain. When you open the fist, you use a different set of tendons in your fingers, the extensors.

If you pump your fist—closed, open, closed, open—you raise the blood pressure in the neighborhood and speed up the circulation of your blood. If you add some deep breathing and hum a mantra as you pump, you change your brain wave pattern. Even the simplest movement has consequences. Just bending over sends blood to your head. Just throwing your shoulders back and standing up straight helps your lungs work more efficiently. Swallowing, squatting, singing, and even smiling have physiological results that can help heal.

The secret of maximum health and healing is to understand these chain reactions and tap their therapeutic power. This book offers customized advice for easy and practical movements that will help you use your body to heal your body. It explores both the highways of health—stretching, strength training, and aerobics—and the quirky back roads of the body. It explains why pulling your tongue can stop hiccups, why conducting an orchestra can banish a headache, why putting stickers on your computer can ease eyestrain, why moving your wallet can stop sciatica, and how laughter can improve your circulation.

The Biggest Ripple of All: Exercise and Your Emotions

For a long time, we've thought of exercise as a way of keeping our bodies well. But recent breakthroughs in brain science suggest that exercise can do much more than keep our blood vessels clear and our posture proud. It can help manage our moods as well. There is powerful evidence that exercise alters our brain chemistry in such a way that it helps with dozens of common emotional problems—anxiety, low self-esteem, shyness, job burnout, midlife crisis, boredom, and even winter blues.

"Our moods create chemical reactions in our bodies, and so do our movements," says Dale L. Anderson, M.D. an urgent-care physician in Minneapolis and author of health humor books such as *The Orchestra Conductor's Secret to Health and Long Life*. "When you exercise, you change your brain chemistry," he says.

Exercise can also ease you through times of stress, which helps all of our relationships. It's no exaggeration to say that exercise can make you a better spouse, parent, worker, and friend. When your body is working at its best, your spirit often is, too.

That's why this book contains such a won-derful multiplicity and variety of advice for healing with motion—all supported by the research and observations of leading experts and doctors who have witnessed the complex interactions of body and mind. You'll discover:

► How guided imagery can strengthen your immune system
► How meditation can improve digestion and ease pain
► How physical challenges can prevent midlife malaise
► How movement can put the fire back in your love life
► How less-mainstream activities like fencing can help you beat boredom

As for the necessary tools for all this healing—well, most of what you need for body repair and maintenance can be seen in a full-length mirror. If you want some further guidance and instruction from experts, we'll tell you how they can help you and give you the flavor of their recommendations. Plus, you might want to invest in a set of hand weights and a few articles of exercise gear, such as walking or running shoes.

The only other thing you'll need is a joy in your body. All of us—no matter what our ages, our weights, or our athletic résumés—have the magic of movement within us. Bruce Springsteen was partly right: We were born to run—and take walks after dinner and lift our arms and tend our gardens—and use the marvelous machinery with which we're blessed.

PREVENTION'S®
HEALING WITH MOTION

ADULT ATTENTION DEFICIT DISORDER

Restoring Order and Calm

Marion, a successful advertising executive, wins praise for her brainstorming talents, but she tends to forget important meetings—a trait that colleagues dismiss as part of her creative personality.

Ben, in his midthirties, can't count the number of jobs he has held since college. Now selling bus tickets at night and delivering newspapers every morning, Ben explains that he becomes bored easily, craves challenge, and never devotes the time necessary to master one career.

At first glance, Marion and Ben appear to be worlds apart, but both are among nearly 8 to 10 million American adults with an often misdiagnosed, untreated condition known as attention deficit disorder (ADD). Until recently, ADD was considered strictly a childhood disorder, and it wasn't until the mid-1980s that the general medical community recognized that ADD symptoms continue through adulthood in some cases.

Although researchers are still developing standards to accurately diagnose ADD, they believe that it is a neurological disorder characterized by three core symptoms: inattention, impulsiveness, and hyperactivity.

To some extent, most of us demonstrate ADD tendencies on occasion. We can't balance our checkbooks, we interrupt others, or we wait until the last minute to write a company report. We find it hard to concentrate while reading a chapter for our microeconomics class or, on impulse, we buy the first sport utility vehicle we spot in a showroom.

"Just about everybody has ADD symptoms to a degree," says Kevin R. Murphy, Ph.D., chief of the adult ADD clinic at the University of Massachusetts Medical Center in Worcester and co-author of *Out of the Fog: Treatment Options and Coping Strategies for Adult Attention Deficit Disorder.* "The major difference is that these symptoms occur far more frequently and cause clearly visible impairment—usually in school, work, and social/family domains—in people diagnosed with ADD.

"If anyone is experiencing significant problems with impulsivity, lack of concentration, inattentiveness, and sustaining effort and motivation, and these symptoms have been evident since childhood, they should seek a professional evaluation. Diagnosing ADD requires a rigorous evaluation from a qualified professional to determine that the problems are indeed due to ADD and not some environmental or situational cause," adds Dr. Murphy.

People with ADD may feel constantly overwhelmed and caught in a tug of demands, adds Kathleen Nadeau, Ph.D., a clinical psychologist, director of Chesapeake Psychological Services of Maryland in Bethesda, and author of *Adventures in Fast Forward: Life, Love, and Work for the ADD Adult.*

Fortunately, ADD is highly treatable and manageable. In most cases, medications such as methylphenidate (Ritalin) or dextroamphetamine (Dexedrine) are an important part of a treatment plan. But exercise and therapy help reduce stress levels and keep hyperactivity and impulsiveness in check, say medical experts who treat and counsel people with ADD.

Sweating Out Excess Energy

For some with ADD, sweat is an ally. Regular, vigorous, and lengthy workouts seem to expel pent-up energy and help people feel calmer, say

Organizing Your Mind

In addition to physical exercise, some easy-to-learn mental strategies can help people channel their attention deficit disorder (ADD) traits into successful skills at home and at work, say experts.

"It's important to recognize how you function and create an ADD-friendly environment at work and at home," says Kathleen Nadeau, Ph.D., a clinical psychologist, director of Chesapeake Psychological Services of Maryland in Bethesda, and author of *Adventures in Fast Forward: Life, Love, and Work for the ADD Adult.*

Here are some mindful ways to sidestep the tendency to be forgetful, overwhelmed, tardy, and disorganized, as offered by Dr. Nadeau and other national ADD experts.

▶Each night, write down a "to do" list for the next day and post these reminders within sight on the refrigerator door, the bathroom mirror, the car dashboard, and your appointment book.

▶If you are unsure of instructions, ask your teacher or boss to repeat them rather than guessing.

▶Break major projects into small, achievable tasks. Set a deadline for each task and reward yourself upon completion.

▶Learn to "underbook" yourself by reducing the number of items on your daily planner. It will reduce stress and bolster your self-esteem.

▶For two days, jot down how long it takes to do everything you do—including showering and brushing your teeth. This should give you a clearer idea of how you can plan your time more effectively in the future.

▶Store similar items in the same place. Keep your canceled checks in one place and your unpaid bills in another.

▶Write phone numbers in one notebook. Never scrawl them on a scrap of paper, which is easily lost.

▶Remember to laugh. It's a healthy way to melt away tension, anxiety, and hostility.

doctors. "Exercise plays a very important role in the management of attention deficit disorder by reducing hyperactivity and stress levels for some," says Dr. Nadeau.

Doctors suspect that moderate to intense exercise, for at least 30 minutes at a time, causes the release of mood-elevating, painkilling hormones called endorphins.

"Some with ADD say that vigorous exercise helps them feel better emotionally and physically," says Dr. Murphy. "At this point, although there are no scientific studies verifying this, there are anecdotal evidence and clinical impressions that seem to suggest that exercise is beneficial to some." Here's what experts recommend.

Seek a sweaty variety. First, get a thorough exam by your family doctor. Then, to tone down your ADD traits, approach your gym as you would an á la carte menu. Doing a variety of activities and mixing up the sweat-producing exercises is the key to maintaining your motivation and preventing boredom.

On the first visit, perhaps focus on yoga positions. The next time, maybe it's you and the stationary cycle. On the third occasion, keep pace on a treadmill. Or you can elect to spend a little bit of time doing each of these exercises during each visit.

"Some people tell me that they go to the gym for two hours and lift weights, run, use the rowing machine, and use the stair machine," says Dr. Murphy.

Walk about. Thirty-minute walks on the same stretch of sidewalk every day do not often inspire people with ADD, says Dr. Nadeau.

"These people have very low boredom tolerance and are less likely to stick with the same healthy but boring exercise routine," she says. "So we advise them to look for places that they enjoy walking and to vary their routes, distances, and times of day."

Try driving to a dream neighborhood, parking the car, and fast-footing it past some handsome mansions and spectacular gardens. Or pace the high school football field a few times and remember your favorite team cheer or imagine that you just caught the game-winning pass in the end zone. Maybe you can spice up your regular mall walk by counting the number of stores or red-headed toddlers you pass.

Spring into cleaning. If the gym doesn't inspire you, try putting on a hip-moving, energizing cassette like the soundtrack from *The Big Chill* and aggressively attack dust bunnies and soil-filled carpets with a thorough vacuuming job. You'll evaporate the surplus energy and clean your floors at the same time, says Dr. Nadeau.

Dismiss distractions. When you're feeling restless, Dr. Murphy offers these five-minute activities to recapture your focusing powers.

▶Stand up, walk away from your desk or the task at hand, and stretch your body from head to toe.

▶Take a brisk walk around the block or around your office building during lunch or breaks. Time yourself.

▶Stand up and perform 10 to 20 jumping jacks, then resume your work.

AEROBIC EXERCISE

Heavy Breathing Helps Hearts

WHAT IT IS

Here are some things that this chapter is *not* about: Wearing a thong leotard and shaking your booty in front of a roomful of strangers, working out so hard that you sweat like a football player and have to gasp to catch your breath, and finally, bouncing up and down for a solid hour, feeling bored and silly.

No one's going to ask you to do any of these things anymore. In fact, here are the easy activities that aerobic exercise is all about: dancing, swimming, walking outside or even in a mall, mowing the lawn with a rotary mower, playing Frisbee, washing your car, and even playing soccer with your kids.

If you've ever driven to the mall, parked far away from the entrance, walked to the front door, and then spent two hours shopping, you already know what aerobic exercise is like. Similarly, if you've ever raced up a flight of stairs while holding a bag of groceries because you heard the phone ringing, you also know what it's like.

"If you're breathing harder, it's aerobic," explains Russ Pate, Ph.D., professor and chairman of the department of exercise science at the University of South Carolina in Columbia. The word *aerobic* actually means "with oxygen," so if you're breathing harder— that is, consuming more oxygen—you're taking part in an aerobic activity.

Of course, there is a difference in intensity between activities such as mall walking and running up stairs. A difference in intensity means a difference in results. "You'll burn calories at a higher rate if you work out more intensely," says Dr. Pate. "That really clouds the issue, however, because walking vigorously and walking at a moderate intensity are both worthwhile activities. As long as you're moving, you're helping yourself."

WHAT IT DOES

"If every benefit of exercise could be known to man, it would easily be the most potent pill we have," says Ken Forsythe, M.D., clinical instructor of sports medicine at the University of California, Davis, San Joaquin Hospital. That's because exercise can reduce blood pressure, decrease total cholesterol, help you lose weight, decrease your risk of diabetes, increase heart function, decrease risk of osteoporosis, and ease the symptoms of anxiety, depression, and tension. Aerobic exercise even improves skin tone.

"Take osteoporosis, for example," says Dr. Forsythe. "Exercise is more beneficial to

Aerobic Activity: A Two-Week Menu

WEEK 1

Sunday	Monday	Tuesday	Wednesday	Thursday	Friday	Saturday
Take the dog and a Frisbee to the park for 45 minutes.	Do 30 minutes of an aerobic exercise video. Walk for 15 minutes on your lunch hour.	Park in the very last space of the office parking lot. Clean part of the house after work, including floors and windows.	Go up and down your stairs for the length of two songs on the radio. Meet your friends and play tennis.	Walk around the block for 20 minutes before going to work. Wash and wax the car.	Do 30 minutes of an aerobic exercise video. Take a leisurely walk after dinner.	Ride your bike for a half-hour. Go dancing with your spouse and some friends.

WEEK 2

Sunday	Monday	Tuesday	Wednesday	Thursday	Friday	Saturday
Go bowling with some friends you haven't seen in a while. Rake (or mow) the lawn.	Do 30 minutes of an aerobic exercise video. Play table tennis with your spouse.	Walk for 15 minutes on your lunch hour. Meet your friends at the gym and go swimming.	Do 30 minutes of an aerobic exercise video. Walk through your neighborhood after dinner.	Turn on the classic rock station and dance in your living room for 15 minutes. Take care of a friend's toddler. Play whatever games she wants to play.	Garden after work. Join kids in a pickup game of basketball.	Play miniature golf with your kids. Walk through the mall before you go to the movies.

women than drinking milk or taking any sort of calcium supplement. A 50-year-old woman who's at risk for osteoporosis will be helped somewhat by drinking milk, but if she started to walk regularly, in one year she would see an increase in bone density of 4 to 5 percent."

If you're looking strictly for a cardiovascular benefit, three times a week for 20 to 30 minutes will work. If you want to lose weight, you'll have to get your heart pumping for at least 45 minutes five times a week, says Dr. Forsythe.

REAPING THE BENEFITS

Now it's time to talk about the hardest part of aerobic exercise—just doing it—because you can only reap the benefits if you actually do the exercise. "The big problem with exercise is compliance," says Dr. Forsythe. "So the most important thing is to find a few activities that you enjoy and alternate doing them."

To create an active lifestyle that works for you, Dr. Pate suggests experimenting with a variety of aerobic activities. "Look for pastimes that give you another reward," he says. "For instance, if you want some privacy, maybe you'll take up walking. If you're a social animal, then you might want to go to aerobics classes or play a regular game of tennis with some friends." Or combine two things that you like to do. Dr. Forsythe listens to books on tape while he jogs. He "reads" two or three books a week that way.

Optimally, strive to be active for at least 45 minutes every day. Your activity does not have to be continuous, however. If you like, break the 45 minutes into shorter time segments throughout the day. "Parking farther away from the entrance to your office, walking up the stairs instead of taking the elevator, and then walking for 15 minutes during your lunch hour can ultimately add up to a moderate amount of activity if you make these consistent habits," says Dr. Forsythe.

One more suggestion: Get support from your spouse or a good friend. "It's extremely important to have partners who want to help you make exercise a consistent habit," says Dr. Pate. For instance, many of Dr. Pate's clients walk the dog every evening with their husbands or meet their girlfriends for a date at the gym. That way, they combine their social lives with their aerobic goals.

TRIAL RUN

If you have never exercised regularly, the best news is that you're the perfect workout candidate. "You really get a big bang for your buck if you're sedentary and you become moderately active," says Dr. Pate. "Those are the people who see the most immediate and important results."

In other words, if you're already a marathon runner, taking up walking isn't going to help your fitness level much. If you're a couch jockey like most of us, though, taking a short walk every evening will make a really big difference in your life. The schedule on page 5 demonstrates how you can work in 45 minutes of aerobic activity every day for two weeks.

AFTERNOON SLUMP

Getting over the Hump

For decades, economists and politicians have campaigned passionately against financial deficits. But there's a bigger deficit that strikes people of all incomes, say health experts: lack of adequate sleep.

Maybe your newborn daughter keeps you up at night. Or maybe you can't stop watching *Late Night with David Letterman*. Or you rise by 6:00 A.M., hit the ground, and don't stop until midnight every day.

Like clockwork, the midday slump typically surfaces between 2:00 and 4:00 P.M.Our eyelids get heavy, our heads start to nod toward our computer keyboards, and our razor-sharp minds seem incapable of comprehending the text on a report page.

"The afternoon slump is extremely common because so many of us are sleep-deprived, and there is a normal little dip in the circadian rhythms in the middle of the afternoon," explains Neil B. Kavey, M.D., director of the sleep disorders center at Columbia Presbyterian Medical Center in New York City.

Circadian rhythms are part of our bodies' built-in 24-hour clock systems that make us sleepy or alert. Even if doctors rule out sleep disorders such as insomnia, it's natural for many of us to be a bit sluggish by midafternoon.

You don't have to let Mr. Sandman get the best of you every afternoon, however. You can cruise through the day energized and focused if you rely on refreshing stretches, easy body movements, and power naps, say experts.

Combating Sluggishness

A big reason that you may feel so lethargic by 3:00 P.M. is that you've barely moved from your desk. After a few hours of little movement, blood does not flow efficiently through the muscles and organs, including the almighty brain, says Dr. Kavey. And when blood flow flags, you can be sure your cells aren't getting all the oxygen they'd like.

Another factor is adrenaline, the body-stimulating hormone that's released when you exercise and that helps get you in gear. "The afternoon slump will hit people who are sedentary a little more because they are not generating a lot of adrenaline to override it," says Dr. Kavey. "That's why it is so important to take frequent breaks in the afternoon and get up and move." Here are some moves that will help lift you from the slump.

Run to nowhere. When you feel the slump coming, try standing up and doing a one-minute jog in place. Make sure that you pump your arms and lift your knees, says Peter Wylie, Ph.D., an organizational psychologist and management consultant in private practice in Washington, D.C.

"It may be inappropriate to jog down your office hallway, so try jogging in place in the privacy of your office or maybe in the restroom," says Dr. Wylie. "Rapid movement gets the motor running and is better than a cup of coffee."

Plant a knockout punch. Imagine you're former heavyweight champion Muhammad Ali during his heyday. Find a place where you won't be observed and punch the air with your right and left fists for 30 seconds to a minute, suggests Dr. Wylie.

"If you train your arms aerobically, you are able to process more oxygen with them," he says. "Shadow boxing with short bursts for 30 seconds or so will definitely energize you."

Solicit fellow slumpers. If a few co-workers gripe about the afternoon slump, why not form a group of slump fighters and spend five minutes together stretching and jogging in place? There is strength—and support—in numbers, says Dr. Wylie.

Act like Rocky. There's no need to shout "Yo!" but you can feel a bit like the movie hero Rocky each time you scale a flight of stairs. After one flight, raise your arms above your head and slowly turn around for one rotation.

"Compliment yourself on doing the first flight of stairs," says Dr. Wylie. And for a second victory, do another.

Keep Moving

Simple stretches can also help you fight the tendency to nod off after lunch, says Dr. Wylie.

If you have a desk job, these stretches can be especially helpful. "As animals, we're not built to sit inside all day long and do sedentary tasks like tapping on a computer keyboard," he says. "We are designed to move, and that's why stretches are so important."

To limber up on your way out of limbo, here are some stretches to oxygenate muscles and get blood flowing in key body parts.

Make a head start. While seated with your feet flat on the floor, tilt your head back so that you eye the ceiling. Move slowly until you feel a stretch to tightness—never pain or dizziness. Breathe naturally as you hold this pose for 15 seconds. Then lower your head until your chin is tucked in to your chest and breathe as you hold that stretch for 15 seconds. Next, tilt your head laterally to your right shoulder and breathe as you hold for 15 seconds. Then tilt your head to your left shoulder and breathe as you hold for 15 seconds.

Try to do one to three sets of each of these neck stretches every midafternoon, says Cam Vuksinich, a certified personal trainer and certified clinical hypnotherapist in Denver.

Rescue your shoulders. Lift both shoulders as if you were shrugging and make sure you breathe naturally as you hold the pose tightly for 15 seconds. Then exhale as you let your shoulders drop naturally to a more relaxed position. Do this three times.

"You're asking your shoulder muscles to contract, and then they have an increased relaxation response when you let them go," explains Vuksinich.

Reach for the world. For a good slump stumper, extend both arms straight in front of you with your palms down. Slowly move both arms in a backward circle. For a change of pace, you can alternate arms as if you were swimming the backstroke. As your arm travels slowly around the circle, imagine that you are pointing to numbers on a clock while you count to 12. Complete three rotations and then reverse and try three forward circles (again, either with both arms or alternating), says Vuksinich. While this exercise can be performed either sitting or standing, if you have lower back problems, you should stand to avoid straining your back.

Listen to your body as you go through these motions and stop if you feel any pain, says Vuksinich. When you are ready to try again, reduce the range of motion you are trying to achieve; see your doctor if you still experience pain.

Make your spine feel fine. Take a break

Afternoon Eye-Openers

Remember how sleepy you felt during a two-hour microeconomics lecture on Friday afternoon in college? You would have done anything to be able to convert your textbook into a pillow. Well, that's a good example of bad timing. The fact is, certain activities are appropriate for certain times of the day, and if your scheduling is awry, well, you just might end up with a textbook for a pillow. Here's what experts recommend for keeping your eyes wide open during the afternoon slump.

Skip afternoon reads. Confine your business reading to the mornings and after 3:00 P.M., when you are most alert, says Peter Wylie, Ph.D., an organizational psychologist and management consultant in private practice in Washington, D.C.

"Reading is a passive activity—just you and the words—that makes it difficult to concentrate or remember what you read during the middle of the afternoon," he says.

Zip off e-mail. Don't just scroll through your screenful of electronic mail messages; instead, use the slump time to fire off e-mail replies and new messages. Finger-tapping action and electronic interaction with others will keep you alert, says Dr. Wylie.

Make the call. Know an energetic buddy or an engaging client? Pick up the phone amd call them in midafternoon, suggests Dr. Wylie.

"The afternoon slump hours are a great time to get on the phone with a live wire," he says. "You're interacting, and sometimes, their energy is contagious."

Flash some color. Add a bright red pillow or a colorful poster to your office decor. Spend a few minutes in the afternoon hugging the pillow or observing that poster to kick up your adrenaline a notch or two, says Neil B. Kavey, M.D., director of the sleep disorders center at Columbia Presbyterian Medical Center in New York City.

"There's no question that bright, cheery colors are more likely to stimulate a certain level of alertness in people than dull, dreary colors," he says. "I recommend decorating your office to make it interesting and attractive to the eye."

from your work and stand up, placing your hands on your hips and moving your feet shoulder-width apart. Leading with your head, slowly rotate your upper body to the right as far as you can while keeping your feet planted. Breathe naturally while you hold this position for 5 to 10 seconds, then alternate by repeating on the other side. Try three sets.

Be cautious about this stretch, however, and stop if it hurts. "It is important to move slowly to protect the spine," cautions Vuksinich.

Dial with your tootsies. Finish your full-body stretch from a seated position by lifting one leg at a time in front of you. Flex your toes and slowly rotate your foot at the ankle three times clockwise and then three times counter-clockwise, says Vuksinich.

"When we've been sitting for a long time, blood flow is inhibited, and a lot of us get swelling in our ankles and feet," she explains. "Toe circles help keep the blood flowing."

Practicing Snooze Control

As odd as it may sound, one of the best ways to fight afternoon sluggishness is with a 10-minute nap. The key is brevity, not a two-hour snooze that will make you feel even more tired after you awaken, says Dr. Kavey. "A short nap in the afternoon makes an individual feel refreshed," he says. "If I get a short nap in at the end of the workday, I do immensely better."

Why a brief nap refreshes us is still being studied by medical experts, but Dr. Kavey says that it may have something to do with being in the first two stages of sleep rather than the heavier levels. Experts agree that a midday nap helps boost creativity, productivity, and mood.

Be a healthy napper. Instead of reaching for your fourth cup of coffee to overcome that listless feeling, experts recommend this step-by-step guide to getting the most out of a 10-minute power nap.

►Close your office door and turn down the volume on your phone so that your voice mail handles all calls.

►Set a timer to go off in 10 minutes.

►If possible, rest your feet flat on the floor and lean back in your chair. Lying horizontal on a couch may make you more likely to fall into a deep sleep.

►Close your eyes. By doing so, you reduce visual distractions and improve concentration.

►Take a minute to take three slow, deep breaths. Put your hand on top of your abdomen and feel it move as you breathe from your diaphragm. This helps slow your heart rate and reduce your blood pressure.

►Visualize a peaceful place that you enjoy. Maybe you see yourself sitting under your backyard maple tree on a summer day. If so, make the feeling and impressions as vivid as possible. Imagine feeling the warm sun on your face, smelling the freshly cut grass, hearing a robin sing, and seeing the out-stretched branches that provide you with shade.

►After you wake up, slowly open your eyes, take a few more diaphragmatic breaths, and give yourself a full-body stretch before resuming your work at full speed.

"After a 10- or 15-minute afternoon nap, I'm up and ready to go," says Dr. Wylie. "Let your bosses know that power napping is a good investment of time. Some of the most creative people, including inventor Thomas Edison, took afternoon naps."

ALEXANDER TECHNIQUE

Posture Skills to Take You Onward and Upward

WHAT IT IS

The Alexander Technique has nothing to do with conquering the known world or tacking the phrase "The Great" onto the end of your name. Developed by an Australian actor at the turn of the century as a cure for vocal problems that plagued his career, it's actually a teaching process that's designed to replace bad posture habits with more economical and graceful use of the head, neck, and spine.

Taking a class in the Alexander Technique is an experience in upward mobility. This doesn't mean that you'll get a promotion at work, but you will learn to change the way you use your body when performing everyday activities.

The technique's goal is to keep the head, neck, and spine flowing in an upward direction. Entire lessons are devoted to keeping this in mind while performing basic activities like sitting, walking, and even lying down, according to Diana Mullman, certified teacher at the New York Center for the Alexander Technique in New York City.

During the one-on-one classes, the teacher assists the student physically and verbally. The teacher encourages upward movement of the spine by directing the student to free the neck of all muscle tension, which then allows the head to go forward and up. As the student sits down or slowly walks across the room, the teacher stands alongside, coaching and sometimes guiding with her hands. "It's difficult to explain the Alexander Technique without experiencing it," says Mullman. "You need to focus on allowing your body to naturally align itself without forcing."

WHAT IT DOES

The Alexander Technique is a posture- and movement-based practice, so the health benefits you'll get from taking lessons in the technique really center on any part of the body affected by posture, including the spine, neck, and chest area.

Lengthening the spine and neck is a powerful way to maintain a healthy back and prevent back trouble in the future, says Bruce I. Kodish, P.T., Ph.D., a physical therapist and certified Alexander Technique teacher in private practice in the Los Angeles area. That's because keeping the spine lengthened protects the posterior ligaments and other supporting structures that run along the back of the spine and hold the vertebral disks in place.

"Habitually collapsing forward, as many of us do, may eventually overstretch the structures in the lower back," says Dr. Kodish. Stretched-out or weakened tissues in that rear position may make so-called slipped disks more likely to happen, according to Dr. Kodish.

The Alexander Technique also has a beneficial effect on breathing, says Michael Gelb, certified Alexander teacher, president of High Performance Learning in Great Falls, Virginia, and author of *Body Learning*. "I used to get a twinge of asthma," he says. "When I noticed trouble coming on, the technique helped me become aware of how I was using my body." Remembering to maintain an erect spine, says Gelb, allows breathing to come more easily.

This makes sense, says Dan Hamner, M.D., a physiatrist and sports medicine specialist in New York City. Keeping the back and neck straight ensures that the bronchi—the tubes leading into the lungs—are not compressed. "Since the symptoms of asthma are mostly due to the collapsing of the bronchi," he says, "maintaining an open airway through better posture could really be helpful."

Here are a few other problems that, according to practitioners, might be eased by the Alexander Technique: scoliosis (curvature of the spine), temporomandibular disorder (TMD), headaches, arthritis, and repetitive strain injury.

REAPING THE BENEFITS

Unless you happen to live near a major city, the Alexander Technique probably isn't one of the classes offered at your local gym. Experience an Alexander Technique session for yourself by contacting the North American Society of Teachers of the Alexander Technique (NA-STAT). For a referral to a certified teacher in your area, write to them at 3010 Hennepin Avenue South, Suite 10, Minneapolis, MN 55408.

TRIAL RUN

A basic Alexander Technique procedure is called the rest position. This simple procedure is very helpful for people with back problems, says Mullman. In fact, she says that her own recovery from a serious back and shoulder injury was greatly helped by the rest position. "It gently helps realign the spine and also allows it to elongate," she says, referring to two of the main goals of the technique.

► Lie on the floor on your back with your knees bent and your feet flat on the floor, about hip-width apart. (Your knees can be separated or touching, whichever is most comfortable.)

► Place your hands on your chest or abdomen and rest your head on a book or a small stack of books (*1*). Experiment with books of different thicknesses. For some, *The Great Gatsby* will do; others may require *War and Peace*. Try to get your face perfectly parallel to the ceiling. If your head is thrown back, you don't have enough books. If your chin is falling toward your chest, you have too many.

<u>1</u>

► Once you are in position, allow your back to fall naturally onto the floor without forcing. Try to release any tension that you may be holding in your body. Lie down in this position for 15 minutes twice a day, Mullman advises.

ANGER

Exercise Restraint

In the 1993 action/misadventure film *Falling Down*, Michael Douglas plays a mild-mannered commuter who runs afoul of one traffic jam too many. Enraged, he abandons his car in a bumper-to-bumper snarl and sets off on a violent rampage through the streets of Los Angeles.

Although that's certainly an extreme example, it is true that anger can drive us to do all manner of outrageous things. It can ruin relationships, careers, and health. When you're angry, your body cranks up production of highly arousing hormones, such as cortisol, that inflate your blood pressure and keep you on Red Alert. Many studies have found that angry types run a higher risk of heart attack than more placid personalities.

Psychologists used to believe that the appropriate way to respond when angered was to vent, to rage like Shakespeare's mad King Lear. The theory was that blowing up and releasing anger was better than repressing it and seething silently. But some research suggests that venting may actually perpetuate anger and be just as unhealthy as swallowing it. Getting yourself to simmer down may require some mental calisthenics—along with some physical exercises.

Get in Shape So You Don't Get Bent Out of Shape

Thought you knew all the reasons to get in shape, didn't you, like lower cholesterol, lower blood pressure, and lower weight? Well, exercise raises some things, too, including your boiling point. A regular exercise program is a good way to prevent getting steamed in the first place. Studies show that people who are aerobically fit are less likely to overreact to all sorts of stressors. They also produce smaller quantities of arousing hormones like cortisol, says Redford Williams, M.D., director of the behavioral medicine research center at Duke University Medical Center in Durham, North Carolina, and co-author of *LifeSkills*. "And they're less responsive to the hormones they do produce," he says. Toning the body is tonic for the mind.

Not only is general fitness a good anger blunter, but working out is a weapon against each flare-up of frustration. Physical exertion helps distract you from your rage, accelerates the burn-off of those incendiary hormones, and prompts your body to produce extra feel-good neurochemicals like serotonin. So here's what to do if you find yourself getting wrought up.

Run around the reservoir. "Try not to obsess over your anger while you're working out," advises Jerry Deffenbacher, Ph.D., professor of psychology at Colorado State University in Fort Collins. "Turning things over and over in your mind angrily only continues to fuel your anger. Spend some time problem-solving or thinking constructively about what you can do about what has angered you," he says. "But when you are exercising, take a break from anger and turn your thoughts to more pacific things, like how good it feels to

The Three Little Questions and Big, Bad Anger

An anger cure often requires a goodly dose of mental medicine. To control anger, it's important to step back and think situations through. The right anger response always varies with the circumstances, says Redford Williams, M.D., director of the behavioral medicine research center at Duke University in Durham, North Carolina, and co-author of *LifeSkills*. The next time your heart starts a-pounding and your teeth begin to grind, Dr. Williams suggests that you pause and ask yourself three questions.

1. *"Is this important?"* So your waiter has muffed your order, and you're just getting your shiitake mushrooms while the rest of the crew is digging into their manicotti. Irritating, yes. But does it matter? Will it matter tomorrow? One week hence? If the answer is yes, proceed to the next question.

2. *"Am I justified in being angry?"* If you answer that your ire is indeed reasonable (or, better yet, if you think a friend would agree that it was justified), move on to the most important question of all.

3. *"Do I have an effective remedy?"* If you do, follow through. But remember, most of the things that make us mad are beyond our control. There's no point in tripping the busboy out of frustration.

If a negative answer to any of these questions doesn't help squelch your anger, there's another effective way to smooth your feathers. You need to challenge the thinking that ruffled them in the first place, says Cary S. Rothstein, Ph.D., director of the Crossroads Center for Psychiatry and Psychology in Doylestown, Pennsylvania.

Learning to re-evaluate anger-provoking thoughts is important. If you tend to have a short fuse, according to Dr. Rothstein, you probably see the world as a pretty hostile place. Get into the habit of challenging enraging thoughts, and you'll start to see the world in a more benign light. Give this cognitive exercise a whirl.

Let's say that the school bus ahead of you is still standing at the traffic light, which turned green a good two seconds ago. Mentally calling on Providence to smite the driver will only make you angrier. So move your mind, shift your perspective, and take a look at the situation from the other guy's point of view. Tell yourself, "Hey, the driver's probably distracted by the kid in the back seat with the boa constrictor." "This kind of self-talk can soften the anger," according to Dr. Rothstein.

How to Cool Out without a Workout

You're in the middle of the most important marketing meeting of your career, and you're ticked off at a colleague. You can't slip away for a long-distance run, but you definitely need to simmer down. Try this do-it-anywhere exercise in which you systematically tense and relax your muscles.

▶ First, tense the muscles in your feet and hold for a moment or two.

▶ Then relax them completely for a few seconds.

▶ Now work your way up to your calf, thigh, stomach, shoulder, and neck muscles, tensing and relaxing each set one at a time.

Relaxation exercises like this will lower your arousal level, says Cary S. Rothstein,

Ph.D., director of the Crossroads Center for Psychiatry and Psychology in Doylestown, Pennsylvania. They'll ease you down from high dudgeon. And no one will know you're doing them. Dr. Rothstein also offers these useful anger-tamping thoughts.

▶ Imagine how the wisest person you know would handle the situation that's got you miffed: Think how Gandhi would handle a traffic gridlock, for example.

▶ Imagine yourself in your favorite peaceful place, such as walking along the Seine at sunset.

▶ Or try humor: Envision the person you're at loggerheads with wearing an oversize diaper and toting a pacifier.

stretch your legs or what the weather and surroundings are like."

Catch some flies. "Competitive sports like softball are a helpful and socially acceptable outlet for blowing off steam," says Cary S. Rothstein, Ph.D., director of the Crossroads Center for Psychiatry and Psychology in Doylestown, Pennsylvania. So organize an impromptu game the next time you're miffed. It's hard to ruminate on what's irked you when you're out in left field and "Home Run" Hallahan has just popped one your way.

Other Rage Assuagers

If you're not in a sporting mood, try these other wrath-diffusing tips. They'll distract you until your arousal level has dropped, and some

will even help your body shift into a low-arousal mode faster.

Tell it to the basset hound. Working your jaw muscles when Fifi's within earshot can soothe the savage beast in you. (*Note:* Don't spend the time telling Fifi the sordid details of what got you in a lather.) In a study at the University of Pennsylvania in Philadelphia, researchers found that pet owners' blood pressures dropped when they chatted with their animals. No dog? Your ferret is a sounding board as well. If your animal isn't much for conversation, try nonverbal communication. "Pet your pet," says Dr. Rothstein. "Unless it's a fish, in which case just watching an aquarium can be very calming."

Beat your bongos. Beating out a heartfelt rendition of "Day-O" will take your mind off the offense at hand. "It'll give you something else to focus on while your arousal level drops," says Dr. Rothstein. Left your bongos at Woodstock? Try these different-drummer variations.

▶Beat a cadence on the countertop.

▶Turn Tupperware into a snare.

▶Use chopsticks as drumsticks.

▶Play "Wipe Out" on your tummy.

Exercise your ears. If you don't feel like making music, listen to it. Research at the Institute of Sport in Warsaw, Poland, shows that listening to music you like can prompt a drop in your body's cortisol levels. Avoid college fight songs, though—a Tony Bennett ballad may be better balm.

Hit the showers. A shower or bath can soothe and refresh you, says Dr. Rothstein. But you already knew that, right? "Women have been taking baths to calm themselves for centuries," he notes.

Build Your Meditation Muscles

Meditation and mad just don't jibe. When you meditate, your heart and breathing rates slow, and your brain waves go from a high-arousal to a low-arousal pattern, says Dr. Williams.

Give it a shot the next time you feel your anger ascending. You can do it at your desk or on the commuter train. Start by breathing deeply and slowly so that your abdomen rises with each inhalation and falls with every exhalation. Focus your eyes on a distant spot—maybe that Bermuda travel poster at the far end of the railroad car. Then repeat a calming word, such as *vacation*, for instance, to yourself each time you exhale.

Practice meditation on a regular basis, and you'll find it easier to start calming yourself when angered, says Dr. Williams, who recommends spending 10 to 20 minutes meditating each day. Make a habit of meditating, and your body will produce smaller quantities of cortisol.

Ankle Pain

Soothing a Sprain

An ankle is a different sort of joint. Why? Because most of the damage that it can sustain happens suddenly. A simple misstep off a curb, and pop goes the ankle. Sprain is the number one cause of ankle pain, doctors say. The ligaments get pulled or torn, leading to swelling, tenderness, and sometimes, a bruise.

For the first few days after an ankle sprain, doctors recommend the RICE (rest, ice, compression, and elevation) treatment plan. But after that, RICE is not enough. You have to gently exercise a sprained ankle back to health. Exercise is important because it reduces the amount of scar tissue that develops after a sprain. Too much scar tissue and you'll lose mobility in the ankle joint. Doctors recommend mild stretching and strengthening exercises.

Stretching the Ankle

To help resolve swelling and soreness, try some gentle stretching exercises to restore flexibility in the ankle's ligaments, tendons, and muscles, suggests Kim Fagan, M.D., a sports medicine physician at the Alabama Sports Medicine and Orthopedic Center in Birmingham.

Wave bye-bye with your toes. Sitting on a chair, extend your legs in front of you, about three inches off the floor. Focus on your toes and try to turn them back toward you as far as you can. Hold for a few seconds. Then tip your toes forward so that they point straight out in front of you. Again, hold for a few seconds. Try 5 repetitions at first and work up to 15 twice a day, suggests Dr. Fagan.

"This back-and-forth motion will get the muscles in the lower leg working and help dissipate any swelling," she says. "You should feel a little discomfort, but don't push to the point of pain."

Try the ankle alphabet. Have some fun and help your ankles at the same time with this toe-spelling alphabet exercise.

Slip off your shoes and sit on a chair or the floor. Using your right big toe as a pointer, rotate your ankle to air-write the letter "A." Be sure you pivot around your ankle. Don't just move your whole leg.

Continue through the alphabet. For variety, write your name in the air.

"You should feel your ankle moving slowly in different ways," says Dr. Fagan. "This stretch improves your lateral (side-to-side) and medial (inside-and-outside) movement."

Play foot tennis. Here's a sprain-recovery stretch recommended by Michael Ciccotti,

Find Out What Gives

Before beginning any exercise program for ankle pain, it is important to be sure that you don't have something more serious than a sprain, such as a ruptured Achilles tendon or a fracture, says Kim Fagan, M.D., sports medicine physician at the Alabama Sports Medicine and Orthopedic Center in Birmingham. So a visit to the doctor may be in order.

"If you have a lot of tenderness right over the bone and it is extremely painful to stand, you may have a fracture," says Dr. Fagan. "Also, if there is a significant amount of swelling in the back of the ankle, it may be a problem with the Achilles tendon and not a typical sprain."

If you feel pain directly on the ankle bone, not merely an ache or tenderness on a muscle or tendon, that's another clear signal to stop exercising and see your doctor, she emphasizes.

M.D., an orthopedic surgeon and director of sports medicine at the Rothman Institute at Thomas Jefferson University in Philadelphia. "Try this as soon as comfort allows," he says. "Some swelling is okay."

Sit on a chair with both feet flat on the floor. Tuck a tennis ball between the arches of your feet. Gripping the ball, raise and extend your legs so that your heels are three to four inches off the floor.

Next, holding the ball between your feet, slowly rotate both ankles clockwise. Try 10 clockwise circles and then 10 circles counter-clockwise. Remember to make the circles by rotating your ankles, not just by moving your legs and feet. Start and end your day with this stretch.

"You are restoring the range of motion to your ankle in a coordinated fashion," explains Dr. Ciccotti. "It stretches the joint capsule and the ligaments."

Table your pain. You can deliver a complete stretch from your foot all the way up your calf muscle with this exercise, recommended by Jeffrey Willson, a certified athletic trainer at K Valley Orthopedics in Kalamazoo, Michigan. Here's what to do if your ankle has been sprained.

► From a standing position, lean forward against a table with your hands palms down on the table edge. Stand far enough away so that you feel a moderate stretch in your calf muscles. Your arms should be straight.

► Lock the leg with the sprained ankle at the knee (*1a*). Slide the other leg forward and bend it so that your knee is directly over your toes. Keep your heels on the floor.

► Press your upper body forward (don't bend at the waist) until you feel a moderate stretch in the calf muscles of your back leg. Hold that stretch for 15 to 20 seconds.

► Keeping both heels on the floor, bend your back leg until you feel a moderate stretch in your Achilles tendon (*1b*). Hold that stretch for 15 to 20 seconds.

► To give your good ankle a healthy stretch, change your position and repeat the steps with your other leg.

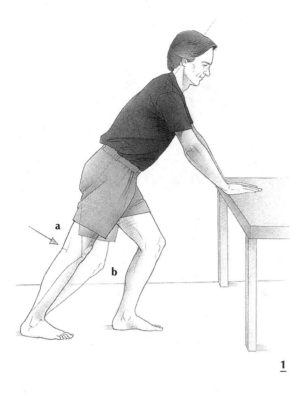

a

b

1

Strengthening the Ankle

Round two of sprain recovery is returning power to the ankle's ligaments, muscles, and tendons, says Dr. Fagan. For the next three exercises, you will need something to pull on that will provide resistance. The ideal equipment is a resistance band, available at sporting goods stores. If that's not handy, a bike inner tube will do.

After you complete these exercises to give your sprained ankle a workout, Dr. Fagan recommends that you repeat them with your good ankle to keep it in top shape.

Move it out and up. Sit barefoot on a chair with your aching foot directly facing the leg of a sturdy table. Loop one end of the resistance band around your foot at the base of your toes (your feet and toes will have to be slightly flexed). Loop the other end around the table leg about two inches above the floor. Keeping your heel on the floor, pull

No Sprain, No Pain: A Question of Balance

Lots of sprains happen because you lose your balance, stumble, and land on one of your feet at an odd angle. Here's a better-balance tip from Kim Fagan, M.D., sports medicine physician at the Alabama Sports Medicine and Orthopedic Center in Birmingham.

Strike the one-legged pose. You may snicker at the flamingo's one-legged pose, but the bird has a brain when it comes to balancing acts.

To improve your balance, try standing on one leg. Keeping your right leg slightly bent, simply lift your left leg off the floor. If you're nervous about being unsteady, put a chair in front of you for support if you wobble. After one minute of standing on your right leg, switch legs.

If you're feeling brave, try the one-legged stance with your eyes closed. "Keeping your eyes closed makes your body adjust without the aid of visual cues," says Dr. Fagan. It will help you achieve better balance.

Oh, in case you're wondering, according to *The Guinness Book of World Records*, in 1992, Girish Sharma of Deori, India, established the world record for standing on one leg, with a time of 55 hours and 35 minutes. Don't try to break this record.

your toes back and slowly turn your foot outward, pulling the band taut (2). Repeat this five times.

Move it in and up. For the second band stretch, start with your foot in the same position as before. With your heel on the floor, slowly turn your ankle inward while flexing your toes as high as you can (3). Feel the resistance from the band. Try five stretches.

Pull straight. For this final strength-building exercise, start with your heel on the floor and your toes facing the table leg. This time, flex your toes straight back as far as you can (4). Feel the tug-of-war from the band. Try five repetitions.

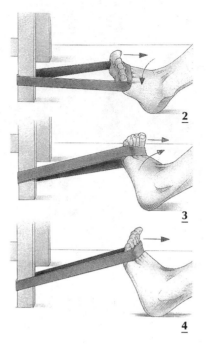

ANXIETY

Cut Your Worries Down to Size

Remember leafing through *Mad Magazine* and chuckling at the grinning Alfred E. Neuman? Well, the "What? Me Worry?" wag is more than just a poster boy for goofing off. In one respect, he's a role model for a healthy life. The man from *Mad* never let anxiety cross his mind.

Research shows that anxiety really takes it out of you. Not only does it undermine your overall health, it can also play a role in myriad maladies, ranging from headaches to wrinkles.

While the not-overly-bright Alfred E. came by his buoyance naturally, you're burdened with a brain, so it isn't as easy for you to be carefree. A powerful anxiety-assuaging tool is at your service, however: Exercise can help you hang loose and stay well. You can relax in your slacks with gentle physical workouts, mental workouts, and a little creative breathing.

Walking and Jogging Away from Anxiety

Healthy movement is a great antidote to anxiety, says Edward M. Hallowell, M.D., senior lecturer at Harvard Medical School and founding director of the Hallowell Center for Cognitive and Emotional Health in Concord, Massachusetts. A daily dose of calorie-burning exercise reduces tension, drains off excess aggression, acts as an antidepressant, enhances well-being, and improves sleep. Exercise makes your body produce chemicals—endorphins, corticosteroids, and neurotrophins—that help ease a worried mind.

Experts agree that if you can spend 20 to 30 minutes a day in some kind of activity, you'll feel less tense. Walking can do the trick. So can easy jogging or biking or dancing. If you have access to a lake or pool, swimming is a superb way to unwind the spring inside you. Try these other suggestions, offered by Dr. Hallowell in his book *Worry: Controlling It and Using It Wisely*.

Take a cold plunge and a quick run. Drown anxiety with a cold shower, or splash water on your face. Then hop into some exercise clothes and take a dash around the block. It's a mental wake-up call that will take the edge off anxiety, says Dr. Hallowell.

Enjoy some rock and roll. Dust off that old rocking chair in your basement and take it up to the living room. Rhythmic motion—such as rocking for 20 minutes—sends serenity signals to the brain, says Dr. Hallowell. If you can sit a spell and set to rocking, you'll be able to leave some anxiety behind.

Love thy partner. Sex is a good anxiety antidote, according to Dr. Hallowell. A loving kiss, caress, or cuddle with someone who cares about you is more effective than any anti-anxiety medication. Lovemaking is an ideal way to change the physical state of your brain from worry to pleasure, says Dr. Hallowell.

Mental Ammunition against Anxiety

The trick to managing anxiety is to keep it from seeping into every corner of your life. Try these ideas for keeping it corraled.

Make a date with anxiety. Set aside a specific time each day to focus on anxious thoughts. Mark it in your appointment book, but give worry only 20 minutes of your precious time. Train yourself to worry about things only during that time frame. It'll take about three weeks to develop the habit. And really worry during this time; don't do it halfheartedly.

"When worries start to enter your thoughts outside this time slot, say, 'Hold it. This is not the time to worry,' " says J. Crit Harley, M.D., a behavioral physician and director of Un-Limited Performance, a stress-management company in Hendersonville, North Carolina. This helps clear your mind to concentrate on other issues. Once you've mastered the 20-minute time slot, work to reduce it to 10 minutes a day, says Dr. Harley.

Wad up your worries. It's downright debilitating to worry about issues that you can't control, but you can actually toss those worries away. Write each of them down on separate pieces of paper, then crumple each worry into a little ball. Squeeze the paper ball into a tossable projectile and aim it toward the nearest trash can, suggests Dr. Harley. "This technique works especially well in helping people get over anxiety about something over which they have no control," he says.

Exercising Tranquillity

In this hurry-scurry world, relaxation can become a rare commodity. It's important to find some methods to make your mind and body go *ahhhhh*. Try these suggestions for dealing with anxiety-producing situations.

Visualize victory over Econ 101. If you're taking a new course at night school and you're worried about an upcoming exam, there's a tried-and-true way to disarm test anxiety. Visualization is a mighty mental warrior, says JoAnne Herman, R.N., Ph.D., a biofeedback expert and associate professor of nursing at the University of South Carolina in Columbia.

"After you've studied for the exam, visualize yourself going through the whole process of taking that economics test ahead of time," she says. "Visualize the classroom in your mind. See yourself walking in, prepared, comfortable, and confident that you will perform at your peak. Practice summoning this image so that you can call it up when you actually take the test."

Take an anxiety aspirin. Whenever you feel the worry monster approaching, think of something that never fails to delight you. Remember a dumb joke that you've loved for years. Think of a cheerful birthday card you received from your child or a friend. Let yourself smile.

Counter anxious thoughts with positive memories or images that make you feel happy and peaceful. Focusing on relaxing images can do wonders to drop your blood pressure and pulse, says Martin Rossman, M.D., clinical associate in the department of medicine at the University of California Medical Center in San Francisco and co-director of the Academy for Guided Imagery in Mill Valley, California.

Breathe like a baby. When anxiety starts rising, we sometimes start taking shallow, stac-

Acupressure vs. Anxiety

Acupressure, the fine art of finger or knuckle pressure, can ease anxiety, says David Nickel, O.M.D., a doctor of Oriental medicine and licensed acupuncturist in Santa Monica. According to Traditional Chinese Medicine, these self-applied finger or knuckle pressings alleviate the emotional toll of anxiety and relieve muscle tension by improving circulation through your body and to your brain. Dr. Nickel offers these acupressure points for anxiety.

Give in to ear pressure. Locate the neurogate acupressure point at the base of the upper

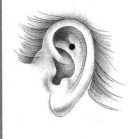

ear triangle (*left*). Using the tip of your forefinger, apply steady pressure on the neurogate point for five seconds and then gradually release the pressure for five seconds, says Dr. Nickel. Time your breathing so that you exhale through your mouth as you apply pressure and inhale through your nose as you ease pressure. Use a deep, rhythmic, relaxed breathing style. Try this pressure-on, pressure-off cycle for one minute.

Become a web-master. Work the web area between your thumb and index finger (*below*), says Dr. Nickel. Hold the web of one hand between the thumb and index finger of your other hand, with the thumb on top and the

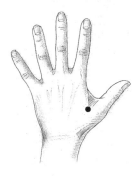

index finger underneath. Squeeze gently but firmly for 5 to 10 seconds while exhaling through your mouth. When you release the squeeze, inhale deeply.

---

cato breaths. To counter rapid-action inhalation, try breathing more like an infant than an adult. "Babies naturally breathe from the lower part of the lungs, where breathing is most efficient," explains Dr. Harley. It's called diaphragmatic breathing.

The next time anxiety ambushes you, close your eyes and concentrate on relaxing each part of your body from head to toe, one section at a time. While you're relaxing, breathe as deeply as you can. Place your right hand on your chest and your left hand on your belly, about three inches above your waistband. If you're breathing deeply, your belly will rise when you inhale.

Once or twice a day, try to spend about 15 minutes breathing deeply. It triggers a relaxation response from your parasympathetic nervous system, says Dr. Harley.

Say a prayer or mantra. If you're religious, get in the habit of praying twice a day. You can break the worry habit by feeling part of something larger than yourself, says Dr. Hallowell.

If you're not religious, perhaps you can use daily meditation by repeating a mantra (words or sounds used as a tool for focusing the attention) to subdue negative emotions and achieve the feeling of being in tune with the rest of the world, he says.

ARTHRITIS

Treating It in Water and on Land

Doyt Conn, M.D., gets out of bed every morning and starts his day with 100 situps. If he were a U.S. Marine, this wouldn't be so remarkable. But he's not a gunnery sergeant, he's a doctor in his late fifties, and he has osteoarthritis of the spine.

"There's a myth that arthritis is something that you can't do anything about," says Dr. Conn, senior vice president of medical affairs for the Arthritis Foundation, based in Atlanta. "I do my situps every morning to maintain my muscles—especially my abdominal and large back muscles."

Dr. Conn is passionate in his belief that you can help all types of arthritis with exercise. You're not powerless against the pain.

Of course, if you have arthritis, the very idea of exercise can seem daunting. You surely won't be able to do demanding activities, says Dr. Conn. But a gentle workout habit is a powerful anti-arthritis weapon. In a study of people 60 and over who had knee pain from osteoarthritis, doctors found that there were a number of exercises that helped reduce the pain and made knee movement easier. For many people in the study, the benefits came from aerobic exercises such as walking. For others, resistance training designed to strengthen major muscle groups was also helpful.

"There's no question that people with any type of arthritis can help themselves with exercise," says Dr. Conn. "If you become sedentary, you'll rapidly lose muscle mass and mobility. The arthritis process may accelerate." If you can handle an easy-does-it exercise plan, you can minimize your pain, increase your range of motion, and keep the arthritis under control.

Doctors and the Arthritis Foundation recommend combining some simple range-of-motion movements, such as reaching, bending, and stretching, with strength-building techniques.

The Morning-and-Evening Exercise Plan

"A daily routine that combines flexibility exercises with strength exercise reduces pain and stiffness," says Dan Hamner, M.D., a physiatrist and sports medicine specialist in New York City. The following 10-minute exercise plan may help people with osteoarthritis. Try these exercises twice a day—right after you wake up and right before you go to bed. Start out by repeating each exercise 3 times and gradually work up to 10 repetitions. The first half of Dr. Hamner's plan focuses on improving flexibility and range of motion.

For ankles: Before you begin, warm up by sitting on a chair and crossing one leg over the other. Slowly rotate the ankle of your top leg so that your foot moves in a circle. Then switch sides and repeat with your other ankle. Now you're ready to begin.

▶Stand facing a wall with your feet flat on the floor and shoulder-width apart.

▶Place both palms against the wall at shoulder height and lean forward from the waist.

▶Keeping your left foot in place and your heel on the floor, take one step forward with your right foot.

▶Keeping your left knee and hip straight, bend your right knee and lean forward a bit at the right hip.

▶Gently stretch forward and hold this position for two to five seconds. You'll feel a little stretch in your left ankle, but stop if you feel any pain.

▶Repeat, reversing the position of your legs.

For knees: Warm up by standing with your feet and legs together and your hands on your knees (if you need more balance, you can keep your feet about six inches apart). Slowly move both knees in a circular motion, first clockwise and then counterclockwise.

▶Next, lie on your back and bend your right leg at the knee; keep your left leg straight.

▶Use both hands to pull your left knee to your chest. Hold for two to five seconds.

▶Push your left leg into the air, as close to vertical as you can manage. Hold this position for a moment and then lower your leg slowly to the floor.

▶Repeat these steps with the right knee. Try doing five sets for each knee.

For hips: You can use the same basic warmup for your hips that you used for your knees. Stand with your legs together, put your hands on your hips, and slowly move your hips in a circle as if you were playing with an imaginary hula hoop, first clockwise and then counterclockwise. Don't try the following exercise, though, if you have had a total hip replacement or if you have lower back problems or osteoporosis.

▶Lie on your back with your legs straight. Place your feet six inches apart with your toes pointed up.

▶Slowly slide your right leg out to the side as far as you can comfortably while keeping your toes pointed up. Hold for two to five seconds and then slowly slide your leg back to its original position.

▶Repeat with your left leg.

For shoulders: Standing straight, raise your left arm over your head, keeping your elbow straight and your arm close to your ear. Hold the stretch for two to five seconds. Then sweep your arm—like the hand of a clock—forward in a clockwise motion and bring it slowly back to your side. Repeat with your right arm.

The second half of Dr. Hamner's daily workout centers on maintaining and increasing muscle strength.

For ankles: Sit in a comfortable chair with both feet flat on the floor. Try to lift the toes of your left foot as high as possible while keeping your heel on the floor. Hold this position for 5 seconds. Then lift the toes on your right foot and hold for 5 seconds. Next, lift the front of your left foot and tap with your toes for 10 seconds, then tap the toes of your right foot for 10 seconds. Then tap both feet for 10 seconds. Try to work up to 60 seconds for each foot.

For knees: Sit in a comfortable chair with both feet on the floor and spread slightly apart.

Move your left foot up and out until your leg is parallel to the floor. Tighten your leg muscles and hold for 10 seconds. Slowly lower your leg, then repeat the same steps with your right leg.

For hips: Stand straight and place your right hand on the back of a chair in front of you. Using your hip muscles, extend your left leg behind you while keeping your right knee fairly straight. Hold for two seconds, then slowly lower your leg. Raise it again for two seconds. Think up, one-two, down, one-two. Repeat with your right leg.

Water Workouts to Fight Pain

Exercising in water can be a great anti-arthritis option. The water not only keeps your aching bones from being jarred, it also offers resistance and thus helps you build strength. "Water is the mainstream alternative for low-impact, soft exercise," says Jane Katz, Ed.D., professor of health and physical education at John Jay College of Criminal Justice at the City University of New York, world Masters champion swimmer, member of the 1964 U.S. Olympic performance synchronized swimming team, and author of *The New W.E.T. Workout*. She suggests these pool exercises.

Follow your toes. Sit on the edge of a pool and dangle your legs up to midcalf in the water, says Dr. Katz. Begin by making foot circles, rotating from the ankle clockwise and then counterclockwise. Next, move your feet forward and back, then left and right, keeping your legs slightly bent and your ankles loose. Repeat each five times.

Try a wet walk. Stand in chest-deep water and simply walk forward normally, making sure to swing your arms as you would on dry land.

Once you feel stable doing this, try walking backward and then sideways, says Dr. Katz. Begin with 1 minute, resting as needed, and progress to 10 minutes.

Do the aqua jog. Step up the pace of your walk. Pretend you're in a race and run in place, lifting your knees and moving your arms with each stride. As you get comfortable with this water jog, try to run faster and faster, says Dr. Katz. Begin with one minute, resting as needed, and progress to five minutes.

Dr. Katz offers the following water exercises for stiffness and pain in your hands and wrists.

Try a wrist roll. Standing in chest-deep water, extend your arms to the sides, keeping them underwater, and slowly close and open your fists six times. Then, to work your wrists, point your fingers up and then down, then keep your wrists straight as you move your hands from side to side. Finish by making clockwise and counterclockwise hand circles from the wrists, with both open hands and closed fists (*1*). Progress to three sets, resting briefly between sets.

1

Deliver water punches. Still standing chest-deep, take a boxing stance, with one foot in front of the other. With your hands underwater, throw a series of punches. First, try the jab, punching your left and then your right hand out in front of your body (2). Then try a hook, swinging your left arm and then your right around from the side and out in front to the torso of your imaginary opponent (3). Finish with a knockout uppercut,

2

3

4

starting from your waist and extending each arm upward through the water's surface (4). Practice each punch four times to complete the set. Rest briefly before repeating a set and work your way up to three sets.

Landlocked Water Exercises

Even outside a pool, warm water is a valuable arthritis-workout asset. Doctors recommend these dry-land H₂O exercises.

Play towel toss. Moist heat helps improve blood flow and relax tight muscles around joints, says Dr. Hamner. The next time your knee throbs or your elbow swells, try wetting a hand towel and then putting it in your dryer at a high temperature for about 10 minutes. Take it out while it is still wet and hot. Place it on your knee or elbow and then lay a dry towel on top to keep the heat from escaping.

Hit the showers. This "no-sweat-but-get-wet" technique can decrease muscle stiffness associated with arthritis in your neck and back, says Jane Sullivan-Durand, M.D., a behavioral medicine physician from Contoocook, New Hampshire. "Stand under a hot shower for 10 minutes with the water beading on the affected area. Get the temperature as hot as you can stand," she says.

Get Rubbed, Think, and Giggle

If you prefer exercising without getting wet, try these moves to ease arthritis ache.

Request some rubbing. Ask your spouse or a friend for a 10-minute massage. Massage can reduce pain, says Dr. Hamner. It can push blood and other body fluids through the bloodstream to reduce swelling and eliminate waste products from tissues (for more details, see "Massage" on page 246).

Play mind games. Try to find at least 15 minutes of quiet time a day. Close your eyes and take deep, deliberate breaths. Inhale and exhale slowly and fully. Fifteen minutes of happy, healthy thoughts can ease muscle tension, say doctors.

Imagery (page 192), meditation (page 249), tai chi (page 386), and yoga (page 426) are effective ways to counter pain and induce relaxation, say doctors.

Laugh the pain away. There's no need to mimic Robin Williams or Jerry Seinfeld. But frivolity can actively help, says Norman Harden, M.D., a neurologist and director of the Center for Pain Studies at the Rehabilitation Institute of Chicago. "Laughter can trigger wonderful endorphins, natural painkillers that also combat depression and anxiety," he says.

Give peas a chance. Pay some mind to your peas—frozen peas, that is. While watching television or reading a book, place a bag of frozen peas on your painful joint, suggests Dr. Harden. "A cold pack is good to use when there is swelling and inflammation; you want to calm this down." Just make sure that you don't leave it on for more than 15 minutes at a time.

Walk and talk. A 30-minute brisk walk—or three 10-minute walks—every day can improve strength and reduce pain in arthritic joints, says Dr. Hamner. "When you are walking, you should be able to carry on a normal conversation, sing a song, or recite a poem. If you get out of breath, you are working too hard and you need to back off and take it easy."

If you are unable to walk comfortably for 10 minutes, he recommends that you consult your doctor before continuing to exercise.

Arnie's Army Fights Arthritis

If you're a golfer with arthritis, the Arthritis Foundation offers these exercise tips to stay limber on the links.

▶ Before you tee off, walk for a few minutes to loosen your joints.

▶ Devote 5 to 10 minutes to gentle stretches like trunk twists and hamstring stretches.

▶ Be sure to spend some time on the practice range. Loosen your muscles with a few dozen swings.

Here are two suggestions to try if your playing partners aren't sticklers for the rules.

▶ Use a tee for every shot that you take, even on the fairway. This will help you avoid striking the ground and jarring your joints.

▶ If you start feeling tired, skip the tee area and begin your pursuit of the hole from the 150-yard marker on the fairway.

BACK PAIN

Put Your Woes behind You

Physical therapist Peggy Anglin, P.T., makes a living from bad backs. But more than 20 years of demonstrating the wrong moves finally caught up with her.

"I spent so much time demonstrating the wrong techniques over and over again to my patients that I became a statistic myself," says Anglin, who practices at Duke University Medical Center in Durham, North Carolina. "I was picking up a paper one morning at home and it just hit me. I nearly fainted from the excruciating, terrifying pain."

Almost all people experience acute lower back problems at some time in their adult lives. One of every two American adults experiences back pain every year. For people under age 45, lower back problems rank as the top disability.

Why is back pain so common?

It's simple, sort of. The back is an amazingly complex piece of anatomical real estate. Our spines are made up of small bones called vertebrae that are stacked on top of one another like dinner plates. Sandwiched between the vertebrae are cushionlike disks. Openings in the vertebrae line up to form a long, hollow canal, through which the spinal cord runs. Nerves from the spinal cord pass through small gaps between the vertebrae. When you add the dozens of back muscles, it's easy to see why backs often go blooey.

Not only is back pain widespread, it's also tough to track down. Pinpointing the cause and even the precise location of back pain often turns doctors into body detectives. "When one muscle in the back becomes inflamed or spasms, it affects a whole bunch of muscles that may have nothing to do with it, structurally," says Patrick Massey, M.D., Ph.D., an internist at Alexian Brothers Medical Center in Elk Grove Village, Illinois. "Sometimes, moving your skull will affect your tailbone."

Just about the only back pain certainty is that no one magic exercise program cures all backs. "Everybody is completely different," says Dr. Massey. "The back pain is specific to each person."

But there's hope. First, much back pain is preventable. Second, 80 to 90 percent of people with back pain recover within six weeks. And third, experts say that even people who have chronic back pain can often manage it—with exercise. Here are a few techniques that have proven successful in strengthening the back and abdominal muscles. And if you practice doing them, you'll be well on your way to coping with a balky back.

Oh, My Achy Back

If you sometimes have back pain, you should try to make some low-stress, low-intensity aerobic exercise part of your routine. Walking, stationary biking (with upright handlebars), and swimming are healthy choices, say physical therapists. In addition, between attacks of pain, these targeted activities and exercises may bolster your back.

Try a pseudo-situp. Patrick Fallon, P.T., a physical therapist with the Texas Back Institute in Plano, suggests this modified situp.

Lie on your back with your knees bent in a comfortable position and your hands on your thighs. Lift your head and shoulders off the floor slowly as you slide your fingertips to reach your knees. Hold for a count of three and gradually ease back down. Start slowly, but aim for three sets of 10 repetitions. Don't attempt this exercise, however, if you are currently experiencing back pain.

Stretch like an elephant. Pretend that you're Dumbo with this back stretch offered by Meir Schneider, Ph.D., a licensed massage therapist, founder of the Center and School for Self-Healing in San Francisco, and creator of the Meir Schneider Self-Healing Method.

Stand up and bend your torso forward as far as is comfortable. Let your arms hang separate and loosely at your sides. Next, swing your arms gently from side to side like the trunk of a contented elephant. This relieves pressure in the sacrum—the lower part of the spine—and loosens your hip and lower back muscles. After a minute of swaying, uncurl your spine and slowly straighten to a standing position. Bring your head up last. Repeat these steps at least six times a day, says Dr. Schneider.

Chart your course. Yoga stretches, called asanas, may help relieve some back pain, says Alice Christensen, founder and executive director of the American Yoga Association in Sarasota, Florida. But be sure to get your doctor's approval before trying this or other yoga poses. With this pose, Christensen suggests, pretend that you're an ocean liner slicing through the deep blue.

▶Lie on your stomach with your arms outstretched in front of you and your forehead on the floor (*1*).
▶Exhale completely, then inhale as you raise your legs, arms, and head all at once, looking up (*2*). Lift yourself up only as far as you can comfortably.
▶Exhale and lower your body. Rest completely, then repeat two more times.

1

2

Try tai chi. Dorothy Garwood had severe lower back pain, and physical therapy didn't completely take it away. She turned to tai chi and has been practicing its movements for 15 years at a park in Los Angeles. "Tai chi is a gentle exercise that has been very helpful to me and my back," says Garwood, who is in her seventies.

This "moving meditation" offers a series of graceful movements that are relaxing and energizing, says Lana Spraker, a Los Angeles master instructor who has taught tai chi for more than 25 years. "Tai chi is successful for older people because it is done slowly and gently, exercising all the muscles of the body in a balanced way," says Spraker, who is also a certified Alexander Technique instructor. (If you want to try this technique for your back, see "Tai Chi" on page 386.)

Slip into the pool. Water provides buoyancy and allows your tender muscles the chance to move freely without a lot of resistance, says Jane Sullivan-Durand, M.D., a behavioral medicine physician in Contoocook, New Hampshire. Here's a tip: Take gentle laps for at least 10 to 12 minutes in a pool with a water temperature of at least 83°F. Warm water helps relax back muscles.

No pool? Head for a hot shower. "Try 10 minutes under a hot shower with the water beating on the affected area," says Dr. Sullivan-Durand. "Moist heat is best, because dry heat from heating pads tends to increase inflammation and stiffness."

Find a hero. Attention all partners of back pain sufferers—here's your chance to be a real hero to your mate. When your partner begs for a gentle massage, rub to the rescue, says Dr. Sullivan-Durand.

Using both your thumbs or the heels of your hands, rub up the center of the spine, starting at the bottom of the back and stopping just below the shoulder blades. Then start over again at the bottom. Massage stimulates the circulation of blood and lymphatic fluid (which tends to accumulate around an injury site). If you do this kind of massage, it reduces swelling and promotes healing, says Dr. Sullivan-Durand.

How to Prevent a Back Attack

There are no guarantees. But doctors and physical therapists agree that you can reduce the

When to See a Doctor

Here are some flares that signal it's time to see a doctor for your back pain, as offered by Patrick Massey, M.D., Ph.D., an internist at Alexian Brothers Medical Center in Elk Grove Village, Illinois, and a former sciatica sufferer.

► The problem persists beyond four days without improvement.
► You have back pain that persists after a fall or accident.
► The pain travels down your leg into the foot.

According to Dr. Massey, if you have any of the following symptoms, see a doctor immediately.

► You have trouble controlling your bowels or bladder.
► You have numbness in the groin or rectal area.
► Your legs feel extremely weak.

chance of getting back pain. Here's how.
►Avoid lifting while twisting, bending forward, or reaching.
►Bend at the knees, not at the waist, when lifting.
►Lose excess weight, especially in the abdomen.
►Improve your posture. Keep your natural back curves in alignment by lifting your breastbone and maintaining a slight hollow in your lower back.
►Don't sleep on your stomach. Lie on your side with your knees bent or on your back with a small pillow under your knees.

▶Make exercise part of your daily routine. Take walks, swim, and tone up your abdominal and back muscles through exercises. Stay flexible.

The Back-Pain-Sufferer's Guide to Gardening, Golf, and Love

Having back pain doesn't mean that you have to give up doing the things you love. But you may have to customize your technique and develop some pain-avoidance strategies. Physical therapists offer these tips to help you keep enjoying your hobbies, sports, and other leisure activities.

Dig

There are enough twists in gardening and yardwork to make Chubby Checker jealous. Keep these thoughts in mind.

Pace yourself. If you're a weekend gardener who tries to get everything done on Saturday, you might want to be a little kinder to your back. "People tend to go from doing nothing during the week to gardening for four hours or more on the weekend," says Wendy Woods, P.T., a physical therapist at K Valley Orthopedics in Kalamazoo, Michigan. Take a break every half-hour, she says.

Get the real hoe down. The smaller the chopping motions, the less strain on your back muscles, says Anglin. "When you're using a hoe, you don't need to swing your arms in wide, big motions," she says. "With hoes, brooms, or mops, keep your upper arms close to your sides. Use your leg muscles for power, not your back."

Weed with your knees. Get up close to those wicked weeds by getting down on your hands and knees, says Anglin. "Crawling around on your hands and knees is a fine position for the back," she says. "It's much better than bending

High and Mighty Laid Low by Backache

Political clout, stardom, or a super dunk shot is no guarantee against back pain. Consider this who's who of the high and mighty whose backs blew.

President John F. Kennedy. His wooden rocker provided a place of comfort in the Oval Office. The blame for the damaging blow to his back lies with a football game back in his Harvard University days, not the PT 109 collision with a Japanese destroyer during World War II.

Basketball megastar Michael Jordan. Muscle spasms in the 1996 NBA playoffs sent His Airness crashing to Earth. But he winced through pain to guide the Chicago Bulls to their fourth NBA championship.

Actress Elizabeth Taylor. She's been married eight times, but her first head-over-heels tumble was the source of her back pain. She injured her back as a child in a fall from her horse during the filming of *National Velvet*.

over and rounding your back to pull weeds. That position puts a lot of stress on your lower back."

Stop rake ache. Woods offers this leaf-raking tip for autumn weekends: When you're raking, be sure to stand really upright. Use your knees and hips so that the power of your legs can help your arms pull the leaves. And take stretch breaks every 15 minutes.

Swing

Golf, because it involves so much twisting, can sometimes cause back pain. "We get a lot of golfers here at the center," says Woods. "They want to keep golfing and don't want to stop at any cost." Try these ideas to deal with hacker's back.

Slow down. In general, Woods recommends that golfers with bum backs slow down their strokes to reduce stress on the lower back. Remember the wisdom of Bobby Jones, the biggest name in golf during the 1920s: A golf club cannot be swung too slowly.

Five Days to a Happier Back

Although there is no one-plan-fits-all cure for back pain, Jane Sullivan-Durand, M.D., a behavioral medicine physician in Contoocook, New Hampshire, offers these general day-to-day guidelines for the first few days after the pain strikes.

Day 1: Pick your pocket. If your back feels good enough for you to drive, here are some pain-cutting tips. Place a rolled-up towel behind your lower back for support and remove your wallet from your back pocket. A fat wallet makes you lopsided and can put pressure on the sciatic nerve, which could send shooting pain up your legs and back.

Day 2: Ice, ice, baby. Frozen peas in a bag, ice cubes wrapped in a washcloth, or a chilled gel-pack will restrict blood flow to the injured spot and cut down on the chance of inflammation.

"It may be uncomfortable or even hurt at first, but tolerate the hurt until the skin feels numb," says Dr. Sullivan-Durand. "Then you know the cold has penetrated. Leaving it on for no more than 20 minutes at a time, two or three times a day or whenever the area feels hot, seems to work the best."

Day 3: Message for a massage. When you feel that the ache is more tolerable, ask your spouse or a friend to lend you a hand. "A muscle that is tight can usually be manually softened and released through massage," Dr. Sullivan-Durand says. "Have them run their fingers along the area that is sore and feel the spasm. Ask them to take their thumbs or the heels of their hands and stroke that area as deeply as possible without causing any discomfort. Don't squeeze the area."

Day 4: Get a knee up. Start the morning with this stretch. Lie on your back on the floor or bed with your feet flat and your knees bent. Use both hands to bring your left knee to your chest. Hold for 10 seconds. Repeat with your right knee. Repeat once more, bringing both knees up at the same time.

This exercise helps to loosen your lower back muscles and can become part of your daily routine even after the back pain subsides, Dr. Sullivan-Durand says.

Day 5: See a doctor or take a walk. This is judgment day. If your back still isn't up to par, see a doctor, advises Dr. Sullivan-Durand. If you feel better, put on your sneakers and do a 30-minute steady walk on level ground or on a treadmill.

Shorten your swing. Everyone's a Tiger Woods wannabe; everyone wants to blast the dimples off the ball. But a long backswing is not just bad golf strategy, it's also bad back medicine. Wendy Woods says that shortening your backswing by not pulling back so far with the club reduces twisting and the strain on your back.

Meet the tee. When it's time to tee up the ball or pull it out of the cup, posture is key, says Woods. "Don't bend from the back," she says. "Try to bend at the knees, keep your back in a neutral position, and slowly lower yourself down to put the ball on the tee or take it out of the hole."

Get symmetrical. If you carry your bag, be sure to use a strap that goes over both shoulders. That way, you distribute the weight evenly and can prevent some back problems caused by habitual overuse of one side of the back.

Cuddle

Your hormones scream yes; your back says no way. You can win this battle by practicing safe sex for backs, say physical therapists. The most important things are to keep the spine

Lifting the Groceries and Little Greg

Many back injuries are caused by lifting something that's too heavy at a tough angle, such as a grocery bag full of cartons of milk and juice or a recalcitrant three-year-old, says Jane Sullivan-Durand, M.D., a behavioral medicine physician from Contoocook, New Hampshire. Here's what she suggests to keep those loads from laying you low.

Hug your groceries. To avoid more than a financial pinch at the supermarket, here's how to load groceries into your car pain-free.
► Position the grocery cart next to your open trunk or car door.
► Bend your knees slightly, keeping your back straight.
► Lift a bag from the cart using both hands underneath it, then bring it close to your body, still holding it at the bottom.
► To place a bag in the car, bend your knees first, keeping your back as straight as possible. Don't bend from the waist or overextend your arms when putting the bag into your trunk or backseat. Remember to keep your arms slightly bent.

"Never use only your arms and upper back to lift groceries," she says. "That puts strain on your lower and middle back."

Train your toddler. Many new mothers suffer from back pain. The culprit may be all the extended arm action used to get a child in and out of a car seat, says Dr. Sullivan-Durand. When your child reaches about 14 months, you should train him to climb into his car seat himself. This alleviates having to hold him at arm's length, which strains the lower back. When you do have to lift your pride and joy, be sure to hold him close to your torso. This shows that you love both your baby and your back. Remember to keep your back straight and use your legs to do the lifting.

aligned properly and to use positions that don't require the partner with back pain to be active or bear weight, says Anglin. Try these back-safe positions.

Spoon your mate. In this position, both partners lie on their sides with their knees bent and the man behind the woman. Nobody bears any body weight, Woods says.

Try a pillow prop. Here, the partner with the bad back lies on his back, using pillows to support the neck and lower back if necessary. This provides a more stable position for the back during sex, says Woods. If the man has a bad back, his partner should straddle him with her knees bent while supporting her upper body weight with her arms. If it's the woman who has a bad back, her partner should support himself with his arms and knees so that little weight falls on her.

Head for the hot tub. Let buoyancy and the warm, massaging bubbles of a hot tub keep both of you in the mood, says Woods. She recommends that the man sit in the tub with the woman sitting on his thighs and facing him. This position will work if either the man or the woman has a bad back, she notes. "The key here is to not make twisting motions," Woods adds.

Say Goodbye to Back Pain

Chung Moo Doe sounds like a delicious Chinese entrée, but it is a soft form of karate (not the board-breaking variety) that helps some people with chronic back pain, says Patrick Massey, M.D., Ph.D., an internist at Alexian Brothers Medical Center in Elk Grove Village, Illinois, sixth-degree black belt holder, and former sciatica sufferer.

In a study that he co-directed, Dr. Massey prescribed a modified version of Chung Moo Doe for 58 men and women ages 25 to 71 who had histories of serious back problems. All had failed to improve with physical therapy, steroid injections, or surgery. But after eight weeks under his guidance, 93 percent were pain-free.

For more information on this program, contact Dr. Massey at Alexian Brothers Medical Center, 850 Biesterfield Road, Suite 4011, Elk Grove Village, IL 60007.

BELCHING

Gulp Less Now, Burp Less Later

Air equals breath. Too much air equals belch.

Often, the burping equation is that simple. "The most common cause of belching is swallowing too much air," says William Ruderman, M.D., a physician at Gastroenterology Associates of Central Florida in Orlando. People who are cursed with burps are inadvertently wolfing down air. If you inhale more than your share of air, the pressure builds up in the esophagus and stomach, and suddenly you're begging everyone's pardon.

In rare cases, chronic belching may signal a serious medical problem, like a hiatal hernia or a peptic ulcer. But more often, the air-guzzling habit can be easily broken by some common-sense steps. The first is to exercise restraint at mealtime, when a lot of excessive air swallowing occurs.

Eat, Drink, and Be Belch-Free

Maybe you've never thought of eating, munching, and swallowing as exercise, but they really are. If you're cursed with burps, give these etiquette exercises a little more attention the next time you sit down to dine.

Do the fork lift—between bites. In general, the faster you eat, the more air you swallow. Put your fork down between bites, suggests William Whitehead, Ph.D., co-director of the Center of Functional Gastrointestinal Disorders at the University of North Carolina at Chapel Hill. You'll eat much more slowly and swallow less air.

Block That Burp

When you're tense and a belch tries to surface, try this relaxing head-tilting exercise, suggests William Whitehead, Ph.D., co-director of the Center of Functional Gastrointestinal Disorders at the University of North Carolina at Chapel Hill.

While sitting comfortably, slowly rock your head back. Take a deep, slow breath, then move your head forward until your chin touches your chest. Exhale slowly. Repeat these steps three times.

"The idea behind this is that after tension, people usually can relax more thoroughly," explains Dr. Whitehead. "Using relaxation is a good way to counteract the effects of stressful situations that may trigger belching."

Work your jaw muscles. Take care to chew your food well. This will not only slow down your eating and inhibit belching, it also has the benefit of easing the whole digestive process.

Get on your feet, soldier. You've just over-dosed on Thanksgiving bird and you're headed for the couch and a quick liedown. Bad idea. It can lead to what's known as reflux belching. That's a belch caused by the backup of stomach acid, says Dr. Ruderman. "Gravity helps to keep the acid where it belongs—in the stomach, not coming back up the esophagus. After a big meal, walk around, then give your body an hour or two to digest the food. Fight the urge to recline, and stand up tall," he says.

What's that? You love to sit around the table after dinner and gab? Don't do it. "People who sit around after a big meal drinking coffee and talking tend to increase the amount of air they swallow," says Dr. Ruderman. For good fellow-ship, ask your dinner companions to take a quick constitutional with you and keep up the conversation.

Put the bite on it. If you can't blame food or drink for your belches, try sticking a pencil or even an ice-pop stick in your mouth and biting down, says Dr. Ruderman. True, you might feel like Black Beauty out for a gallop, but it could block belches. "If you concentrate on biting down on something, you can break the air-swallowing habit," he suggests. "When something is keeping your mouth open, it is hard to swallow air." Try this technique for an hour or two at a time.

Hypnosis Stops Belching Rampage

"Hello (burp), Dr. Spiegel? I (burp) really need (burp) your help. I can't (burp) seem to (burp) stop belching. I'm (burp) worried. Please (burp) help me."

This desperate plea was left on an answering machine by an active 71-year-old Maryland woman whose uncontrollable belching was making her a social recluse.

Sharon Spiegel, Ph.D., a clinical psychol-ogist in private practice in Bethesda, Maryland, and former president of the Society of Clinical Hypnosis in Washington, D.C., was aware of studies using hypnosis to cure hiccups, but she could not find anything linking hypnosis and belching. "I knew, though, that anything that would promote a sense of relaxation would have a calming ef-fect," she says. "My goal was to teach her to relax and focus away from the symptom."

Dr. Speigel treated the woman with psychotherapy using hypnosis. After the first session, the woman went 24 hours without belching. Although the therapy didn't relieve the belching for good, gradually, over several weeks, the woman's symptoms were almost entirely eliminated. Eventually, the patient learned self-hypnosis and was able to image pleasant events from her past that could ease her into a trance and control the belches that had controlled her.

Dr. Spiegel's case was published in the *American Journal of Clinical Hypnosis* in what she says may be the first documented medical case of stopping belching with hypnosis.

Tension Suspension

Ever notice how you get a case of the croaks right before your speech to the school board? Although tension isn't a primary cause of belching, it can be an aggravating factor. Try these pre-speech warmups to help prevent embarrassing eruptions during your introductory remarks.

Breathe away stress. To combat stress-related air gulping, Dr. Whitehead suggests taking slow, deep, relaxing breaths. "People unconsciously swallow more frequently when they are experiencing stress or tension," he explains. "Breathing deeply, focusing on breathing, and focusing on relaxing with each breath should help." Try yoga breathing techniques and meditation. Serenity is an anti-belching medication. (For details, see "Yoga" on page 426.)

Exercise your optimism muscles. Dr. Whitehead says that you can head off stress—and burps—by using mental imagery. "Identifying patterns of thinking that make you tense and replacing those thoughts with positive expectations often help control stress," he explains. "When people are less tense, they don't swallow as much air." So before you go in for your next audition, substitute thoughts of worry and panic with an image of you taking home the Oscar.

Belay These Burp Boosters

Here are a few common habits that can contribute to belching, says William Ruderman, M.D., a physician at Gastroenterology Associates of Central Florida in Orlando.

Smoking. Lung cancer doesn't convince you? Smoking cigarettes can cause air pockets to form in the esophagus and build up burp-producing pressure.

Chewing gum. Working a wad can overload you with air.

Frequent snacking. More munchies, more air.

Hard candy. Constantly sucking on mints can make you swallow air.

Bubbly beverages. Carbonated drinks are loaded with air bubbles.

Straws. If you can't resist the effervescence of root beer, at least ditch the straw. "You take in more air when you sip carbonated beverages through straws," says Dr. Ruderman.

BICYCLING

The Adventurous Exercise

WHAT IT IS

Marcia Andrews rarely knows where she's going when she leaves home each morning. She ventures where wanderlust, her muscular 60-year-old legs, and her bicycle take her. Sometimes, she rides around the canals that crisscross her neck of southern Florida and sizes up the yachts. Other times, she pedals over to the new developments, spies on the construction crews, and watches the houses take shape. In spring, she might steer her bike past the spots where burrowing owls make their underground nests.

"Biking brings out the curious, expectant kid in me," says Andrews, who hadn't been on a bike in decades when she moved to Florida and began her early-morning bicycling adventures.

Cycling is the ideal pursuit for the expectant kid in *you*. It combines adventure and exercise. On a bike, innumerable destinations and possibilities await. On a bike, you can go places that are too distant to reach on foot and enjoy them in a way that you can't when you're sealed up inside a car.

"One of the great things about cycling is that it gets you out there," says Edmund Burke, Ph.D., director of the exercise science program at the University of Colorado at Colorado Springs, a staff member of the 1980 and 1984 Olympic cycling teams, and author of *Serious Cycling*.

WHAT IT DOES

Among the surprises that Andrews discovered while cycling: After a few months of pedaling and exploration, she'd lost the weight that she'd been trying to lose. And she felt better than she had in years.

"I think I'm healthier than I've ever been," says Andrews, former executive editor of *Metropolitan Home* magazine.

A cycling habit *will* help you leave excess weight by the wayside, confirms Dr. Burke. But that's not the only benefit. Because it helps you lose weight, cycling can help bring your blood pressure down as well. It'll give your heart a good workout, making that most important muscle stronger and more efficient. And it will lower the levels of bad LDL cholesterol lurking in your bloodstream while boosting levels of the good stuff, HDL cholesterol.

Habitual cycling also strengthens your other muscles, particularly your buttock and leg muscles. And since it's good for your muscles, it's also good for your joints. Strong muscles do a better job of holding joints in their proper places and preventing the kind of joint wear and tear that contributes to osteoarthritis.

Regular cycling may deliver other health benefits, too. A Japanese study found that bicycling helps people with diabetes regulate insulin levels and even heightens natural immunity.

Because the bike frame supports much of your weight when you're cycling, bicycling isn't quite as good at building bones as weight-bearing exercises like jogging or aerobic dance. But your bones do have to carry some of your bulk when you're biking, especially when you're climbing uphill, says Dr. Burke. Because building muscle helps protect against fractures, a cycling habit also can offer some protection against osteoporosis.

While so-called weight-bearing exercises are good for some conditions, the advantage of cycling is its lightweight action. Since the bike bears most of your weight, cycling is easy on your joints and vertebrae. It's a particularly good exercise if you have back injuries or arthritis or are overweight, Dr. Burke says.

REAPING THE BENEFITS

To get the most from your bike workout, you need the right equipment.

Mount a mountain bike. Mountain bikes, with their big, cushy tires, are your best bet for comfort, says Dr. Burke.

Buy in a bike shop. Even if it's a bit more pricey than your local EverythingMart, Dr. Burke suggests that you buy your bike at a reputable bicycle specialty shop. The advantage? An experienced salesperson can adjust the bike, fitting the cycle frame to your frame. "Fitting a bike is fairly technical," he explains.

In fact, even if you already have a bike, it's still worth a trip to the bike shop, particularly if you have an old Schwinn that you haven't taken out in a while. Have one of the shop's mechanics check to make sure that everything is in working order and that the bike is adjusted to fit you properly, says Dr. Burke. He recommends adjusting the saddle and handlebars so that:

►Your elbows are bent only slightly—roughly 5 to 10 degrees—when your hands are on the handlebars. Of course, this angle would change if you were leaning forward or standing up while riding.

►Your knee is bent slightly—again, about 5 to 10 degrees—when your foot is on the pedal and your leg is fully extended. (*Note:* Your knees should never lock when you're cycling.)

Suit yourself. Once you have the bike, you don't need a lot of other stuff. Just remember a helmet. "If you get no other piece of equipment, get this," says Iona Passik, a certified personal trainer, certified master Spinning trainer, and certified group fitness instructor at Chelsea Piers, a fitness center in New York City.

Another essential accessory when you ride is a water bottle. "When you're cycling, you lose a lot of fluid," Dr. Burke explains. "You can never drink too much water."

TRIAL RUN

You already know how to ride a bike. Learning cost you lots of skinned knees when you were a kid. Here are some ideas for starting the adult phase of your cycling career.

Warm up before you burn rubber. A warmup may seem tedious when you're itching to go, but do it anyway. Warmed-up muscles are less injury-prone than cold ones, says Passik. To warm yours sufficiently, spend 5 to 10 minutes cycling nice and slowly over relatively flat ground, she suggests.

Extend yourself—slowly. Don't try to become a Tour de France contender overnight, especially if you haven't been on a bike in decades. Build up distance and speed slowly. On Day 1, your 5-minute warmup should be your whole bike workout, Dr. Burke says. Each day, add another minute or two. Your goal should be three 30-minute sessions a week, he says.

Sweat and talk at the same time. To get all the health benefits that cycling has to offer, you should work up a sweat while pedaling. But you shouldn't pedal yourself to the brink of exhaustion. "If you can't talk, you're working too hard," Dr. Burke says.

Adventures in Indoor Cycling

When the weather gets surly, an outdoor cycling adventure can turn into a misadventure, replete with slips and falls and cold-water drenchings. That's no reason, though, to forgo your usual exploratory jaunt. Just hop on a stationary bike. If you don't have one in your attic, your local YMCA is bound to have a half-dozen.

If that doesn't sound like much of an adventure, though, consider Spinning, suggests Iona Passik, a certified personal trainer and certified group fitness instructor at Chelsea Piers, a fitness center in New York City, who is also certified to train instructors in Spinning.

Spinning is a workout designed for a special stationary bike—a Schwinn "Johnny G Spinner" by name. The bike has a 40-pound flywheel where you'd expect to find a front tire.

In the course of a Spinning class, an instructor takes the group on what Passik calls a Spinning journey. The journey is set to inspiring music, which can be anything from New Age to opera. The instructor guides you through different types of imaginary terrain—up hills and down. The pedaling is linked with breathing and relaxation exercises, such as closing your eyes and breathing deeply as you imagine yourself cruising through a mountain pass, for example.

Passik explains that although the workout can be tough, you can take Spinning classes with people of all fitness levels because you adjust the resistance knob on your bike to create your own level of workout intensity. "An elite athlete can Spin next to my mother," she says. Classes are typically 45 to 90 minutes long.

Roughly 3,000 fitness centers in the United States now offer Spinning classes. To find classes where you live, write to Mad Dogg Athletics, 1119 Colorado Avenue, Suite 23, Santa Monica, CA 90401. Include your telephone number, and someone from the company will contact you to let you know which fitness centers in your area offer Spinning.

Add resistance. Been cycling for 30 minutes without even a damp brow? Then add some resistance, suggests Passik. Start pedaling up steeper inclines, or raise the resistance level if you're riding a stationary bike. One caveat: If your knees start to hurt, you're probably going up hills that are too precipitous (either that, or your saddle is too high or too low). Scale back to more gradual slopes.

Other areas of strain to watch for include your lower back and your shoulders and neck. If your back starts aching while you ride, you may have the resistance set too high, Dr. Burke says. Sore shoulders and a stiff neck are indications that you aren't positioned properly: You're probably leaning too far forward and putting too much weight on your arms. Dr. Burke explains, however, that a little discomfort is normal until you've gotten used to your cycling muscles.

Get rubbery. Stretching after a workout keeps you flexible, notes Passik. Don't skip it. (For a complete routine, see "Stretching" on page 380.)

BIOFEEDBACK

Your Best Body Language

WHAT IT IS

Every day, unspoken signals tell you exactly where you stand with others. That glare from your teenager when you dare to tie up the phone line. The wag from your cocker spaniel when you walk in the front door. The head-shaking from your boss when the report is late. The thumbs-up sign from your best friend when you run your first mile.

People spend so much time worrying about or anticipating the actions of others that they often ignore feedback from their closest, most trusted allies—their bodies.

"A lot of us have lost touch with our bodies," says Michael McKee, Ph.D., a psychologist and vice chairman of psychiatry and psychology at the Cleveland Clinic Foundation in Ohio. Biofeedback empowers you by giving you control over your body to some extent. Using instruments that monitor skin temperature, blood pressure, heart rate, and muscle tension, you can learn to relax your muscles and blood vessels to regulate your body in a healthy way with the help of breathing exercises, guided imagery, and relaxation methods, says Dr. McKee.

Biofeedback is safe and noninvasive, but body self-control takes some time to learn.

Practicing it daily delivers the most effective results, experts add.

On average, it takes up to eight sessions with a trained expert operating a biofeedback machine to learn how to accurately read and control your body's physiological responses to stress, pain, and other intruders. Once you know your body cues, you can use deep diaphragmatic breathing or guided imagery to put yourself into a relaxed, healthy state without the help of a machine or a professional.

Biofeedback instruments can monitor your brain waves, blood pressure, heart rate, skin temperature, and muscle tension through electrodes or temperature-sensitive probes attached to your skin. Depending on whether the instrument operates with sounds or lights, you can "hear" or "see" when you become relaxed or stressed. Relaxation warms your fingers and releases tension in your muscles.

"The more relaxed you are, the more your arteries dilate and bring more blood to your hands and feet, and your body temperature rises," explains JoAnne Herman, R.N., Ph.D., a biofeedback expert and associate professor of nursing at the University of South Carolina in Columbia.

Some instruments signal a rise in temperature by increasing the pace or volume of beeps. Others emit light signals that become brighter the more relaxed you become. In some cases, a person's finger temperature can rise by as much as 20 degrees once they shed a highly stressful state and plunge into total relaxation and control.

"Biofeedback teaches you how to relax," says Dennis Turk, Ph.D., professor of anesthesiology and pain research at the University of Washington in Seattle. "If you are more relaxed, you will sleep better and deal with stress better. You learn a sense of control over your body so you don't feel helpless."

The appeal of biofeedback is that it offers immediate information about tension in your body and produces concrete results. This differs from other relaxation methods such as yoga or meditation, where objective indications of relaxation are not present.

"I've been doing relaxation training for 15 years, and I've found that some Americans who love instruments feel more comfortable with biofeedback than with, say, yoga because they can get readouts immediately to see how they're doing," says Dr. Herman.

WHAT IT DOES

When you become captain of your body, you can steer yourself on a healthy course through life. That's the goal behind learning biofeedback.

This technique, when practiced daily, has been shown in studies to help people fight stress, control Raynaud's disease, regulate blood pressure, reduce muscle tension, erase insomnia, ease tension headaches and migraines, sharpen memory, eliminate facial pain,

and even stop smoking. So you can depend on biofeedback-assisted relaxation the next time a stressful situation surfaces—like receiving an unexpected letter from the IRS or getting stuck for 20 minutes at a railroad crossing on the way to your office.

"Sometimes, your body mobilizes for what it perceives to be World War III when most of the time, it is not a life-threatening situation but just a daily deadline that you're facing," says Dr. McKee. "No one can escape stress, but biofeedback helps you deal with it better." When you consciously relax, your breathing slows, your muscles loosen, and your blood vessels widen to allow more oxygen throughout your entire body, Dr. McKee says.

In addition, biofeedback gets high marks for helping people with the "cold hands–cold feet" condition known as Raynaud's disease. "For Raynaud's disease, biofeedback is the treatment of choice," says Dr. Herman. "We use thermobiofeedback to warm the hands."

Other studies have documented success using biofeedback to combat writer's cramp and even constipation. Some experts also use biofeedback to re-educate the sphincter muscles to work better if you're incontinent.

REAPING THE BENEFITS

To get the most out of biofeedback, doctors offer the following tips for first-timers looking for a way to relax and fight stress and aches.

Scout for a qualified expert. Find a medical doctor, psychologist, psychotherapist, or nurse who is certified in biofeedback training, says Dr. Turk. To find a biofeedback professional near you, send a self-addressed, stamped envelope to the Association for Applied Psychophysiology and Biofeedback, 10200 West 44th Avenue, Suite 304, Wheat Ridge, CO 80033. Request the free information packet and the list of contacts in your state.

Practice patience. Babe Ruth wasn't nick-named the Sultan of Swat in his first professional season. Mary Baker Eddy was on this earth nearly 90 years before she founded the *Christian Science Monitor*. Learning biofeedback also takes some time, says Dr. McKee, but most people can do it.

"Essentially, you're teaching your body a new language, and it takes about six months on average to internalize it as a new language," he says. "If you're stuck behind a slow driver and you're late to give a lecture, biofeedback can help you practice a new response. Instead of getting frustrated, use the time in traffic to think of a joke to offset your tardiness."

Invest in yourself. Biofeedback training varies in price from $60 to $125 per session, and most people need about eight sessions to master the technique, says Dr. Herman.

Think about how well you are spending that money, she suggests. "When you think about getting into a daily habit of performing a relax-ation technique that works, think of what it will save you in terms of cardiac medicines, high blood pressure medicines, pain, and suf-fering. Biofeedback can be a bargain," she says.

Convert threats into challenges. Biofeed-back is a new bodyspeak that welcomes chal-lenges, says Dr. McKee. "Think of the world as full of challenges and opportunities instead of threats, and you can channel your physiolog-ical response into a healthy response," he says.

Don't discount mood rings. Remember the mood rings of the 1970s? In a primitive way, those finger fads worked like portable biofeedback aids, say doctors. "Mood rings recorded skin temperature," says Dr. Turk. "A blue tone meant you were relaxed."

Dr. Herman doesn't hand out mood rings to her clients, but she does give them Biodots—heat-sensitive dots the size of a thumbtack. Her patients place them on the fleshy parts of their hands between their thumbs and forefingers as biofeedback "homework." The dots change color depending on the skin temperature of the wearer. The more stressed the person is, the blacker the dot becomes.

Biodot creator Bob Grabhorn, an Indi-anapolis inventor, relies on these dots. "I learned a long time ago to do deep, relaxing breathing and get my blood flowing," he says. "By practicing relaxation, I've learned how to keep the dot from turning black."

TRIAL RUN

The use of mental imagery is one effective way to harness the benefits of biofeedback in dealing with life's stresses, says Dr. Turk.

"Biofeedback teaches people how to control their imaginations," he says. "For some people, a relaxing place would be a beach in Hawaii. But for others, it may be skiing down a moun-tain. It is important for the practitioner to work with each client on an individual basis."

Let's say that you've just walked in your front door from a 10-hour workday to a house rattled by your twins arguing over ownership of a blue wool sweater, your beagle chewing your favorite black socks, and a ringing tele-phone. Feel the stress soaring?

Now let's fight this stress by selecting an image that is most pleasing and relaxing for you. Here's a step-by-step guide to raising your body temperature and slowing your heart rate, as offered by Dr. Turk.

►Wear loose-fitting clothes.

►Sit in a comfortable, relaxed position in a quiet part of the house. A cushy recliner is better than a straight-backed kitchen chair.

►Inhale slowly and deeply through your nose.

►Slowly count to five before you slowly exhale. Or, before you exhale, try to see a single word such as "c-a-l-m" or "p-e-a-c-e" to help free your mind of distracting or stressful thoughts. Seeing the word appear in your mind's eye, letter by letter, will take about five seconds. Then exhale.

►While exhaling, try to let your chest and stomach muscles relax and your shoulders drop.

►Repeat this inhale-exhale cycle for about five minutes.

"What this breathing exercise does is help make you aware of your body and slowly release tension from every major body part," says Dr. Turk.

BIRTHDAY BLUES

Don't Let Those Milestones Weigh You Down

Off-key and cheery, your family and friends belt out "Happy Birthday" lyrics and present a cake with enough candles to light a small city. You should feel as jubilant as they do. Instead, you dwell on past mistakes and dread the future.

Birthday blues can hit at any age. In a country that worships youth, each passing year after age 25 is an indictment. It's just another step toward the day when your television-viewing habits are of absolutely no interest to network executives. In our youth-centric culture, the so-called milestone birthdays—the big 4-0 and the even bigger 5-0—can be perfect excuses for brooding.

But that's no fun! It's your birthday! The trick to dodging despondence is to get some new sights, sounds, scents, and stimuli into your life on or around your birthday.

"I encourage people to use occasions such as birthdays not to mourn the time lost but to look forward to the time ahead of them," says C. R. Snyder, Ph.D., professor and director of the graduate training program in clinical psychology at the University of Kansas in Lawrence.

The Sky's the Limit

You've got to throw open the doors and windows and let in some fresh air. You may just

> ### Rejoice in Numbers
>
> You never have to feel alone on your birthday. You have plenty of company. According to research by Hallmark Cards, 675,000 people in the United States, on average, celebrate their birthdays each day. And here's a news flash for baby boomers: 11,000 of you turn 50 each day. That's a trend that's expected to continue for the next 18 years, according to Hallmark officials.

need an attitude adjustment. Try these exercise suggestions for new highs—or lows.

Take a flying leap. According to Caleb Esmiol, owner of Air Adventures in Clewiston, Florida, milestone birthdays often prompt people to try skydiving for the first time. "When you pull your own ripcord and land safely, it gives you a whole new confidence level and puts things into perspective. It's a birthday you'll never forget," he says.

Is skydiving more adventurous than you'd like to get? Well, then . . .

Climb every mountain. Or one, anyway.

And there's no alpine mountaineering equipment required. Your trek can be a gentle hike to the top of your local highlands. Not only will you burn a few calories but a peek from the peak may lift your spirits, says Bruce Tuckman, Ph.D., educational research professor at Florida State University in Tallahassee.

And if you're not much for heights . . .

Go chase a kite. Kite flying is just the kind of novel activity to help people feel young at heart. "For someone reaching a landmark birthday, kite flying makes you feel like a kid again," says Lisa Heath Fraser, manager of Once upon a Breeze, a kite shop in Cannon Beach, Oregon.

Although there are more than 100 styles of kites, the two main types are single-line and stunt, says Fraser. "With single-line kites, you let the string out, watch it soar, and feel relaxed. Stunt kites have dual hand controls, giving you a definite aerobic workout. You concentrate so hard to get the kite up that hours go by and you don't even realize it," she explains.

Flying stunt kites offers an aerobic workout for both the upper and lower extremities, says Dan Hamner, M.D., a physiatrist and sports medicine specialist in New York City. "You do a lot of bending and dipping and use lateral movements to catch the wind," he says.

If you're not inclined to seek new highs, how about some new depths?

Take up scuba diving. Remember Lloyd Bridges in *Sea Hunt*? Well, there's no reason that you shouldn't try to salvage that old pirate ship or get a close-up look at brain coral. New sights, sounds, and sharks from the sea may give you the jolt you need, says Dr. Tuckman.

"Part of the reason that people get depressed

Vintage Victors

You are never too old to achieve. Need some motivation? Take a look at some of these late-in-life champions.

▶ Harlan Sanders, the man known as the Colonel, founded Kentucky Fried Chicken in 1956 at the age of 66—no spring chicken.

▶ Mary Baker Eddy founded the *Christian Science Monitor* at age 87.

▶ Kathrine Robinson Everett practiced law in North Carolina at age 96.

▶ Anna Mary Moses, a/k/a Grandma Moses, didn't start painting until she was in her seventies. Now her work is in museums.

▶ George Foreman became the oldest boxing champ in history at 46 when he knocked out Michael Moorer in 1994.

---

on their birthdays is that it feels like any other day of the week," he says. "Make this a special day. Try something new."

Start spelunking. Exploring caves, that is, suggests Dr. Tuckman. Caverns are wonderful, exotic places—unless, of course, you're claustrophobic or you've got a thing about bats. In which case, try a different way to get down.

Change the music. Exercise your ears with some fresh sounds. If you've been a folkie since Joni Mitchell was a pup, try some jazz CDs. New input often helps your outlook, says Dr. Tuckman. You may learn something about the joy of sax.

BOREDOM

Boost Your Ego and Beat the Blahs

Alan Caruba, a writer from Maplewood, New Jersey, founded The Boring Institute in 1984 as a media spoof. The institute has since evolved into a clearinghouse for information about boredom. It also spoofs celebrities, films, and popular culture. Caruba's research has made him an expert on tedium and given him material to write the booklet "Beating Boredom."

"I get letters from people all over the world who tell me they're bored," Caruba says. Unfortunately, the run-off-to-Tahiti fantasy is, at best, a short-term boredom cure. Even Tahitians get the blues, reports Caruba. "Boredom is a universal experience."

The big misconception about boredom is that we get bored with routine. But, according to Neil Fiore, Ph.D., a psychologist in Berkeley, California, and author of *The Road Back to Health*, that's not exactly right. "Boredom arises when you don't have a sense of ego gratification, when you don't feel effective," he says. Consequently, one of the best antidotes to monotony is challenge, and the sense of engagement and accomplishment that comes with it.

"Because it offers limitless opportunity for challenge and accomplishment, exercise is a great boredom breaker. It's a handy way to work yourself out of a rut," says Dr. Fiore.

In Search of a Challenge

Here are a few suggestions on using exercise to fight the funk.

Be your own coach. To shake off inertia, take on a leadership and coaching role with yourself, suggests Dr. Fiore. "Tell yourself, 'I'm choosing to overcome this inertia. I'm choosing to exercise so that I can feel better,'" he says. A pep talk can get you through the hardest part—getting started. After five minutes or so of jogging, dancing, or aerobics, your body is warmed up. You will have broken through inertia and will have some momentum to help carry you forward, he says.

Get strange to stay motivated. Consider an unusual exercise avenue. There's more to the exercise world than walking, jogging, and weight lifting. Karate? Aero-boxing? Fencing? An exotic undertaking adds freshness and will help you stick with it.

Pick up a book. Whatever exercise you choose, buy a book or subscribe to a magazine about the discipline. Lots of exercise pastimes are intriguing little cultures unto themselves. To get drawn into the world of scuba diving or cross-country skiing, you need a glimpse behind the scenes that will give you a sense of the sport's particular challenges and benefits.

Unending Boredom: A Cautionary, Contrarian Tale

Chronic boredom isn't simply tedious, it's dangerous, says Sam Keen, Ph.D., a philosopher and author of several books, including *To Love and Be Loved* and *Inward Bound*, which deals with boredom and depression.

Dr. Keen believes that if you don't deal with garden-variety boredom, it can grow into depression. "If you keep ignoring your boredom, that's a sure recipe for getting seriously depressed," he says. And he has an unconventional view of the right medicine for monotony.

Contrary to popular belief, Dr. Keen believes that the last thing you should do if you're chronically bored is try to keep busy and distract yourself from your boredom. If you do that, you'll never get to the heart of the matter and figure out why you're bored. "Instead, stop and do nothing—except think and feel," he says. Here's what he suggests.

First, meditate. "Meditation means stopping and thinking about your life, paying attention to your thoughts and feelings," says Dr. Keen. Every day, go someplace where there are no distractions, sit quietly, and pay attention to the thoughts and feelings that crop up. "It's time to ask yourself, 'If what I'm doing is not engaging me, if it's boring me, what is it that I really want to be doing?'" Dr. Keen says.

Look for clues. When you're feeling particularly bored or, on the other hand, when you find some momentary relief from boredom, take note. "When you're excited, that's a clue that you're doing something you want to be doing," says Dr. Keen. And when you're really bored? That's a clue that you're heading down the wrong road.

Daydream. Allow yourself time to daydream and fantasize. If you could do anything, what would it be? Where would you live? With whom? Let your mind wander. "Your dreams will surface," says Dr. Keen.

Play matchmaker. If your spouse isn't interested in taking synchronized swimming lessons, fix yourself up with a workout buddy. If you make a standing date to exercise together, you'll find that it's easier to start (and keep) moving. "A commitment to another person provides powerful motivation," Dr. Fiore says.

Remember, you're an acorn. If you're taking up an exercise regimen for the first time, start with small challenges. If you set unreasonable goals, you'll get discouraged and never grow into the oak you can be.

"Map out a series of small, achievable goals that will continue to provide a sense of accomplishment and reward," suggests Dr. Fiore.

Have a goal line. It's important that you move the ball down the field, so choose a target. Make competing in the hit-and-giggle tennis tournament or taking part in a fund-raising run your goal. Don't be afraid to take baby steps toward it.

Work out with Ben Hur. The next time you head out for a boredom-dispelling walk through the neighborhood moors, take Jane Eyre along.

You can pick up a tape-recorded version of the Gothic classic—and hundreds of other titles—at libraries and bookstores.

"While you exercise your body, you can exercise your mind by listening to recorded books on a portable cassette player," suggests Caruba. "Think of it as a way to master two challenges—one physical, the other mental—at the same time.

When shopping for a book on tape, look for something that intrigues and excites you. Don't pick something up simply because you think you *should* read it.

Revel in reinforcement. When you reach a goal—say, doing the box step without having to stare at your feet—enjoy your victory. Savor a sense of accomplishment. Here are a few ways to give yourself a gold star.

►Chart your course. Dr. Fiore suggests making a visual record of your achievement. Every day that you exercise for at least 15 minutes—whether you boogaloo or jump rope—record it on a chart. "It's tangible," says Dr. Fiore. "You can look at it. It's evidence that you did what you set out to do, and it will show your improvement from month to month."

►Be like Miss Crabtree. Remember your beloved fourth-grade teacher and those sticky stars she used to put on your homework back at Riverside Elementary? Go to the store and buy a box of them. Slap one down on your chart with pride each time you exercise. Go ahead. You deserve it. You'll be surprised by how good it feels.

►Make your computer enthusiastic. Your exercise log can be electronic. Enter your jogging miles into your computer every day. If you can program it to play the theme from *Rocky* every time you pass another 10-mile mark, you'll get a musical pat on the back each time you do.

BREAST SAGGING

Gaining Support

For centuries, women have been trying to fight time and gravity with the brassiere. But there's a much better weapon, and it's close to their hearts. The best defense against a downward trend is strengthening your pectoral muscles.

You can't actually build up the breasts themselves. They're composed of fat and other soft tissue, which doesn't respond to exercise. But you can work the muscles that support your breasts—your pecs, the muscular mats that cover the upper chest and the ribs. Firmer, more substantial pecs will bring your bust line both up and out.

Pump Up Your Pecs

Here are three good exercises that work your pecs.

Churn for your chest. Swimming's great for boosting your bust, says Dan Hamner, M.D., a physiatrist and sports medicine specialist in New York City. The crawl, the butterfly, and— what else?—the breast stroke all exercise the important supportive pectoral muscles.

Try pushups. The pushup is the classic pec improver. If you can do the old-fashioned Marine Corps–style pushup—lying on your stomach and using your arms to push your body straight up—go to it. But they're very hard to do. You might want to try this easier, modified wall pushup suggested by Robyn Stuhr, exercise physiologist at the Women's Sports Medicine Center at the Hospital for Special Surgery in New York City. (See page 336 for an illustration of this exercise.)

If you do these three or four times a week for two months, you should start to notice results.

▶Stand facing a wall with your toes touching the wall, then take one or two steps back. Place your palms flat against the wall at chest height and shoulder-width apart.

▶Bend your elbows toward the outside to move your body toward the wall as one unit. Be careful not to bend at the waist.

▶Then straighten your arms by pushing back off the wall, feeling it work in your chest muscles. Start with one set of 10 to 15 pushups. Gradually, over the course of two months, try to build up to three sets.

▶As you get stronger, stand a little farther back from the wall. This will make the wall pushups a little more difficult.

If you should find yourself looking for a bigger challenge, give this tougher version a try.

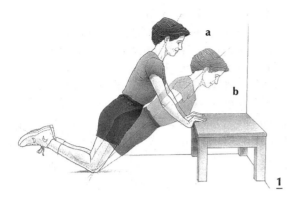

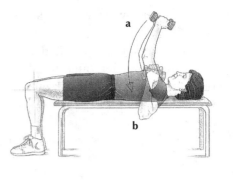

▶Kneel a few feet away from a sturdy coffee table (it's a good idea to move the table against a wall so it won't move) and place your hands on the edge, shoulder-width apart. Bend your knees and raise your feet behind you (*1a*).

▶Hold your body in a straight line from your knees to your neck as you bend your elbows outward and lower yourself toward the table (*1b*).

▶Hold there for a beat, then slowly push yourself back up to the starting position.

Lift a little weight. Weight lifting is another way to speed development of firmer pectoral muscles. Try this basic weight workout recommended by Dr. Hamner.

Ideally, you should do this fly exercise on a weight bench, but if you don't have a bench, a sturdy coffee table padded with a towel or blanket makes a handy stand-in. You can also do this exercise while lying on the floor, but it won't be quite as effective because your arms won't have the optimum range of motion.

▶Lie on your back with both feet flat on the floor.

▶Grasp a five-pound dumbbell in each hand and raise your arms above your chest, with your palms facing each other. Bend your elbows

slightly and press your lower back against the bench (*2a*).

▶Slowly lower the dumbbells to either side, keeping your elbows flexed and your wrists strong.

▶When the weights reach about chest level, hold them there for a beat or two (*2b*). Then slowly raise them to the starting position. Start with one set of 8 to 10 repetitions three times a week and build up to three sets as your strength increases.

Home and Garden Workouts

Here are three common home maintenance tasks that—with some slight symmetry modifications—can actually improve your pecs, which can help keep your breasts out in front, says Dr. Hamner.

Vacuum voluptuously. The back-and-forth arm stretch of vacuuming actually pumps your pectorals a bit. For maximum benefit, extend your normal reach as you clean the carpet. And here's the modification—use both arms equally. You want the pecs on both sides of your chest to get the same workout.

Rake leaves with a lift. The annual autumnal burden can actually be a bust-maintaining activity. The motion of extending the rake and pulling it toward you works the vital chest muscles.

Here, too, extend your reach, and keep symmetry in mind. It may be tough to rake leaves with your dominant hand placed lower on the shaft of the rake, but give it a try. Even if you rake a little more slowly, so be it. Remember, you're not just raking leaves, you're building up your body, too.

Get good chest panes. Washing windows may be the single worst task of home maintenance. Maybe this thought will keep you going from pane to pane: The movement required to wash windows is a pectoral plus. For figure enhancement, extend your arms farther than you normally might. Remember, you're not just letting the sun shine in, you're also maintaining your figure.

BREATHING DISORDERS

Get Your Fair Share of Air

One blustery day in 1997, Susie Maroney dove into the churning surf off Havana, Cuba, and started swimming north. She didn't stop until she reached Key West, Florida, 112 white-capped, shark-infested miles later. The swim earned the 22-year-old Australian a spot in the record books as the first swimmer to cross the Florida Straits.

Not bad for someone with asthma.

Of course, if you have any of the three most common breathing disorders—asthma, chronic bronchitis, or emphysema—exercise may be the last thing that you feel like doing. Hey, it's tough to dance till dawn if you can't draw a breath. In fact, if you have asthma, exercise may be one of the things that triggers your attacks. But exercise is now considered a key part of treatment for many breathing disorders, says Linda Ford, M.D., president of the American Lung Association, a practicing allergist in Omaha, Nebraska, and chair of the National Allergy Bureau. Although working out won't cure breathing disorders, it can help you manage them and improve your life. "If you exercise regularly, you'll feel breathless less often," Dr. Ford promises. Here's why.

Although exercise increases your body's need for oxygen, over time, it also makes your heart a more forceful oxygen pump and makes your chest muscles more efficient and stronger for breathing. Toned muscles require less oxygen to do their jobs than flabby ones do. A more powerful heart and more efficient muscles mean that you'll feel breathless less often, whether you're doing the crawl from Cuba to Cocoa Beach, walking to the Winn Dixie, or just climbing the stairs, explains Dr. Ford.

Caution: If you have a breathing disorder, *always* consult your doctor before starting an exercise program or trying the exercises below. There are risks involved. The advice that he'll give you will vary, depending on the severity and type of lung disorder you have.

Once you've gotten clearance from the doctor, here are some important training tips to keep in mind.

Become an Environmentalist: Crave Clean Air

For people with breathing disorders, air pollutants are public enemy number one. They can irritate your airways and lungs and aggravate breathing problems, says Dr. Ford. So always heed the following advice when you exercise.

Steer clear of traffic. Tailpipes are spewers

of carbon monoxide. Exercise far from the madding crowds, away from well-traveled roads.

Avoid the afternoon. Ozone levels are highest in midafternoon. If you're going to exercise outside, it's best to do it in the late morning.

Be a pollen dodger. Lots of people with asthma are allergic to the male spores of seed plants—that is, pollen. Keep tabs on the pollen count, says Sally Wenzel, M.D. associate professor of medicine at the University of Colorado in Denver. You can get information about a daily tally by checking with your local chapter of the American Lung Association.

Most plants release their downy barrages in early morning, says Dr. Ford, so exercising later in the day is a better option. When the pollen is

---

Breath Busters: The Big Three

Although asthma, chronic bronchitis, and emphysema are all conditions that affect your airways, their causes and symptoms are different. Here are the basics of the big three.

Asthma. Sudden but reversible attacks of breathlessness, wheezing, and coughing are hallmarks of asthma. With this disorder, the problem lies deep in your bronchi and bronchioles, the delicate airways leading to your lungs. For many reasons, including genetic factors, the linings of these airways are chronically inflamed. Unfortunately, this inflammation makes the airways hypersensitive to all sorts of attack-inducing triggers—irritants that further inflame and constrict the airways, limiting air flow.

Cold and flu viruses and allergens such as dust mite and cockroach feces, animal hair, pollens, molds, smoke, and air pollution can all trigger attacks. So can the rapid drying and cooling of the airways that accompanies certain types of exercise.

If you have asthma, your doctor will probably prescribe two types of medications, an anti-inflammatory drug to control the underlying inflammation and a bronchodilator to open your bronchioles if needed.

Chronic bronchitis. If you're short of breath and coughing up mucus, chronic bronchitis may be the problem. With this condition, the larger bronchi become inflamed and narrow. They also excrete excess mucus, which further blocks air flow. Almost exclusively a result of smoking, chronic bronchitis can lead to emphysema if it goes untreated.

Emphysema. A life-threatening disorder, emphysema can cause near-constant shortness of breath, coughing, and wheezing. With emphysema, the tiny air sacs deep in your lungs actually start to break down. This is bad news because these air sacs transfer oxygen to your bloodstream, and the breakdown leaves big air spaces where there once were millions of sacs. Smoking is usually the culprit.

Other health problems, such as anemia, hormone imbalances, and heart disease, can also cause shortness of breath during exercise. See your doctor for a diagnosis if you're feeling winded.

Caution!

Neither emphysema nor chronic bronchitis nor asthma should keep you from your appointed rounds at the gym or the walking trail or pool. Just remember to follow your doctor's advice. And keep two caveats in mind.

Be on the lookout for signs of hypoxemia. This a dangerous drop in oxygen levels in your blood. Warning signs are extreme shortness of breath, dizziness, rapid heartbeat, fatigue, and lips and fingers that turn an unflattering shade of blue. If you have any of these symptoms, stop whatever you're doing. If your doctor has prescribed medication for times like these, take it. See your doctor if your symptoms don't improve within minutes, says Stanley Wolf, M.D., an allergy specialist in Silver Spring, Maryland, and head of the Quality of Care for Asthma Committee of the American Academy of Allergy, Asthma, and Immunology.

Stop for an attack. Don't try to "run through" or "play through" an asthma attack, warns Norman H. Edelman, M.D., dean of the School of Medicine at the State University of New York at Stony Brook and consultant for medical affairs for the American Lung Association. Stop, take whatever medication your doctor has prescribed for such incidents, and take it easy until you can breathe easily. If that doesn't help, call your doctor. If you have a severe attack— one that leaves you so breathless that you have trouble talking and your lips and fingernails start turning blue or gray—you may become hypoxic. Take your medication and see your doctor or head for the emergency room right away.

flying high, exercise inside. Go mall walking, head for the Y, or do some stair-climbing.

Take It Slow and Easy

It's important that you don't take your lungs by surprise and overtax them suddenly. Approach exercise with the word *gradually* in your mind.

Think tortoise, not hare. Start modestly. In the beginning, you may not be able to exercise for more than 5 or 10 minutes before you start feeling out of breath. Don't give up. It'll get easier over time. "After a while, 10 minutes won't leave you short of breath anymore, so you can go longer and longer," says Dr. Ford. Add 5 or 10 minutes to your workout every week until you're exercising for at least 30 minutes at a time three times a week.

Walk, don't run; dance, don't rebound. Choose a reasonably gentle exercise to start with. Walking is perfect. Want a little more action? Try country line dancing, suggests Barbara Alexander, P.T., a physical therapist at the National Jewish Center of Immunology and Respiratory Medicine in Denver. Biking can be terrific, too.

If you have chronic bronchitis or emphysema, steer clear of high-intensity sports like basketball and rope-jumping, which require barrels of oxygen. If you have asthma, however, you should be able to work your way up to just about any kind of exercise, says Allan Weinstein, M.D., an allergist in Washington, D. C., and author of *Asthma: The Complete Guide to Self-Management*. Just gradually lengthen the duration of your workouts.

Ease on down the road. Think "nice and easy" when you're starting a workout. According to Dr. Weinstein, a 10-minute warmup session gives your airways and lungs a chance to adapt to the coming changes in exertion level. "When you warm up beforehand, your body releases various chemicals that help prevent an asthma attack," he adds. You know that you've warmed up enough when you just break a sweat, says Dr. Wenzel.

Ease out of your workout, too. Don't go instantly from a brisk walk to the recliner in front of the tube. Let your lungs adjust slowly. A gradual cooldown can help prevent a postexercise asthma attack, says Stuart Stoloff, M.D., an asthma specialist and clinical associate professor of family and community medicine at the University of Nevada School of Medicine in Reno.

Maximize Your Breathing Muscles

If you have chronic bronchitis or emphysema, exercises that strengthen the respiratory muscles in your chest and abdomen may help you breathe easier, says Norman H. Edelman, M.D., dean of the School of Medicine at the State University of New York at Stony Brook and consultant for medical affairs for the American Lung Association. Here's what experts recommend.

Crunch your tummy. Abdominal crunches can strengthen the muscles that control exhalation, says Cindy Raymond, an exercise physiologist at the National Jewish Center in Denver.

Picking the Proper Place

If you have a breathing disorder, be careful where you put your home gym.

Beware your basement. If you're allergic to mold (and lots of people with asthma are), a damp cellar is a bad bet for the new treadmill, especially if there's carpet on the concrete floor.

Watch out for carpet creatures. If you're allergic to dust mites (another common trigger for asthma attacks), don't work out in rooms with carpets or that Victorian sofa you inherited from Aunt Blanche. Rugs and plush fabric are favorite hangouts for these microscopic spiderlike pests. They also love humidity.

If you have a lung disorder like emphysema, crunching regularly will help you exhale more deeply. (For tips on how to do crunches correctly, see "Resistance Training" on page 331.) Your ultimate goal: strength workouts twice a week, says Raymond.

Pump your chest. Exercises with a special device to train respiratory muscles, called a Threshold IMT, can strengthen the chest muscles that you use to inhale, says Raymond. The device, intended for people with lung disorders like emphysema, looks a bit like an oversize pipe—without the tobacco, of course. To use it, you inhale through the mouthpiece. An adjustable valve in the "bowl" supplies resistance that your chest muscles work against each time you suck air. The higher the resistance, the harder your muscles have to work, and the stronger they get.

To work out with this device, you'll need a prescription and help from your doctor or therapist.

Exercise-Induced Asthma: The Blessing of Warm and Wet

If you have asthma attacks when you exercise, odds are that you've been diagnosed with exercise-induced asthma. The truth is, it's not exercise itself that causes asthma attacks. It's the cold, dry air that you suck down your airways when you exert yourself, explains Dr. Edelman. When you're breathing at a normal pace, your nose warms and humidifies the air you inhale, but when you're exercising, you breathe harder and faster than usual, so your nose can't keep up with warming and humidifying. Cool, dry air irritates your airways and can provoke an attack. Here's how to keep your airways moist and placid.

Do a swan dive. Swimming, water aerobics, and water polo are all good bets. The water humidifies the air, so your risk of an attack drops considerably, says Dr. Edelman.

Choose stop-and-go sports. Tennis and other stop-and-go games such as softball, football, badminton, and golf are smart picks if you have asthma attacks when you exercise, says Dr. Weinstein. Because they all allow you to take breaks between bouts of exertion, stop-and-go sports afford your airways some valuable protection. Remember: It's the cooling and drying of your airways during exertion that trigger this type of asthma attack. Take periodic breaks, and your airways won't get as dry, irritated, and ready to revolt. By contrast, duration sports like long-distance running and cross-country skiing aren't so good, Dr. Weinstein says.

Stroll in the surf. The humid shore air makes the beach a good place to walk, run, or toss a sea-soaked football around. If you're in Kansas, look around for a lake, a pond, or a river and do your exercising nearby, suggests Dr. Weinstein.

Sprinklercize. No beach, lake, or pond nearby? On arid days, try exercising on the lawn with the sprinkler going, says Dr. Weinstein. The sprinkling will add a bit of humidity and make life somewhat easier for your airways.

Give your nose a workout. Your nose does a far better job of warming and humidifying the air en route to your airways than your mouth does, so close your mouth and leave the breathing to your proboscis, says Dr. Ford.

Take a fashion tip from Wild West outlaws. If you're exercising outside during winter, when the air is particularly dry and cold, try covering your face from the nose down. There's no need to pull your turtleneck up to your eyelashes (unless you really like that look). Just wrap a nice, cozy scarf around your face. This way, you can breathe the slightly warmer and more humid air that gets trapped between your face and the fabric, says Dr. Ford.

Pump saline. When exercising in an arid locale, spritz your hardworking nose with a saltwater nasal spray. It will help keep your airways moist and contented, Dr. Weinstein says. You can find saline nasal spray in most drugstores.

Humidify your home gym. Use a humidifier in the room where you exercise, and you'll give yourself another hedge against attacks, says Dr. Weinstein. Just remember: An unkempt humidifier will quickly become a breeding ground for allergy- and attack-inducing mold. So wash your humidifier daily with a 1:20 solution of bleach and water, or follow the manufacturer's recommendations.

Don't Be Shallow: Breathe Deep

Most of us are shallow, at least when it comes to breathing. But you'll need to master some deep-breathing techniques if you have asthma, chronic bronchitis, or emphysema. Like any kind of exercise, deep breathing takes practice. Here's how to drink deep of nourishing air.

Develop your diaphragm. Diaphragmatic breathing exercises train you to breathe deeply and more effectively by using your diaphragm, the dome-shaped muscle below your rib cage, says Dr. Edelman. When you breathe with your diaphragm, your lower abdomen rises with each inhalation and sinks with each exhalation.

While practicing diaphragmatic breathing, imagine that your body is a giant, upside-down eyedropper, suggests Nancy Zi, the inventor of the chi yi system of breathing exercises and the author and producer of the book and videotape, *The Art of Breathing* (available from the Vivi Company, P.O. Box 750, Glendale, CA 91209-0750). Imagine that your mouth and nose are the opening of the eyedropper and your stomach is the bulb. With your hands on your stomach, breathe in deeply, imagining air filling the bulb. Your stomach should rise as you do this. Then exhale, tightening your abdominal muscles as though you're squeezing the eyedropper bulb.

Try visualization exercises. In a study at Mount Sinai Medical Center in New York City, folks with asthma reported relief from symptoms when they used mental imagery. When breathing got difficult, they used controlled deep breathing and envisioned a variety of breath-easing images. Some imagined new sets of lungs in their chests. Others saw themselves in the woods, inhaling the refreshing scent of pine.

Make like an arch. At least once a day, arch

Inspiring Words

The words for "breath" and "spirit" are identical in various languages, notes Andrew Weil, M.D., author of *Spontaneous Healing*. In Sanskrit, the word is *prana*, in Greek, it's *pneuma*, and in Latin, *Spiritus*. We make the same connection in English, with words like *respiration* and *inspire*.

your back. Lie on the floor on your back. At first, use a rolled bath towel beneath your shoulder blades to make a gentle arch. If you feel strain in your neck, support your head with a pillow. Let your muscles relax. Gradually increase the thickness of what you lie on, but make sure it's comfortable. This stretch will help open your chest cavity and relax your chest muscles. "This makes more room in your chest so you can breathe more deeply," says Carrie Demers, M.D., medical director of the Center for Health and Healing and the Himalayan International Institute of Yoga Science and Philosophy in Honesdale, Pennsylvania. Dr. Demers suggests that you spend five minutes a day in this stretch position.

Take Lauren Bacall's advice. Put your lips together and blow the next time you feel short of breath. Inhaling through your nose, then exhaling through pursed lips may make breathing easier, says Dr. Wenzel. That's because you have to exhale more slowly when you do it this way. When you exhale more slowly, the air pressure that keeps your airway open decreases more gradually. Your airways stay open longer, and you can breathe more deeply.

BREATHING EXERCISES

The Easiest Way to Ease Stress

WHAT IT IS

Inside the classroom, the lighting is dim, but no one seems to notice. Silhouetted in the semi-darkness, the 30 or so students standing around the room are supplying their own illumination—small balls of light that hover waist-high.

The spheres of light aren't visible to the naked eye, of course. But they're quite visible in the mind's eyes of the men and women who have come to the Turtle Island Health Center in St. Paul, Minnesota, to study the ancient art of qigong. A 5,000-year-old Chinese healing practice, qigong combines gentle movements, visualization, and breathing exercises.

"Breathe deeply through your nose and visualize the air you're inhaling and the qi, or energy, that you're inhaling along with it," instructs the center's founder, Russell Desmarais, D.C., a chiropractor and certified acupuncturist, who is teaching the class. "Then visualize the qi as a ball of energy or light gathering just below your belly button."

The goal of qigong (pronounced *chee-gong*) is to cultivate qi, or "vital energy," and move it where it's needed in your body. In keeping with the tenets of Traditional Chinese Medicine, health depends on a sufficient and balanced flow of qi throughout the body. If you have too little qi in general, or too much qi here and not enough there, your health suffers.

Breathing exercises are a key part of qigong because it's primarily through breathing that you gather qi from the environment, explains Dr. Desmarais.

"You breathe out old qi and breathe in fresh, vital qi," he says. Once you've done that, you use visualization exercises and movements to get the qi where it's needed.

WHAT IT DOES

There is plenty of evidence that a certain type of breathing—deep breathing—can help relieve stress and the health problems that can accompany it, according to Laurence Smolley, M.D., a pulmonologist with the Cleveland Clinic–Florida in Fort Lauderdale and co-author of *Breathe Right Now*.

Research suggests that stress can contribute to a whole litany of health problems, from heart disease, insomnia, and headaches to lowered immunity, asthma, and digestive troubles. "So by helping to relieve stress, deep breathing can offer a wide variety of benefits," says Dr. Smolley.

Because it's really a form of meditation, deep breathing can help defuse stress by helping you shift your attention from whatever it is that's

Original coupon must be redeemed at time of purchase. Offer is not redeemable for cash. May not be combined with other offers or program discounts (i.e., classroom, corporate, nonprofit, institutional). Offer is valid only at Cafe Espresso. One offer per customer. Coupon is valid only January 1-31, 2001.

Bring this coupon into your favorite Borders store anytime between January 1 and January 31 for a 50¢ double cappuccino. Relax in the café, and enjoy.

BORDERS®

Borders moment no. 100

When the holidays are finally over, relax in our café with a 50¢ double cappuccino.

BORDERS

upsetting you. And it can trigger stress-relieving physiological changes, says Robert M. Sapolsky, Ph.D., professor of biological sciences and neuroscience at Stanford University in California and the author of *Why Zebras Don't Get Ulcers: A Guide to Stress, Stress-Related Diseases, and Coping.* Deep, relaxed breathing can shut down the release of the stress hormones epinephrine and norepinephrine, slow your heart rate, and lower your blood pressure. It also relaxes your muscles. says Dr. Sapolsky.

REAPING THE BENEFITS

Disciplines such as qigong and yoga use a variety of breathing techniques, but one of the most frequently employed is deep, diaphragmatic breathing. When you breathe deeply, your diaphragm (the muscle that separates your chest from your stomach) moves down with each inhalation, forcing your abdomen outward and making more room for your lungs to expand. You'll know that you're breathing deeply and diaphragmatically because your belly will rise with each inhalation and fall with each exhalation.

You don't have to breathe this way all the time, says Dr. Smolley. Even if you breathe shallowly, you still get all the oxygen you need. But it's worth giving deep breathing a try when you're under stress. And you'll need to breathe this way when you're practicing many kinds of qigong, yoga, and tai chi exercises.

TRIAL RUN

A good way to practice diaphragmatic breathing is to lie down and put your hand or a book on your lower abdomen. When you breathe in, make sure that your hand or the book rises. When you exhale, it should fall.

The best way to learn qigong and to combine qigong movements and visualization techniques with deep breathing exercises is to study with a certified instructor, says Dr. Desmarais (for a free list of instructors, write to the National Qigong Association, 571 Selby Avenue, St. Paul, MN 55102). To get a feel for the ancient art right now, you can try the Orbit Meditation movement offered by Dr. Desmarais, which is designed to generate qi and distribute it evenly throughout the body.

▶Stand with your feet shoulder-width apart and parallel to one another. Your knees should be slightly bent, your pelvis tucked in, and your chin parallel to the floor.

▶Relax your shoulders and let your arms hang at your sides with your elbows slightly bent and your fingers relaxed. Hold your tongue gently against the roof of your mouth, right on the ridge just behind your front teeth. Close your eyes.

▶Now that you're in position, visualize a ball of energy gathering just below your belly button. You might try visualizing a ball of white or yellow light taking shape at that spot.

▶Imagine the ball of energy moving slowly straight down between your legs and up to your lower back. Visualize it moving up your back to your neck and over the top of your head to your forehead. Then imagine it moving down your face, chest, and stomach and back to your navel.

▶Throughout this exercise, breathe naturally but deeply, so that your stomach pouches out when you inhale and sinks back down when you exhale. Keep your mouth closed and breathe through your nose.

BURSITIS AND TENDINITIS

Making Smooth Moves

Think of a garden hose bulging with water in one spot. That's bursitis. Now picture a clothesline stretched beyond its comfortable limits. That's tendinitis.

Every day, our bodies rely on dozens of bursae and taut tendons to keep our muscles and joints moving smoothly. Bursae are sacs or saclike cavities filled with lubricating fluid and sandwiched in where muscles or tendons pass over bony places. Tendons are taffylike cords that attach muscles to bones.

Problems develop from overuse. Sometimes, the soreness and stiffness is caused by that nonstop weekend tennis tournament. And sometimes, it's an occupational hazard of living, a result of years of wear and tear—from swinging a golf club, raking leaves, or just reaching for the pickle jar on the top shelf. As we get older, our bursae and tendons often become more vulnerable.

Shoulders are choice spots for flare-ups, but both bursitis and tendinitis can develop in the knees, hips, and elbows as well. It can be difficult to tell whether you have bursitis, tendinitis, or both. The signs and symptoms can be very similar, including pain and irritation when you move and when you're trying to sleep. Bursitis may come with more noticeable inflammation and can seem to be more severe. If pain or inflammation is persistent for more than two to three days, you should see a doctor for an accurate diagnosis.

Bursitis and tendinitis are very seldom life-threatening, says John Cianca, M.D., assistant professor of physical medicine and rehabilitation and director of the sports and human performance medicine program at Baylor College of Medicine in Houston. He adds, however, that some cases are more serious than others and may require surgical correction. Chronic joint instability or a torn rotator cuff tendon are such examples.

Although both bursitis and tendinitis are inflammatory conditions, doctors differ on the kind of treatment—be it heat or ice—that is best. Talk with your doctor to decide on the treatment that is right for you. "With bursitis, the main thing is to get the swelling down through rest and applying ice. Immediate exercises can be irritating to the bursa and cause it to swell even more," says Dr. Cianca. "But with tendinitis, stretching a tendon can make it feel better. Once the inflammation is down, you can work to restore range of motion and make the tendons stronger."

Babying Bursitis

Step one for dealing with bursitis is to *not* exercise. If you have a flare-up, the only exercise you should do is lift an ice pack onto the sore spot. Once the pain and swelling subside, you should do some mild stretching exercises, says Dan Hamner, M.D., a physiatrist and sports medicine specialist in New York City.

"With bursitis, you have to be careful and hold back a bit on activity," he says. "But if you totally rest an extremity for too long, you'll lose strength and range of motion. This is particularly true for the shoulder."

Try some shoulder soothers. For shoulder bursitis, Dr. Hamner offers the following easy movements to restore range of motion and keep the swelling down. Before all of these exercises, he recommends doing five minutes of ice treatment (simply wrap some ice in a towel and hold it on the painful area). He also suggests controlled, slow movements for best results.

►Stand and bend forward at the waist at a 45-degree angle.
►Let both arms hang loosely straight down, parallel to your legs, and bend your knees slightly to keep pressure off your back.
►Move one arm clockwise in a small circle; in 10 rotations, increase the size of the circle from about a foot to about three feet in diameter, allowing your entire arm, from your shoulder to your hand, to rotate (*1*).

_1_

►Wind down in another 10 rotations back to the smallest circle.
►Then try the rotations counterclockwise, using the same arm movement.
►Repeat the sequence on the other side. Try this exercise three to five times a day.

For variety—and only if your shoulders can tolerate it—try this exercise with a one- or two-pound weight in each hand, says Dr. Hamner.

Remember how those toy soldiers from the movie *Babes in Toyland* swung their arms dramatically back and forth? Well, that kind of arm movement can help beat bursitis, says Dr. Hamner. Here's how to do it.

►Stand with your feet shoulder-width apart and your arms at your sides.
►Raise your injured arm, with your elbow locked and the palm of your hand facing forward, above your head about 10 degrees short of vertical, as if you were raising your hand in class.

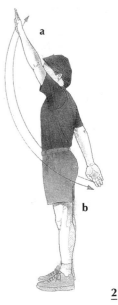

_2_

►At the same time, extend your other arm, with your elbow locked and the palm of your hand facing your leg, behind you about 10 degrees. Hold this stretch for one to two seconds (*2a*).
►In a controlled manner, gently swing your injured arm down and back and your other arm forward and up. Each arm should mirror the previous position of the opposite arm (*2b*).
►Repeat 10 times, alternating arms.

"Doing these motions slowly should help bring relief to the bursae," says Dr. Hamner.

Help your hips. The hips are another common spot for bursitis, says Dr. Hamner. Here's a stretch to ease the pain. Ice the spot for five minutes before giving this a try.

►Lie flat on your back with your knees bent and your feet flat on the floor (*3a*).

►Using your abdominal and quadriceps (front of thigh) muscles, and your hands if you need to, raise your legs (*3b*).

►In a slow, controlled movement, bring your knees toward your chest as far as you can (*3c*).

►Hold the stretch for two to three seconds.

►Relax and slowly lower your legs back to the original position.

►Repeat this stretch 10 times.

<u>3</u>

Taming Tendinitis

Tendons that have been inflamed prefer gentle massages and some stretches to return to full flexibility, says Dr. Cianca. Once full flexibility is restored and the pain is alleviated, strengthening exercises should be added.

Drop your heels to ease your Achilles. The Achilles tendon, the thick band that attaches your calf muscles to your heel bone, often becomes irritated. According to doctors, these exercises will help ease the pain and repair the tendon.

The first exercise, a stair stretch, is fairly vigorous, but the Achilles is a tough, long tendon, and it takes a fair amount of effort to overcome tightness, according to Dr. Cianca.

►Stand on a step with your heels hanging over the edge. To bolster your balance, hold the railing.

►Gradually drop your heels until you feel a stretch, but not pain. Try to hold the stretch for 20 to 30 seconds.

►Then step off and let your feet rest on the floor in a neutral position. Repeat this exercise two more times, says Dr. Cianca.

Write with your feet. Heat helps with tendinitis, notes Dr. Hamner. Here's an exercise that gives your Achilles tendon a little bit of stretch but doesn't overstretch it.

►Sit in a hot tub or whirlpool or in a bathtub with your feet under the stream of warm water from the faucet.

►Pretend to write the letters of the alphabet with your feet. Be sure to include letters that involve a lot of pointing and flexing, recommends Dr. Hamner. (If making letters doesn't appeal to you, try doing foot circles—25 one way and 25 the other.)

If you're a landlubber, you can do the same exercises while seated.

Pick a pair of elbow aids. For elbow tendinitis relief, you'll need support from a sturdy tabletop, suggests Carl Fried, P.T., a physical therapist at K Valley Orthopedics in Kalamazoo, Michigan.

►Sit facing the long end of a narrow table.

►Stretch the painful arm across to the far side of the table so that your wrist hangs completely over the edge and your fingers are extended. Keep your arm straight, with your elbow locked and your palm facing down.

In Pain from Painting? Hit the Showers

Uh-oh, you're two days into the ambitious plan to paint the downstairs when your shoulders stage a protest. Now what?

Head for the showers with your tendinitis, but not in defeat, says Dan Hamner, M.D., a physiatrist and sports medicine expert in New York City. Before you start painting again, warm your shoulders with this shower exercise.

► Face away from the spray toward a sturdy wall in your shower.

► Make sure the jets of warm water are hitting your shoulders.

► With your elbows slightly bent, place both palms on the shower wall.

► Let the fingers of your good arm slowly crawl like a spider as high as you can reach on the wall, then stop and lower your arm to the starting position.

► Repeat the finger crawl using the arm on your sore side. Try to go as high as you did with your good arm.

► Let the water warm your shoulder tendons for about 10 to 20 seconds.

► Try four more finger marches up the wall, trying to go up farther each time.

Now your shoulders are ready to take on the painting job. When you finish painting for the day, head for a second shower and repeat the finger marches. "Two showers a day can keep the doctor away—at least for some tendinitis in the shoulder," says Dr. Hamner.

► Grasp the back of that hand with your other hand and slowly try to bend the wrist down as far as possible (4).

► Hold for 10 seconds and then release.

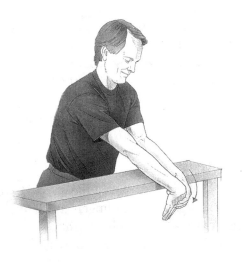

4

► Repeat these steps up to 10 times twice a day.

Or try a variation on this exercise. Head for the refrigerator or pantry and grab a can that fits your grip. Put your arm in the same position along the edge of the tabletop—palm down and wrist over the edge—but hold the can in your hand. Gently move your wrist up and down 25 times. Pause for a few seconds and do 25 more. Try this twice a day, says Fried.

Try a shoulder solution. A door frame can be a true ally against shoulder tendinitis, either as a good preventive measure or whenever pain strikes, says Dr. Hamner. He recommends doing at least four stretches twice a day. "We would be in great shape if we got into the habit of doing stretches every day. Just look at our cats and how often they stretch," he says.

▶Step inside a door opening.

▶Place your palms and forearms flat on the jamb on each side of the door frame, with your hands slightly above shoulder level.

▶Gently lean forward through the doorway while your lower arms bear the weight. You should feel a stretch in both shoulders.

▶Hold the stretch for at least six seconds.

Knead your knees. To relieve tendinitis in your knees, Dr. Hamner recommends doing what he calls cross-friction finger massage.

▶Sit in a chair.

▶If your right knee is red and swollen, cup your right hand across the front of that knee so that your index finger is on top of the painful area and lying across the tendon running down the front of the knee.

▶Place your middle finger on top of your index finger.

▶Press down and gently massage the painful spot, working your fingers firmly for five to seven minutes. This could be painful at first, but it should ease up. If your fingers get tired, knead the spot with your thumb.

CABIN FEVER

Working Out When You're Stuck Inside

Energy-packed muscle rippling under striped flesh, the caged tiger paces endlessly behind iron bars. With every double-back maneuver, the frustration in his green eyes grows. The sensitive tail, quivering with tension, is the fuse for this living time bomb.

That caged-up tiger is you. Snowbound in the winter, stuck at home with sick kids, waiting around for the cable guy—you're housebound against your will. But where there's a will—any will—there's a way. A way to work out within four walls. A way to burn off some of that excess energy that's got you eating everything in sight. A way to stick to your new exercise regimen in spite of the snowstorm that arrived last night.

Do-It-Youself Calorie Crunchers

Nobody actually plans to be stranded at home—it's the kind of thing that just sort of happens without warning. And when it does, we're frequently unprepared. Counter that closed-in feeling by getting active without accessories. The possibilities for equipment-free exercise are vast, but try these suggestions for starters.

Do some classic moves. Good old calisthenics can be a great choice for temporarily homebound folk, says Gordon Stewart, certified athletic fitness trainer at Gold's Gym in Anchorage, Alaska. Sure, they're not as flashy as kick-boxing or Spinning classes, but a sincere session of abdominal crunches, pushups, leg raises, and other classic exercises can really fit the bill when you're surprised by a snow day.

Stretching can be a blessing. That confined tiger lacks the room to really stretch his long limbs. But you don't, even if your house starts to feel like a cage to you. A regimen of total-body stretching or gentle yoga will relieve the boxed-in feeling that can come from inactivity, says Stewart (check out "Stretching" on page 380 for specifics).

Set a pace in place. Put on your running shoes and hit the carpet, says George McGlynn, Ph.D., professor of exercise physiology at the University of San Francisco and author of *The Dynamics of Fitness*. "You need that aerobic workout, even if you can't get outside," he says. Running in place gives a surprisingly strenuous cardiovascular workout. And if you run while watching your favorite TV show, you can be entertained while your feet eat up the stationary miles.

Dance your ants away. Music and movement go together like, well, like Fred and

Ginger. Turn up some fast tunes and let your body follow, says Dr. McGlynn. "It's just like an aerobics class, but without the choreography," he says. You'll discover the fun of making up your own moves as you go along.

The Minimal Moves Toolbox

Maybe you've already experienced cabin fever and didn't care for it much. If you'd like to prepare ahead for future "rainy days," consider the following suggestions. The bare-bones equipment recommended below will make your home an instant, in-a-pinch (but temporary) health club.

Get a jump on it. Jumping rope is a great aerobic exercise, but can you really do it indoors? Sure, says Stewart, if your ceiling is high enough. Oh, and be sure to keep away from any heirloom china that would be hard to glue together again. "Rope jumping is really intense, simple, and cheap—it's great," he says. Purchase a quality rope at any fitness store and you'll always have an instant cure for cabin fever. Start with two to five minutes of light jumping to warm up, then do a stretching routine before continuing with your jumping workout.

Revive with videos. The television sings a siren song, drawing you in when you can't get outside. Resistance may be futile. Instead, heed the call and burn calories at the same time. Stock a small selection of fitness videos and there'll be no excuse to plow the couch-potato field ever again, says Dr. McGlynn. With so many types to choose from, you can work up a sweat with an aerobics session or go the peaceful route with a video class in yoga or tai chi.

Start stepping. Step aerobics seems right at home in the gym, but you can go up and down at home as well. Either purchase a plastic exercise step from a sporting goods store or be creative and make your own. Or you can try a low, sturdy stepstool about 2 feet long by 1½ feet wide. The ideal height for an exercise step is no more than 8 to 10 inches, depending on your height, says Dr. McGlynn. If you are shorter than average, your step should be lower as well.

Bring out the bike. If it's not spring or summer, your trusty bicycle is probably languishing in cold storage. Haul out that Huffy and get riding with a bike trainer. No, we don't mean a burly guy riding alongside you as you pedal through the rain. A bike trainer is a device that lets you turn your road or mountain bike into a stationary exercise bike. You can find them at most sporting goods stores for around $100.

CAFFEINE WITHDRAWAL

How to Survive Week One

Science can't yet pinpoint exactly why we love it the way we do. But ask loyal café crawlers what gets them out of bed in the morning, and there's no doubt about it. Caffeine—the active and activating ingredient in coffee—is the most popular drug in the world.

Once you're hooked, junking the java habit may be tough. But you may have good reasons for doing so: If your doctor says you should, for instance, or if you are anxious and tense or have trouble sleeping.

For the majority of those who try to quit, swearing off the coffeepot means a week's worth of annoying side effects. Headaches are the worst and most common symptom, says Suzette Evans, Ph.D., assistant professor of psychiatry at Columbia University in New York City. Next in line are fatigue and trouble concentrating.

These exercise tips will help you endure the travails of week one and get on with your caffeine-free life.

Rub Out Headaches

Caffeine-withdrawal headaches can often be tamed with over-the-counter painkillers, says Dr. Evans. But if you need more relief than that, try these mini-massages to quiet your banging brain.

Try an eye-socket soother. Start with this pressure-point maneuver, says Allison King, a licensed massage therapist on staff at a five-star hotel in New York City.

Find the orbital notches—the spots at the inner edge of each eyebrow where the eye sockets meet the bridge of the nose. Place the pad of one forefinger at each notch and press firmly, using slow, circular movements, for about a minute. "I know from personal experience that quitting coffee can bring on headaches that feel like you've been stabbed right behind your eyes," says King. "This is something you can do anytime, all by yourself, to get some quick relief."

Tame your scalp. Massage can gently work the scalp, releasing muscle tension in the head and neck, says King. But if your head is really pounding, a hairline rub may be more welcome than a vigorous skull scrub. "The hairline, right where the hair and skin meet, is particularly sensitive to touch," she says.

To begin, place the first two fingers of each hand behind your ear on the same side. Rub in gentle circles as you follow your hairline up and around your ears to the sideburn area. Pay special attention to your temples, says King. You can even work your way up to your forehead. Just do what feels best to you.

Finish by bringing your fingers behind your head to the place where the skull meets the spinal cord. Circle your fingers here for a few extra seconds, letting go of tension.

If these techniques just don't do it for you, Dr. Evans suggests allowing yourself a small amount of coffee—about two tablespoons—to stop the withdrawal headache without harming your progress in kicking the habit.

An Alternative Jump-Start

Without your morning cup (or two), you may find that you've lost all ability to sit up straight, operate a pencil, or even hold your eyelids open. You need an alternative energizer to get you going. Try this energy-enhancing yoga technique in place of the usual caffeine jump-start.

Do a super morning sniff. This yoga breathing exercise is great for generating morning energy, says yoga instructor Lilias Folan, whose video titles include *Forever Flexible* and *Energize with Yoga*. "It's a wonderful breath to do right in bed, to help you wake up," she says.

Begin either lying on your back or seated. Exhale to empty your lungs completely, then inhale by taking small consecutive sniffs through your nose—it should take 8 or 10 tiny sips of air to fill your lungs. Don't struggle at the top of the breath, says Folan. "Breathing this way should feel gentle and effortless without causing any pressure inside." Hold the breath comfortably for a second or two, and then slowly exhale completely. Repeat four or five times.

Concentration Crisis: Try a Quick Workout

Another sign that your body misses its usual caffeine fix is the cloud right behind your eyes. You know the one—the gray mist that won't let you remember how to start the car, much less how to drive it. Here are some suggestions to help you find your mind.

Aerobicize. Increased oxygen can help clear your mind, according to Dan Hamner, M.D., a physiatrist and sports medicine expert in New York City. "If you're already exercising regularly," he says, "you know that a good workout always brings the world into focus." So when you find your concentration waning without warning, even a quick 20-minute walk or other simple aerobic activity can do the trick.

Kid around. Want some unusual activities to help you tune in? Try using some pieces of exercise equipment that the kids may have left in the garage.

▶Try a session of rope skipping. It will not only help lift the veil of caffeine deprivation, it will also make you feel either young again or as though you're the heavyweight champ.

▶Grab that old pogo stick and bounce around the driveway a bit. It will require some concentration and get you puffing.

▶Grab a bike and cruise the neighborhood.

"I'd say to keep moving for at least 20 minutes," says Dr. Hamner. "And do as much deep breathing as you can while you're at it."

CARPAL TUNNEL SYNDROME

Stop Squeezing the Nerve

Picture New York City's Lincoln Tunnel during a Friday afternoon rush hour. Thousands of cars, trucks, and buses stream into the tunnel. Horns sound and tempers flare as congestion builds and traffic comes to a halt. Well, some people have a similar tunnel of torment in their hands and wrists.

Your carpal tunnel is a passageway of bone and ligament in your wrist. The traffic through it consists of nine tendons and the median nerve, the major nerve between your arm and your fingers. If the tendons in the tunnel become inflamed or fill with fluid, they can swell, pinching the median nerve. This often causes pain, tingling, and numbness in the fingers.

Often, carpal tunnel syndrome is caused by repetitive strain injury, which is the result of over-and-over-again wrist or finger motion, especially if it's combined with gripping vibrating tools or using instruments that put pressure at the base of the palm. People who type or sew or hammer or paint for many hours every day often get the tendon swelling that causes the nerve pain.

Severe cases cannot be healed by exercise, and doctors caution that you should stop any wrist or finger exercise if there is constant pain or total numbness. But moderate levels of carpal tunnel syndrome can be treated with gentle exercises. The trick is to strengthen and flex the tendons and stimulate the movement of fluid out of the crowded tunnel, thus taking pressure off the median nerve.

Pain-Fighting Toys and Games

"In carpal tunnel syndrome, it is important to have some wrist extension, wrist flexion, and side-to-side movement," says Dan Hamner, M.D., a physiatrist and sports medicine specialist in New York City. He offers these gentle—even playful—exercises.

Do the Slinky shimmy. Remember that coiled toy from your childhood that gave such delight as it slinked down your stairway? Well, that toy of yesteryear can be an exercise tool for today, says Dr. Hamner. Simply cradle the ends in the open palms of both hands and gently move your wrists up and down as you flop the Slinky from hand to hand.

Deal yourself a great hand. You don't have to be Amarillo Slim to ease carpal tunnel pain, says Dr. Hamner. The simple act of card shuffling is an ideal exercise to work the wrist muscles from side to side. When your bridge club meets, volunteer to be the steady dealer.

Do the yo-yo. The up-and-down glide of a yo-yo is a great exercise to flex and extend the wrists, says Dr. Hamner. Advanced yo-yo performers can try the cradle and walk-the-dog moves.

Gentle and Varied Exercises to Quiet the Carpal

After your symptoms are under control, exercise can help prevent the return of carpal tunnel pain. The key is to use gentle strengthening movements and to vary the motions, say doctors and physical therapists. Here are a few simple exercises that you can do while watching television or even relaxing in a bathtub.

Mimic a mime. For this wrist exercise, stand facing a wall. Extend your arms out at shoulder level directly in front of you with your palms facing the wall. Without touching the wall, pretend that you are a mime trying to find an exit from a box. Slowly move your palms and shoulders up as you reach for the imaginary ceiling, says Dr. Hamner.

Try pretend window washing. While standing or sitting, extend your arms in front of you and gently move your hands clockwise as if you were rubbing smudges off a window. Then change directions and move counter-clockwise, says Dr. Hamner.

Offer some high-pressure prayer. Put your hands together as though to pray, then press the palms together. This helps to stretch the wrist muscles, says Susan Isernhagen, P.T., a physical

---

Punching Your Way Out of Carpal Tunnel

Marketing director Marcia Miller dodged hand surgery for carpal tunnel syndrome by throwing punches with former world middleweight contender Michael Olajide Jr.

Years of repetitive computer keyboard work caught up with her a few years ago, causing loss of most of the movement in her right hand and some in her left. Since aspirin and other painkillers couldn't ease the throbbing, doctors recommended hand surgery to open up the carpal tunnel. But she searched for a more natural alternative.

Dan Hamner, M.D., a physiatrist and sports medicine specialist in New York City, introduced Miller to Aerobox. This highly aerobic program incorporates the jabs, uppercuts, hooks, head bobs, and body weaves of a boxing match without actual hitting.

Under Dr. Hamner's medical supervision and with the benefit of Olajide's boxing expertise, Miller says that her wrist became pain-free within a few months.

"With this type of shadow boxing, you are not hitting anyone, so you don't hurt your muscles or tendons," says Olajide. "Marcia gained both physical and psychological benefits from Aerobox."

Miller cites two major benefits of this exercise program. "I solved the problems with my hands and also learned more about fitness than I ever thought I would in my life," she says. "It puts you in a frame of mind that you can do anything if you set your mind to it."

therapist and director of Isernhagen Work Systems in Duluth, Minnesota. Hold the pressure
for a count of five and relax. As a variation,
keep your fingertips touching as you move your
palms apart.

Play sweatless tennis. There's no need for
a racquet or net for this strengthening exercise
offered by Dr. Hamner. Simply extend your
arms straight out in front of you with your
palms up and hold a tennis ball in each hand.
Squeeze the balls, then relax. Do this 10 times,
shake out your hands, and repeat.

Reach for new heights. Here is an exercise
to get the fluid out of your wrists and fight the
numbness, tingling, and jabbing pain it causes
by pressing against the median nerve, says Dr.
Hamner.

While watching television in your favorite
chair, rest your hands and elbows on top of the
backrest so that your hands are behind your
head—and most important, above the level of
your heart. Hold that pose during an entire
commercial and gradually build your way up to
a half-hour sitcom.

A Keyboard Survival Guide

Hours at the computer keyboard are a common
cause of carpal crisis. You can avoid the
shooting pain through your hand and wrist by
using proper computer poses and office exercises, say doctors.

"We often put ourselves in a head-forward,
limb-forward posture at a computer keyboard,
and that puts strain on the neck, shoulders, elbows, arms, wrists, and fingers," says Robert
Markison, M.D., a hand surgeon and associate
clinical professor of surgery at the University of
California, San Francisco. He offers this fight-
back survival guide.

►Maintain proper posture. Keep your head
resting on the center between your shoulders
and avoid leaning forward.

►Avoid extreme wrist positions by keeping your
wrists almost straight on the keyboard, not
below it or arched upward.

►When typing, move your little fingers over to
reach the keys on the edges to type the letters
Q, A, Z, and punctuation symbols. Avoid
straining your tendons by stretching the little
finger out.

►Don't pound the keyboard like a jackhammer.
Try a kinder, gentler touch.

►Alternate tapping the spacebar with your right
and left thumbs to avoid overusing one
thumb.

►Every 30 minutes or so, stop typing or
working the mouse and stand up and stretch.

CELLULITE

Getting Rid of French Fat

French, the so-called language of love—the language that makes even cuss words sound poetic—made a contribution to the English vocabulary that is now universally recognized. In 1968, the French gave us the word *cellulite*.

True, it's a beautiful, sophisticated-sounding word. But all it means is well, er . . . fat.

Cellulite definitely looks a little different from regular fat. It's more dimpled and lumpier, and it's found mostly on thighs and buttocks. But deep down, fat it is. You can't loofah, massage, wrap, pound, or vibrate it away. And despite what some cosmetics companies wish that you'd believe, there isn't a wonder cream that will expunge it.

There is a way to get rid of it, however—exercise. While there's no such thing as spot reducing (exercising one area to lose weight only in that area), toning your problem areas can help smooth out cellulite while you start working on the overall solution: losing the extra pounds.

"Resistance training is one of the best things you can do to get rid of cellulite," says Karen Harkaway, M.D., clinical instructor of dermatology at the University of Pennsylvania School of Medicine in Philadelphia. "Building up the muscle helps fill out dimples, giving the skin a much smoother appearance."

"Just 20 minutes of strength training two times a week can dramatically improve the look of areas with cellulite by increasing muscle definition," says Joan Price, a certified fitness instructor in Sebastopol, California, and author of *Joan Price Says You Can Get in Shape*.

Toward a Better Bottom Line

To help get that defined look, Price suggests that you space your exercise sessions over the week. And always leave a day of rest between sessions, she advises. Here are the movements she recommends.

Assert your squatter's rights. "Squats are a great all-purpose lower body exercise," says Price. They work the fronts and backs of your thighs and your buttocks, she says. But if you want to really work to define the muscles and improve the appearance of the areas with cellulite, she recommends using some extra equipment—either latex bands or free weights—to provide resistance.

To do squats, stand with your feet shoulder-width apart, bend your knees, and slowly squat down as if sitting on a chair. As you lower yourself, be careful to keep your weight over your heels and make sure that your knees don't go past your toes. Lower yourself until your thighs are about parallel to the floor—no lower. Rise slowly, squeezing your buttocks until you reach full height before repeating.

Using latex bands such as Dyna-Bands (available at most sporting goods stores) makes this exercise more challenging. Here's how to bring them into the picture.

▶Choose two bands that are three feet long.

▶Place one end of each band under the heel of each foot, using your full weight to hold the ends down.

▶Wrap the free end of each band around each hand and hold them in your fists.

▶While you're in a squatting position and the bands are slack, bring both ends behind your back and over your shoulders. Hold the bands in position with your fists in front of your shoulders.

▶Rise, stretching the bands and feeling the resistance. If they're too taut, just adjust your hands for more slack.

▶Now do the squats, lowering and raising your upper body against the resistance of the bands. Do a set of 8 to 12 slow repetitions.

If you use free weights such as dumbbells instead of bands, you'll get a similar amount of resistance (for information on selecting dumbbells, see "How to Lift without Straining Your Budget"). Lift the weights and hold them in a strongman position—that is, with your arms curled up and the weights directly over your shoulders, palms facing your body. Keep them in this position as you do the squats.

Whether you're using bands or weights, do 8 to 12 repetitions. If the squats begin to seem easy after several sessions, you can increase resistance by shortening the slack on the bands or using heavier weights.

Go ahead—take the lunge. Lunges can be done either with or without using dumbbells.

▶Stand upright with your feet shoulder-width apart. If you are using dumbbells, your arms should be fully extended at your sides with your palms facing in.

▶To get into the starting position, take a long step forward, planting your right foot ahead of your left.

▶Keeping your right knee aligned over your right foot, bend your left knee to lower your center of gravity until your right thigh is parallel with the floor. Resist the urge to lean forward as you do the exercise. Keep your back and neck upright so that your torso descends in a straight line toward the floor.

▶While keeping your right foot stationary, straighten your left leg and press your left heel to the floor.

▶Repeat by bending your left leg and again lowering your center of gravity.

▶Do 8 to 12 repetitions, then return to the starting position, switch legs, and repeat.

Test your push-aways. Don't let memories of straight-legged pushups from high school gym class scare you. There are many different kinds of pushups with various levels of difficulty, and any of them can help as cellulite-fighting exercises.

The least difficult pushup is what Price calls a push-away. "People of any fitness level can do push-aways," she says. Here's how she describes them.

▶Stand facing a wall and about an arm's length away, with your feet slightly apart.

▶Place your hands on the wall at chest level, a little more than shoulder-width apart.

▶Make your body very rigid (pretend you're in a body cast). Only your heels and elbows should move. Rise on the balls of your feet as you bend your elbows and lead with your chest, moving your body toward the wall. Be careful not to hunch your shoulders.

▶Do 6 to 15 repetitions. When 15 become easy, move to the next level.

Rise to the occasion. The bent-leg pushup is next in terms of difficulty, says Price. While doing this exercise, it helps to look at the floor a few feet ahead of you, not directly beneath you, Price says, to help keep your head and neck in

line with your spine. If you can't lower your chest to a few inches from the floor, just go as far as you can, she says. Gradually work toward going as far as is comfortable for you.

▶Get down on your hands and knees and lift your feet a few inches off the floor so that your weight is balanced on the padded area just above your knees.

▶Keeping your head, neck, and spine in a straight line, slowly lower yourself until your chest is a few inches from the floor, then

How to Lift without Straining Your Budget

Equipment for resistance training doesn't have to be expensive. Either latex resistance bands or free weights will do the trick, and they're very reasonably priced.

Resistance bands, such as Dyna-Bands, are six-inch-wide latex strips that come in three- or four-foot lengths. The elastic quality of these bands, along with your body, provides the resistance. You can purchase them at most sporting goods stores or by writing to Unconventional Moves, 13091 Bodega Highway, Sebastopol, CA 95472.

If you decide to go with free weights such as dumbbells, you'll get the best deal on them at secondhand sporting goods stores such as Play It Again Sports. "To really work all your muscles, you need three sets of dumbbells, varying in heaviness," says Joan Price, a certified fitness instructor in Sebastopol, California, and author of *Joan Price Says You* Can *Get in Shape*. But you can get away with two—a heavy set and a lighter set, she says.

To choose the proper dumbbells to serve as your heavy set, pick up two of the same weight. Hold one in each hand over your shoulders and do 8 to 12 squats. If you can't do 8 repetitions, choose lighter weights. If 12 repetitions seem easy, you need heavier ones.

Select your lighter dumbbells by doing deltoid raises. Select two that are lighter than your heavy set. Hold one in each hand in front of your hip bones with your arms slightly bent, then arc your arms out to the side and bring them up to shoulder level. Be careful not to lift your shoulders during the movement, which would use your upper back instead of the intended shoulder muscles. If 8 to 12 repetitions of this exercise are challenging but not impossible, you have the right dumbbells.

If you want a third set of dumbbells to use for arm curls and other exercises, choose a weight that's between your heavy and light set.

come back up. Be careful to lead with your chest, not your chin or nose.

►Do 6 to 15 repetitions. When 15 become easy, either lower yourself more slowly or do more repetitions. Or you can upgrade to the gym-class variety of pushup, where you straighten your legs, balance on your toes, and do the pushups with your legs and back in a straight line.

The Clincher

Squats, lunges, and pushups can give your muscles the definition they need to smooth out cellulite's lumps, but your trouble areas are unlikely to get much smaller. You need an aerobic workout to banish the fat lying on top of your newly toned muscles, says Price. That entails any activity that raises your heart rate

by using large muscle groups such as your legs, hips, and buttocks. It can be separate 10-minute bouts of exercise that add up to 30 to 60 minutes of activity on most days, or it can be 30 to 60 minutes of continuous blood-pumping activity at least three days a week, she says.

The key is to pick something you like so that you'll stick with it. If you prefer solo activities, you can go for brisk walks or take up running. Other people prefer to walk up and down the stairs in their houses while listening to music. Remember to check with your doctor before starting an exercise program.

"Just get out there," says Price. "Do whatever it is that pleases you most." To make sure that you're working at your proper fitness level, you should be able to talk comfortably while you're doing the activity.

CLUMSINESS

Moving toward a State of Grace

Under the big top, up above the crowd, David Dimitri is dancing on a wire. He is jigging and hopping—tempting fate and gravity in increasingly outrageous ways.

Suddenly, Dimitri bounces, and both his feet leave the wire. He swings up, his heels arcing over his head. He appears to hang upside down in midair. Just when it seems that he'll crash to Earth, he rights himself on the wire. Below, the crowd gasps, then breaks into applause.

A wire walker and acrobat with New York's Big Apple Circus, Dimitri can't afford to be klutzy. One clumsy move, and he can be badly hurt—and poorly reviewed.

The secret to his commendable coordination? Movement—lots of it, and lots of different kinds.

"I practice my act a lot, and I work out on the wire, running back and forth and doing somersaults on it," says Dimitri. "I also ride a bike everywhere. And I stretch."

No matter how clumsy you are, no matter how inclined to trip over carpets and spill the milk, movement can help you achieve a state of relative grace, says Fay Horak, P.T., Ph.D., a physical therapist and professor of neurology and physiology at Oregon Health Sciences University in Portland.

To an extent, the problem with us clumsy types is that we get stuck in a movement rut, Dr. Horak explains. We can move in routine, familiar ways with no problem. But when situations require us to move in different and novel ways, to adapt to changes in our surroundings, we trip up. Sure, we can walk through the dining room without a hitch when the chairs are in their usual pushed-in places. But let someone leave a chair pulled out a bit and . . . Oof! We're perfectly competent drinking our coffee out of our usual mug, but hand us that Wedgwood teacup, and . . . Oh, dear.

To become less clumsy, you have to practice moving your body—all parts of your body—in a multitude of different ways and different settings, says Dr. Horak. "You need practice in adapting the way you move, " she says. "The more you practice moving and adapting, the more coordinated you'll be."

Toward More Random—And Surefooted—Movement

One caveat before you get moving. Vision problems, such as trouble with depth perception, and snafus in the part of your inner ear that tells your brain which way is up, can contribute to clumsiness. A loss of sensation in

Fine Motor Skill Drill: Toot and Fold

Doing the tango, boxing, fencing, and the like will help you coordinate your arms and legs and develop what physiologists call gross motor skills. These skills will enable you to negotiate the trickiest of obstacle courses without a pratfall.

What they won't do, however, is keep you from dropping Aunt Mary's crystal wine goblets. To avoid such mishaps, you need fine motor skills, which are those involving your fingers. The key to refining your fine motor skills is practicing exercises that involve your digits, says Fay Horak, P.T., Ph.D., a physical therapist and professor of neurology and physiology at Oregon Health Sciences University in Portland.

Just about anything that requires dexterity will help. Playing the piano, knitting, and crocheting are good choices. And here are some others.

Fiddle with a flute. You can pick up a recorder at a music store for about $10. Easy to learn, most recorders come with a fingering chart and some music. If your old elementary-school recorder is still lying around somewhere, you can start playing right away.

Fold with your fingers. Practice the ancient Japanese art of origami, or paper folding, and you'll get plenty of practice coordinating your fingers. The odds are that your local library has a number of books on origami that will show you how to start.

your skin, joints, and muscles can also make you clumsy. So if you suspect that your coordination isn't up to par, see your doctor and have your eyes, inner ears, skin, and joints checked. If everything checks out, then move—all over, in all sorts of ways. Consider these possibilities.

Walk on the wild side. When you walk, vary the terrain. Skip the indoor track at the Y and take a hike through the woods up the side of Blue Mountain. Or walk around the old part of town, with the cobblestone streets and meandering sidewalks. Practice moving over different types of turf, on different inclines, around different obstacles, and you'll get good practice in adapting your movements to different environmental demands, says Dr. Horak.

Ride on the wild side, too. Ride a mountain bike over uneven, bumpy terrain, and you'll get plenty of practice in moving and adjusting your movement to changes in your surroundings. If you tend to be on the clumsy side, start your mountain-biking career on a path that's relatively level and smooth, suggests Edmund Burke, Ph.D., director of the exercise science program at the University of Colorado at Colorado Springs, a staff member of the 1980 and 1984 Olympic cycling teams, and author of *Serious Cycling.* "Start on a basic dirt road that's kind of bumpy but not rocky," says Dr. Burke. "Then progress up to a trail with some turns in it, then a trail with hills and maybe a few bumps." To protect yourself while you're out biking and improving your coordination, always wear a helmet.

Come and join the dance. You don't need to be a good dancer, just a willing one. Dance is a great exercise option for clumsy people, says Dr. Horak. Dance and the music it's set to helps everyone coordinate their movements. All dance is good for curing clumsiness, but some, like improvisational and modern dance, are better, since they call for more variation in movement.

You can learn to dance from videos, but it's easier with a teacher. To find dance classes in your area, call your local community college and ask about adult education dancing lessons. Another option: Check the yellow pages under "Dancing Instruction."

Step into the ring. Boxing is a good antidote to clumsiness because it requires responsive, adaptive movement. When you're sparring, you have to adapt your moves to your partner's or wind up getting socked. "Boxing is the *best* sport to help you get coordinated; it isn't the same old thing all the time," says Ron Casella, an undefeated professional boxer and three-time Golden Gloves champ, who is the boxaerobics instructor at the Greenwich, Connecticut, YMCA. Boxaerobics, a still more diverse variation on boxing, offers aerobic exercise and anti-klutz workouts all in one, incorporating stretching, shadow boxing, sparring, and rope skipping. To

Mental Moves: Have Graceful Thoughts

Imagine that you're the hoofer Donald O'Connor and the world is a dance floor. The truth is, your thoughts influence your movements. Here's how to think—and move—with ease.

Create an obstacle course. One reason that some clumsy people bump into things is that they often try to fade into the walls or the furniture, says Louise Cash, professor and chair of the department of speech, communication, and theater arts at Emanuel College in Boston. Instead, think of yourself as a performer. Actors and actresses train their peripheral vision so that they don't have to look at objects directly, which prevents them from walking into things on stage. "As an exercise, try positioning furniture or boxes in a room and wandering around without looking at or bumping into the pieces. This instantly gives you the feeling of being in control and builds confidence and grace," says Cash.

See yourself as the personification of grace. A wire walker and acrobat with New York City's Big Apple Circus, David Dimitri imagines himself doing his entire routine perfectly before he ever touches his foot to the wire. "You have to convince yourself that you can do it," he says. Studies show that imagining yourself moving flawlessly and gracefully will actually help you move more adroitly, says Fay Horak, P.T., Ph.D., a physical therapist and professor of neurology and physiology at Oregon Health Sciences University in Portland.

So borrow a tai chi video from the library and watch it, paying close attention to how the teacher moves. Then get up and do a few moves yourself. Studies show that watching videos and imagining a movement just before actually doing it can improve performance, explains Dr. Horak. Keep the image in mind the next time your coordination is being taxed. "It's amazing how this helps," she says.

find boxaerobics programs near you, call your local Y or neighborhood gyms.

Make like a musketeer. Fencing is like boxing, except that your partner has a three-foot-long sword (albeit with a safety bead on the end). Fortunately, you have a sword, too, so you can defend yourself. To do that adequately, you have to respond to your opponent's moves and adjust your own accordingly.

"Fencing is very good for developing balance, agility, and coordination," says Michael Marx, a former Olympic fencer, a fencing master, and coach at the Rochester Fencing Center in New York. For a referral to a fencing school or teacher near you, write to The United States Fencing Association, 1 Olympic Plaza, Colorado Springs, CO 80909.

Try tai chi. A saving grace for clumsy people, tai chi is the ancient Chinese art of practicing precisely choreographed movements (more than 150 in all) in very slow motion.

"Tai chi takes these coordinated movements and slows them down to a speed you can manage, " says Alan Tillotson, director of the Chrysalis Natural Medicine Clinic and a tai chi teacher in Wilmington, Delaware. "As you practice and learn to do the movements perfectly, you'll become more coordinated."

If you want to learn tai chi, your best bet is to find a good instructor, says Tillotson. Call community college adult learning centers or martial arts centers in your area and ask whether they offer classes. To get a feel for tai chi in the meantime, try this posture.

▶Stand with your feet 12 to 15 inches apart, your toes pointing outward about 45 degrees, and your arms at your sides with your palms facing up.

▶Bend your elbows slightly and slowly raise your arms until your hands are at chin level.

▶Slowly rotate your hands so that your palms are facing down and lower your arms to waist level.

▶As you lower your arms, gently bend your knees and sink down toward the ground, exhaling as you drop down. Stop your downward movement when it becomes uncomfortable.

▶From your semi-squatting position, rotate your palms upward and slowly stand, inhaling as you rise.

▶Repeat 10 to 15 times, slowly rising and sinking, breathing slowly and deeply from your diaphragm as you move up and down.

"Gradually, you'll be able to go lower and lower without any strain until you get to the lowest possible position, with your elbows touching your knees," says Tillotson.

COLDS AND FLU

Effortless Remedies

If you're sniffling, sneezing, coughing, and just plain tuckered out, you probably won't welcome the sight of a fitness trainer shouting, "Hup, hup, hup, hup, one, two three, four." But for some people with the all-too-common cold, a modest workout might be just the thing to kick the germs.

When can you hit the trail, and when should you hit the hay? To help guide your exercise decisions when you're under the weather, you should do a quick symptom check, suggests William A. Primos Jr., M.D., who practices primary-care sports medicine in Charlotte, North Carolina. If your symptoms are above the neck, such as a stuffy or runny nose or a sore throat, exercising is probably all right, advises Dr. Primos. But start at half speed, he cautions. If after 10 minutes you feel okay, you can increase the intensity and finish your workout. But if you feel horrible after 10 minutes, stop.

When your symptoms are below the neck, avoid exercise completely, recommends Dr. Primos. Some common symptoms that fall into this category are muscle aches, a hacking cough, a fever of 100°F or higher, chills, diarrhea, or vomiting. Many of these are indicators of flu, and if you work out in this condition, you're likely to feel even weaker and certainly

become dehydrated. If after five or six days, your symptoms haven't improved or have worsened, you should see a doctor.

Some Soothing Moves

Whether you have a cold or flu, though, you can try some modest tactics to give you relief from symptoms. Here are some ways to put healing into motion, even if you're feeling more like lounging than doing lunges.

Lean on your thumb joint. To help clear up a headache, use the thumb and forefinger of one hand to press and hold the top of the first joint below your thumbnail on the other hand, says Mary Muryn, a certified teacher of polarity (energy healing) and reflexology (a type of acupressure that focuses on the hands and feet) in Westport, Connecticut. "But never press for more than five minutes," she advises, since you can actually make your headache worse if you lean on that spot too long.

Fight that fever. When Victorian maidens were in danger of swooning, someone invariably applied a cool compress. Should your brow become feverish, try that maidens' cure. Wet a washcloth in cool water and wipe down your face and neck, says Don Beckstead, M.D., a family-practice physician in Altoona, Pennsylvania. As the water evaporates, it takes some of the heat from your skin. If it makes you feel better, you can leave the cloth on your forehead until it warms up, he says. Reapply a freshly cooled cloth as many times as you like. But if your temperature reaches 101°F, acetaminophen is generally the drug of choice, advises Dr. Beckstead.

Nudge your nostrils. An acupressure technique can help relieve a clogged nose, says Michael Reed Gach, Ph.D., director of the Acupressure Institute in Berkeley, California. Place the tips of your middle and index fingers on either side of your nostrils directly under your cheekbones. Gently press each side of your nostrils, applying pressure upward and inward while your index fingers push the area alongside your middle fingers. Hold for two minutes.

Press the eye area. There's another pair of acupressure points to know about, especially if you're suffering from a stuffy nose. Place your index fingers directly underneath each cheekbone again, but this time, they should be under the center of each eye rather than near the nostrils. Gently press and hold for two minutes.

Toe, toe, toe your throat. There's an acupressure point on your big toe that can help relieve sore throat pain, according to Muryn. The point is on the underside of each toe—the area underneath where the toe curls. To reach that area, lift your foot to the opposite knee and press the underside of your big toe with your thumb, or use both thumbs for extra pressure. "You need a good deal of pressure so you really feel it," says Muryn, but don't press so hard that you're in pain. If it's hard to get the pressure in that point, you can use the

Preparing for Quicker Relief

In one study, researchers found that moderate exercise can help reduce the number of days that you experience symptoms once you get a cold or the flu. In the study, conducted at Loma Linda University's department of health science in California, researchers divided 50 overweight, sedentary women, ages 25 to 45, into two groups. For 15 weeks, one group walked briskly for 45 minutes five days a week. The women in the other group went about their normal routine.

During those weeks, researchers discovered that the exercise group had half as many days with cold symptoms as the nonexercisers. And when the scientists did blood tests, they found that the gladiators of the immune system, called natural killer cells, were more active in the exercise group. Since these killer cells help people fight off colds more quickly, their prevalence helps to explain why the exercise group had more resistance. Researchers concluded that the steady exercise provided preventive power.

So if you want to be less miserable with cold symptoms this winter, start walking, riding, swimming, dancing, or doing any other kind of heart-pumping activity before the cold season descends on your neighborhood.

eraser end of a pencil, she says. Hold for up to five minutes.

Oil out the mucus. To help clear mucus from your chest, put one or two drops of eucalyptus oil, available in health food stores, on your thumbs and massage the upper portion of the soles of your feet, says Muryn. (If you have sensitive skin, mix the oil with your favorite body cream.) Hold your foot in the palm of one hand while using the other to knead the flat area just below the toes. "Always massage up toward the toes," Muryn says.

Mollify your muscles. If you don't have a fever, take a nice warm soak in the tub to help ease the tired, achy muscles that are especially prevalent with the flu, says Dr. Beckstead. Twenty to 30 minutes is all it takes to start feeling better. Plus, the steam might help open up a clogged nose.

Use the snooze treatment. Try to sleep at least eight hours during the night and get as much rest as possible, even if you don't actually sleep, during the day, says Dr. Beckstead.

There's no better way to relax your body and recoup some of the energy that will help you fight viral invaders.

If you have trouble getting to sleep because you're really congested, raise your head with a few extra pillows to help you breathe more easily. Lying on your back with your head elevated, you might prevent mucus from draining down your throat and disrupting your breathing. As for when you should sleep, "your body will let you know when it's tired," says Dr. Beckstead. Your job is to listen.

Take it slow when you start again. If you've had flu or the below-the-neck symptoms of a bad cold (like a hacking cough), you can resume exercising when those symptoms subside, advises Dr. Primos. But ease back into it, or you could end up sick again. For every day that you were sick, exercise at a lower intensity for two days upon your return. If you were sick for three days, for instance, take it easy for the first six days that you work out after your illness.

COMMUTER'S BACK

How to Veer from Pain

First, the bad news: Since 1987, traffic on American roads has increased by 40 percent. Now, the worse news: The number of miles of new roads has increased by only 1 percent. Translation: All of us—especially folks who commute by car—are going to be spending more time trapped behind the wheel and thus be more at risk for the aches and pains that are an occupational hazard of driving.

But fear not. There's good news as well. You can avoid the pothole of back pain by turning your Honda Civic into a workout gym on wheels.

Mapping Out a Back Strategy

Here are some stress busters and posture promoters to help you steer clear of back attacks.

Strike the right pose. Proper posture is the key to guaranteeing the long ride doesn't charge tolls on your back, says Jane Sullivan-Durand, M.D., a behavioral medicine physician in Contoocook, New Hampshire.

"When you climb into a car, you lose the stabilizing forces of the buttocks and legs to hold you up," says Dr. Sullivan-Durand. "You must count on your abdomen, back, and vertebrae. That's where good posture comes in." Here is her plan for perfect driving posture.

►Sit in an upright position.
►Make sure that the small of your back is flush with the back of the car seat. Use a small pillow or cushion if you need extra support.
►Use your stomach muscles to hold in your torso.

To help maintain your posture, Dr. Sullivan-Durand suggests that you adjust your rearview and side mirrors so that they are best viewed when you are sitting up straight.

Here are some more techniques for saving your back the next time you hit the road.

Stay in motion. No time to exercise, what with the job and playing super-parent? Simply plug in Karkicks. The 60-minute audiotape, created by Natalie Manor of Merrimack, New Hampshire, and endorsed by the New Hampshire Safety Council, lets you and a carload of kids exercise your backs and the rest of your bodies while traveling from here to there. The tape can be purchased by writing to Karkicks, P.O. Box 1508, Merrimack, NH 03054, or you can call the national Toll-Free Directory (800-555-1212) for their phone number.

"One of the great things about Karkicks is that it is always reminding you to tuck in your tummy," says Dr. Sullivan-Durand. "When you do that, you take strain off your lower back."

Shrug off traffic. To relieve tension in the upper back, Dr. Sullivan-Durand suggests this shoulder shimmy from Karkicks.

▶Keep both hands on the steering wheel.

▶Lift both shoulders toward your ears as if you were shrugging, then roll them backward. Hold for a few seconds.

▶Try 10 shoulder shrugs every hour during your commute.

Hit the ceiling. Give your total back a lift and increase blood flow with this upside-down pushup offered by Karkicks, says Dr. Sullivan-Durand (don't forget to close the sunroof first).

▶Keep your right hand on the steering wheel.

▶Place your left hand, palm side up, flat on the car's ceiling (be sure that you can still see through the windshield and in the mirrors).

▶Push your left hand into the ceiling.

▶Hold the pressure for a few seconds, then relax. Do 10 of these pushups.

▶Repeat these steps with your right hand while you hold the wheel with your left hand.

Turn your Subaru into a stretch limo. Kimbra Kimball, a licensed massage therapist and co-owner of Massage Therapeutics in Allentown, Pennsylvania, suggests this stretch-while-you-drive exercise.

▶Extend your left hand toward the windshield.

▶Lean toward the steering wheel just until you feel a stretch in your back and shoulders. Don't touch the wheel with your chest.

▶Hold this stretch for 30 seconds and repeat three or four times. Be sure to keep your head up and your eyes on the road while doing this stretch.

Trying Some Kicks on Route 66

They drove from southern California to eastern Pennsylvania without a drop of coffee. More important, Kimbra Kimball and a friend covered the 4,000-mile trek in eight days—without a single back twinge, ache, or spasm.

Their secret? The pair kept their backs and bodies limber by doing exercises in the car, at rest stops, at gas stations, and along the shoulder of the road.

"Sometimes, we would pull way off on the shoulder, lean against the car, and do pushups or jog in place," says Kimball, a licensed massage therapist who relocated from Los Angeles to Allentown, Pennsylvania. "People did give us strange looks, but so what? We felt a ton better, and hopefully, we set an example for others to follow."

At roadside stops, she and her friend took turns lying on the front seat with their feet dangling out the open passenger door. Kimball would do leg lifts, which stretch the quadriceps and the abdomen, then do curl-ups, which help stretch the back muscles and relieve tension. She recommends doing two or three leg lifts and two or three curl-ups, holding each for 30 seconds. (For directions on how to do leg lifts and curl-ups, see pages 336 and 337.)

"The key thing for anyone who spends a lot of time in the car is to get out and do stretches," says Kimball. "The stretching movements bring oxygen into the body and brain. They clear your head, making you feel energized and revived."

►Switch hands and repeat three or four times with your other arm. "You're elongating the back, stretching those very long muscles on both sides of the spine," explains Kimball.

Unclench that steering wheel. Have you ever felt aches in your fingers, wrists, and back after a two-hour plow through traffic? Maybe it's because you were unconsciously putting a death grip on the steering wheel.

"When we drive, we increase our muscle tension because we maintain this vigilant state of driving defensively," explains Dr. Sullivan-Durand. To fight it, she suggests this exercise.

Place both hands on the steering wheel. Curl your hands over the top (but don't let go). Squeeze your hands for a few seconds, then relax them. Remember to breathe in through your nose and out through your mouth as you relax your muscles to deepen the relaxation. Repeat this movement 10 times every hour.

Shift into rear. An important rule: Never neglect your buttocks on a long car ride. Dr. Sullivan-Durand and Rodger Koppa, Ph.D., transportation researcher at the Texas Transportation Institute and industrial engineering professor at Texas A&M University in College Station, say this exercise from Karkicks works well. The payoff is both a relaxed back and a firmer backside.

Lift your rear slightly from the seat by squeezing the buttock muscles. Hold for 10 seconds, then drop it down. If you are truly talented, try lifting your right buttock slightly and then letting it drop down. Do 10 lifts, alternating your right and left buttocks. Keep your stomach tucked in and sit up straight.

Let the luggage wait. Don't be in a big hurry to leap out of your car, scoot around to

Pocket the Watch

Stress goes right to your back. So Rodger Koppa, Ph.D., transportation researcher at the Texas Transportation Institute and industrial engineering professor at Texas A&M University in College Station, offers this timeless tip: Put your wristwatch in your pocket or purse.

"In rush hour, you're not going to make up any time. Your watch or dashboard clock is an instrument of torture, telling you that you're late," he says. You can cover your dashboard clock with a Post-It Note.

the trunk, and hoist that heavy suitcase out after a long trip, says Dr. Sullivan-Durand. "Your spinal column needs a little flexing before it can take the stress of lifting something heavy," she says. She recommends doing a full-body stretch first, like an exaggerated yawn. In a standing position, raise your arms over your head, bend to one side, and hold for 10 seconds, then repeat on the other side.

Another good stretch for your neck, shoulders, and back is to stand with your knees bent and roll forward slowly. Start with your neck and progress with each vertebra of your spine until your waist is bent and your head is almost touching your knees. You can place your hands on your thighs for support if necessary.

Now you're ready to deal with that suitcase. First, move your suitcase to the front of the trunk. Then, with your back straight, bend your knees and lift it out.

CONSTIPATION

Graham Crackers, Cornflakes, and Easing the Squeeze

Around the same time that plumbing innovations made shoes and a lantern unnecessary for visits to the bathroom, Americans became obsessed with their bowels. Back in the late nineteenth century, the popular belief was that good health meant a daily bowel movement. In fact, two of the most famous food products in our history were developed to help folks meet that schedule. In the 1830s, Dr. Sylvester Graham claimed that his graham flour, the key ingredient in graham crackers, would help folks stay regular. Dr. John Harvey Kellogg made the same claim for his cornflakes, developed in the late 1800s.

We now know that the once-a-day ideal was misbegotten. There's a wide variation in "normal" bowel habits, according to Joyann Kroser, M.D., a gastroenterologist at Presbyterian Medical Center and assistant professor of medicine at the University of Pennsylvania, both in Philadelphia. Some folks move their bowels three times a day; others go three times a week. There is no "right" number of bowel movements, according to the National Institute of Diabetes and Digestive and Kidney Disease. Your routine is just that—your routine. As long as things move smoothly without pain—whatever the pace—there's rarely reason to be concerned.

But when you get clobbered by constipation, the simplest sit-down becomes a painful trial. Most often, the best strategy for fighting constipation is adjusting your diet, says Benjamin Krevsky, M.D., a gastroenterologist and professor of medicine at Temple University School of Medicine in Philadelphia. Often, we get jammed up because we don't have enough fiber in our diets or because we're not drinking enough fluids, says Dr. Krevsky. Boosting your intake of fiber and drinking more fluids are great constipation remedies.

Keep Things Moving

If you find yourself struggling to move your bowels, talk with your doctor about your specific symptoms and try these exercise tips to ease the strain and the pain.

Move for a movement. Physical activity speeds up bowel transit time—the time between eating and eliminating undigested food. One study at Colorado State University found that transit time decreased by almost 23 percent in men who were put on a six-week running program.

While it isn't clear exactly how exercise peps up the passage, one possibility is that changes

in blood flow during exercise accelerate the process. Another theory is that increased nerve activity in the bowels, which occurs during exercise, speeds up the movement of food. Any kind of exercise that gets you up and moving can help relieve constipation, though, says Dr. Kroser.

And the more intense the exercise, the faster waste moves through your bowels, says Dr. Krevsky.

Mow to go. "Mowing the lawn or running a vacuum cleaner can both help relieve constipation," says Dr. Kroser. Just walking around tends to loosen things up. For a more aerobic remedy, take that old Schwinn for a spin around the neighborhood.

Drink up. Aerobic exercise not only stimulates your bowels, it might also make you thirsty enough to drink more liquids. And when you drink more liquids, you help curb constipation by making stools softer, says Dr. Kroser.

Pump up to pump it out. Strength training has also been shown to speed up bowel transit times, according to a study supervised by Ben F. Hurley, Ph.D., director of the exercise science laboratory at the University of Maryland College of Health and Human Performance in College Park. The seven men in Dr. Hurley's study worked out on weight-training machines three times a week for 13 weeks. Their transit times improved by an average of 56 percent. One possible explanation is that exercise raises the level of a gastrointestinal hormone that helps speed up the movement of stool through the intestine and the bowel.

Steer clear of strenuous, heavy lifting. Gentle weight training is good, but don't push the limits. When you exert yourself at weight lifting, you might aggravate hemorrhoids by increasing the pressure on your rectal area, says Dr. Krevsky.

Sit up for soothing. It's possible that abdominal exercises that work the muscles closest to the intestinal tract can also unblock the dam, says Dr. Hurley. The following simple crunch is easy and effective. He recommends doing it at least three times a week.

▶ Lie flat on your back with your knees bent at about a 45-degree angle and your hands crossed over your chest or cupped behind your ears, with your elbows out.

▶ Position your feet together flat on the floor, about six inches from your buttocks. Keep your legs slightly apart.

▶ Curl your upper torso up toward your knees, raising your shoulder blades as high off the floor as you can. Only your shoulders should lift, not your lower back.

▶ Hold your torso up for a second, feeling your abdominal muscles contract.

▶ Return to the starting position, then continue without relaxing between repetitions. Repeat about 30 times or until you feel fatigued.

Put some spine into it. The spinal twist, a yoga position, can relieve constipation, says Richard Miller, Ph.D., a clinical psychologist and co-founder of the International Association of Yoga Therapists in Mill Valley, California. He recommends practicing the pose every day if you are frequently constipated.

Sit up straight in a chair with your feet flat on the floor. Gently twist your torso to the right, then take two or three deep, even breaths with long exhalations while holding the position. Gently twist to the left and again take two or three deep breaths. Doing this once is fine, but to enhance the results, do it two or three

times. This will help the stool move down toward the rectum, says Dr. Miller.

Shoulder the burden. A yoga position called the shoulder stand can increase peristalsis, the tiny contractions that force stool through the bowels, Dr. Miller says.

Begin by lying on your back on the floor, perpendicular to a wall, with the top of your head close to the wall. Supporting your lower back with your hands, raise your legs and move them past your head so that they lightly touch the wall. Be sure to put the weight on your

---

Winning the Battle of the Backside Bulge

Unless you're creative with mirrors, you've probably never seen the bulging veins called hemorrhoids. But they're hardly uncommon, says Benjamin Krevsky, M.D., a gastroenterologist and professor of medicine at Temple University School of Medicine in Philadelphia. About half of all Americans have them by age 50.

Pregnant women, people who strain at having bowel movements, and those who sit a lot are the primary targets. "Truck drivers have the number one problem with hemorrhoids because they sit all day long," he says.

Unfortunately, once you have hemorrhoids, they usually stick around for life, Dr. Krevsky says. Luckily, though, most remain inside the rectal canal and do not cause the itching, pain, or bleeding that are sometimes associated with them. You don't have to do anything about these silent hemorrhoids, he says; you probably won't even know they are there.

But if these varicose veins of the anal area either inflate outside the rectal canal or become so big that they cause difficulty with your bowel movements, you'll have to act. Here are a few tips.

Fidget. If you have a job in which you sit all day, get up and walk around every hour or two, even just to visit the water fountain, says Dr. Krevsky.

Crank up some Kegels. You can strengthen the rectal muscles and even suck painful exterior hemorrhoids back into the canal by doing Kegels, simple exercises that yoga instructors often teach pregnant women to strengthen the vaginal walls before and after delivery, says Richard Miller, Ph.D., a clinical psychologist and co-founder of the International Association of Yoga Therapists in Mill Valley, California. Best of all, this move is subtle enough to do anywhere. Dr. Miller advises doing it in the car. (For a quick training session in Kegel exercises, see "Incontinence" on page 196.)

Soak. Fill the tub with three to four inches of warm water, find a comfortable position, and read that new Michael Crichton novel for about 20 minutes, Dr. Krevsky says. And don't add anything to the water, such as bubble bath or Epsom salts. "Just warm water soothes most hemorrhoids and helps relieve the itching and burning symptoms," he says. But bathing isn't as therapeutic for internal hemorrhoids that bleed.

upper back and shoulders, not on your neck, Dr. Miller advises. Breathe evenly with long exhalations. Hold the pose for 30 seconds at first, then work up to one to three minutes.

Just breathe. Regular, deep breathing works the diaphragm and relieves constipation, says Dr. Miller. Here's how to do it.

Lie flat on your back with the hardcover edition of *War and Peace*, *Webster's Dictionary* or some other heavy book on your stomach at the navel. Move the book up and down with your breathing for about 10 minutes a day, says Dr. Miller. Try to replicate that deep belly breathing while standing (without the book, of course) and take long, slow exhalations. After you can do that, try it while sitting, he says.

Your goal is to make a habit of that kind of deep breathing. Throughout the day, check yourself now and then to make sure that you're breathing deeply from your diaphragm rather than from your chest, says Dr. Miller.

CREATIVITY BLOCK

Exercises in Innovation

One of the great thinkers of all time, Aristotle, got some of his best ideas while walking. The brainy Greek liked to ponder and teach while ambling around the grounds of his school, the famous Lyceum. And he wasn't the last creative type to find inspiration on the move.

"Many people pace or walk or move around when they're stuck for ideas," says Hannah Steinberg, Ph.D., visiting research professor in psychology at Middlesex University in Enfield, England. "Moving the body helps people get through blocks."

Need to think up a fund-raiser for the high school band? A way to get your partner to do the dishes? A new ad campaign for the local tourist bureau? What you need to solve these and other daily problems is creativity.

What's that? You say you're not creative? Nonsense!

"It's impossible *not* to be creative," says Robert Epstein, Ph.D., founder of the Cambridge Center for Behavioral Studies in Cambridge, Massachusetts, and a psychology professor at San Diego State University, who has spent 20 years studying creativity. We're all born creative, he says. It's just that years of conformist schooling and habitual patterns of thought sometimes get in the path of our clev-

erness. Simple exercises and games can help break down walls and light a creative spot.

Moving toward Inspiration

Obviously, exercise that drains you to the point of exhaustion won't light up your idea lightbulb, or at least not right away. But simpler strategies might break through whatever is blocking your creativity. Try these blockbuster tips.

Chase down your Muse. Still struggling to come up with that fund-raising idea? Take a break and go for a walk. Amble over to Lookout Pond, or if it's frozen, go skating. If you don't have time for a long stroll or a leisurely skate, simply walk up and down the stairs in your office building or go down to your basement and give the exercise bike a quick workout, suggests Dr. Steinberg.

No one knows exactly why exercise seems to spur creativity. Some researchers speculate that it helps because it puts us in a good mood. In a study by Dr. Steinberg and others, though, aerobic exercise made 63 people more creative independently of improving their moods.

"It might be that exercise improves creativity because it gets more blood to the brain, or be-

Daydreaming with Dali: A Surreal Creativity Exercise

When Thomas Edison and Salvador Dali needed creative ideas, they headed for the same destination—sleep. Nodding off can be an exercise in creativity.

Both the Wizard of Menlo Park and the mustachioed surrealist painter sought innovative ideas and images in the semi-sleep called the hypnagogic state. A way-station between wakefulness and true sleep, the hypnagogic state is where dreams and conscious thought mix, giving rise to wild images and wildly innovative ideas.

"Everyone experiences the hypnagogic state, not just Dali," says Robert Epstein, Ph.D., founder of the Cambridge Center for Behavioral Studies in Cambridge, Massachusetts, a psychology professor at San Diego State University, and an expert in creativity. "If you're looking for an interesting image, there's no richer source."

How do you remember these images and ideas so that you can put them to use in the bright light of consciousness? Do as Dali and Edison did and wake yourself up in midstream, suggests Dr. Epstein.

To capture ideas and images from the hypnagogic state, Edison used to nod off in a chair with his arms on the armrests and a ball bearing in each upturned palm. On the floor beneath each hand, he placed a pie tin. As soon as he drifted toward sleep, his hands would relax, and the ball bearings would drop onto the plates. The noise awakened him, and with the return to full consciousness, he immediately scribbled down the ideas that had come to him. Dali did the same thing, Dr. Epstein notes, but the painter held spoons over glasses instead of ball bearings over pie tins.

Both men were prodigiously inventive and creative. Edison created more than a thousand patented products, including the lightbulb that changed our lives. And Dali not only painted canvases during his lifetime, he also wrote several books and made films.

cause exercise releases endorphins, which may stimulate creativity directly," she speculates. "Or it might help because people expect it to help."

Do the twist. Still blocked after your walk? Put on your favorite dance music and pretend you're on *American Bandstand*. (If you're shy, close the curtains first.) People are even more creative after "free" exercise such as improvisational dance than after a session of prescribed exercise like follow-the-teacher aerobics, says Dr. Steinberg.

Recruit some friends. Sometimes, the best way to get through a creative block is to solicit others' ideas. Here's how to blend the lubricant of exercise with the stimulant of verbal play.

Start out jogging with two buddies. As you jog gently, air out possible ideas for that fundraiser. Listen closely to the ideas of the other two people and see if there's a seed that might be nourished into a full-blown idea. Enjoy yourself. Before you know it, you'll have jogged right past the boring idea of a bake sale to the delightful idea of the first annual Moms, Dads, and Teachers Comic Shakespeare Festival.

Dancing

Find Happiness with Some Fancy Footwork

WHAT IT IS

There is, perhaps, only one thing in life that is more romantic and more intimate than dance—which is why, of course, dancing so often leads to love. Fred Astaire always won the heart of Ginger Rogers (and Cyd Charisse and Rita Hayworth and Audrey Hepburn) with a dance. Dancing may simply be moving your feet and swaying your body to a rhythm, but it is also so much more than that. Letting go. Being swept away. Moving from your pelvis. Dance has rocked our world since the beginning of time.

Dancing gives us connection, and connection cures one of our most destructive diseases—loneliness.

"When I was first working as a psychiatrist, I noticed that a lot of my patients had one thing in common: They were lonely," says Barry Sultanoff, M.D., a holistic psychiatrist in private practice in Kensington, Maryland. "It was just them and the television. The flip side of loneliness is connection. When people find community—and they will when they start dancing—it deepens their own soul and creates more radiant health."

"Ballroom dancing is, for some people, a competitive sport that may soon be an Olympic event," says Jackie Rogers of Westfield, New Jersey, public relations director of the National Dance Council of America.

Several dance styles are featured in competitions. There is the International Standard Competition, which features men in white tie and tails and women in elegant, flowing evening gowns. For Latin competitions, dancers wear sexy dresses and suits to slink their way through sultry dance moves. But don't worry if you're not up for gliding gracefully in front of the glaring light of television cameras. For most people, dancing is simply a great way to have fun, be active, and meet people.

WHAT IT DOES

Dancing helps your body, but it also helps your soul by giving you a connection to your partner and, further than that, by connecting your body to the music. "Dance helps flexibility, balance, and bone density, because it is weight-bearing," says Ken Forsythe, M.D., clinical professor of sports medicine at the University of California, Davis, San Joaquin Hospital.

"But what you can get mentally from dance is really special," he says. "It's elegant and graceful and allows you to be soothed by music. Talk about a stress reliever! If you can find something that you like to do with your spouse that combines activity with the closeness associated with dance, boy, that's got to be pretty good."

Taking a few twirls around the dance floor will do more than just help you make friends and perhaps stir up some passions. Done for just an hour, the leisurely box step will burn just over 200 calories for a 150-pound woman, and if you kick that up a notch and start doing the twist or lambada, you can burn three times that amount.

Don't have a partner? That doesn't have to stop you. "I'm 53, and I started doing the Argentine tango a year ago on my own," says Dr. Sultanoff. "The physical benefits include strengthening my legs, an increased sense of balance, and a new feeling of connecting with my body. More than that, however, is how much I enjoy it. I meet so many new people that I would never otherwise meet. The tango community is a diverse and multicultural one. Everyone is there to dance and it's nothing like a pickup. Nobody drinks while they dance."

REAPING THE BENEFITS

Dance presents all dancers with two challenges: Inviting (or being invited by) someone to dance, and then actually dancing. In public. In front of people. Who are watching. And who know how to dance better than you do.

"Most dance studios have introductory offers, such as a free first lesson. So the first thing you should do is call a bunch of dance studios, then take one lesson at each of them to find the right studio and instructor for you," says Del Bradford, a dance instructor and owner of Fred Astaire Dance Studio in Boise, Idaho. "Most studios offer both private lessons and group lessons with an instructor. If you're a beginner and you can afford it, you'll want to start with private lessons so you can develop the leading and following skills. Once you're comfortable with your skills, as a single or part of a couple, you can advance into group lessons."

If you're single and you advance to group lessons, don't be intimidated because you don't have a partner. Eventually, whether you are part of a couple or not, you'll be dancing with everyone. Almost all instructors encourage their students to learn how to dance comfortably with all other dancers, not just their romantic partners. Most dance studios host dance parties as well. Likewise, many communities have swing dance, country line dance, or folk dancing organizations that sponsor public dance activities. These evenings usually begin with free public lessons.

In terms of the male-female ratio, classes are usually split 50-50. "I've been doing this for 19 years," says Bradford. "I've known several couples who have met through dancing, both in class and at dance social events."

TRIAL RUN

The first step you'll learn in most dance studios is the box step. "The box step works on the natural opposition principle," says Bradford. "When the man's left foot is moving, the woman's right foot is moving. People step on each other's feet when they move on the same foot."

For this example, we'll assume that you're the woman. The man's instructions are given in parentheses.

▶Start with your feet together, but keep your weight on your left foot so your right foot is free to move. Step back with your right foot. (The man has his weight on his right foot, with his left foot free. He steps forward with his left.)

▶Second step: Step to the left with your left foot and bring your right foot to meet your left foot. Put your right foot down, shifting your weight from your left foot to your right foot. (The gentleman steps to his right with his right foot, then brings his left foot to meet his right. He shifts his weight to his left foot.)

▶Step forward on your left foot and with your right foot, step to the right. Then bring your left foot to your right and put your weight on your left foot. (The man now steps back with his right foot and brings his left foot to the left side. He then brings his right foot to meet his left, shifting his weight to his right side.)

"Eventually, you'll be making a big box and maneuvering around the room," says Bradford. After you master the box step, you'll go on to learn the fox trot and waltz, then progress to swing and cha-cha. One of the most difficult aspects of dance is learning how to lead and how to follow. Instructors spend time teaching people how to do these things with grace and ease.

Depression

Step Outside of Your Blues

Mild depression is the common cold of mental woes. Almost everyone has a brief bout with the blues now and then, and most often, we just ride out the gloom. But there's no reason to simply wait for the fog to lift. You can put the spring back in your step with exercise. Physical activity can help speed you back to your old self.

A note of caution, however: Although exercise is a powerful weapon against the plain old, everyday blues, you should seek help for a deep depression, even though exercise can benefit you then, too. Signs of serious depression can include despondency that lasts two weeks or more, a change in your sleeping patterns, an inability to take pleasure in anything, and persistent dark or suicidal thoughts. If you have any of these symptoms, it's important that you reach out to family, friends, and mental health professionals.

The Exercise Effect

Exercise is good medicine for lots of reasons. To begin with, it's a healthy distraction and helps take your mind off your troubles. It also shines up the old self-esteem by giving you a sense of mastery and accomplishment. Intense aerobic exercise kicks up the level of those feel-good hormones called endorphins. Finally, exercise has the ability to reduce any stress that may be contributing to your low mood.

When it comes to battling the blues, it doesn't even matter what type of exercise you do. "The literature so far supports the opinion that all kinds of exercise can work against clinical depression," says Maria Fiatarone Singh, M.D., associate professor at Tufts University School of Nutrition Science and Policy in Boston. It has been shown to alleviate the blues, too, she says. Dr. Fiatarone Singh's research has found anti-depression benefits from weight training, and other research indicates that aerobics and relaxation-based exercises like yoga have a mood-boosting effect as well.

Try Nature's Balm

While it doesn't matter what kind of exercise you do, when it comes to stoking your sagging spirits, it may matter *where* you do it. A field of study called ecopsychology suggests that spending time in the great outdoors is your best bet for mood management. Getting in touch with nature is often a spirit lifter, according to Michael Cohen, Ed.D., director of Project Nature Connect in Friday Harbor, Washington, and author of *Reconnecting with Nature*.

We live the majority of our lives indoors, says Dr. Cohen. Life without trees, grass, wind, and sun can make us sensorily deprived in certain ways. "Without the natural stimulation that comes from the out-of-doors, we can begin to feel sad and out of touch," he says.

Since exercise is good for us and nature's good for us, too, it stands to reason that exercising outdoors may give us a double-barreled mood boost—an endorphin rush as well as a Rocky Mountain natural high.

Any weather is good weather for lifting your mood and challenging your senses. Although rainy days may seem intrinsically dreary, with the right attitude, they can actually be invigorating to the downtrodden spirit, says Philip Chard, psychotherapist and author of *The Healing Earth*. In the winter, make a snowball. In the drizzle, savor the wetness on your skin.

It's a joy to fight bad moods with both sweat and an awareness of beauty. You can visit Mother Earth while walking, jogging, cross-country skiing, cycling, or swimming in a lake or river. "Nature stops us from feeling alone," says Dr. Cohen, "because, in reality, we're not."

The Walking and Wonder Cure

Try combining a walk in the wild with periodic "nature pitstops," Dr. Cohen suggests. Both your body and your mind will benefit. Here are some steps on a mood-boosting walking tour.

Set out. If you live in the country, consider yourself lucky and start out right from your front door. But don't despair if you live far from farmland. A municipal park, a garden, or even a tree-lined city street can all give you a dose of nature, says Dr. Cohen. The key is to be outside, away from four walls and a ceiling.

Start walking slowly. Give your body a chance to warm up, then gradually pick up the pace until you're walking briskly. After about five minutes, take your first nature pitstop.

Appreciate a tree. Take a few moments to look closely at a tree. Choose one that appeals to you and get near enough to appreciate its beauty. Reach out and rub a hand over its bark. Take a close look at the symmetry of its leaves.

Get a second wind. Start walking again. Put some pep in your step to get your heart rate up a bit. Try to listen for the sounds that you're usually too busy to hear—birdsong and wind-whisper. This will heighten senses that have been asleep, says Dr. Cohen, and leave you feeling refreshed. After you've walked for five more minutes, take another nature break.

Have a natural seat. Maybe there's a big boulder nearby. A tree stump or grassy hill will do just as well. Take a seat and let your hands examine the texture of the landscape around you. Notice the small things like plants, insects, and pebbles. Appreciate the warmth or coolness of the surface on which you're seated.

Head for home. Walking at a nice, crisp pace, start strutting home. On your return journey, maintain the sense of connection that you've developed. Stay in touch with the color of the sky, the feel of the breeze, the smell of autumn leaves or spring blooms.

Let Fido Help Fight the Funk

Exercising among trees, rocks, and streams is one way to lift your spirits. But you may have another "antidepressant" lying right on your living room rug. Pets—especially dogs—can be mood brighteners, according to Nancy Bradley, R.N., chief evaluator for Therapy

Dogs International, based in Flanders, New Jersey. (Cats make good companions, too, Bradley says, but dogs are in a class of their own because you have to be active just to keep up with them.)

Dogs lift our spirits because they give us un-conditional love. "Dogs don't care where you've been or what you've done," says Bradley. "They love you no matter what." This total acceptance goes a long way toward repairing any damage done by a sometimes-unkind world. At the end of the day, when no one else seems to understand, you can be sure that your dog does.

Getting some exercise with your Airedale will build your body and soothe your soul at the same time. Here are a few fun suggestions for starters.

Jog your dog. Walking or running with your dog is one great way to get in shape while giving your pal a chance to play. Walk or jog for at least 10 minutes at a time, making a total of 30 minutes per day your goal if you want to consider it a workout, says Dr. Fiatarone Singh.

Let Benji fetch it. Fetching can be a good exercise, but only if you do some running, too. To get your own heart rate up, try racing your pet to the pickup. You say you can beat him? Then jog behind him as he runs.

Leap with Lassie. Can your dog catch a Frisbee? Give it a try. It doesn't take a trained stage dog to do these kinds of tricks, just a dime-store throwing disk and a little encour-agement. Make sure that when you throw the Frisbee, you chase after it, too.

Teach him a new trick. Take your canine to obedience classes, where you'll both get a double benefit. The classes will get you moving, but they'll also increase the bond that you share with your pet. Plus, Bradley says, you'll be par-

Tippecanoe and Don't Be Blue

Walking is only one way to get closer to nature. Boating is another great means of combining exercise and elation.

Of course, you'll want to leave the motor behind, recommends Philip Chard, psychotherapist and author of *The Healing Earth*. Try a rowboat or a canoe. The silence, as well as the rowing or paddling action, will do you good.

Start by selecting a placid body of water—a small lake is ideal. If you don't own your own craft, many state parks offer inexpensive rowboat rentals right on site.

As you push off, imagine yourself a modern-day Huckleberry Finn. Paddle or row for a while to leave the shore behind and get your muscles warmed up. As you head into the open water, let the fresh air fill your lungs and mind. Start stroking. Get your heart rate up a bit.

Then take a break and just drift. Ponder the pond below you. There's life on the surface and in the depths. Check out the sky while you're at it. There's beauty there, too.

When you're through lazing for the day, hold on to the sense of calmness that you found on the water. And remember, you can always return.

ticipating in an activity in which you get to talk to people with the same interests. That in itself can boost your spirits.

DIABETES

Controlling a Too-Sweet Bloodstream

You could have a sweet tooth, a sweetheart, even sweet cheeks—but sweet blood? Too much sugar circulating through your vessels is a hallmark of diabetes. While this may sound like a simple enough problem, the long-term complications of uncontrolled diabetes are not simple at all. Infection, vision problems, and heart disease are just a few of the health worries that diabetes brings.

Keeping blood sugar levels close to normal goes a long way toward preventing the worst complications. And exercise is an important tool for regaining control.

The Inside Scoop on Insulin

The star of the diabetes drama is a hormone called insulin, which is produced by special cells in the pancreas. The role of insulin is to help the body's cells burn blood sugar (also called glucose).

A cell taking in glucose is something like a change machine taking in dollar bills. For glucose to pass muster, insulin must first bind to a special receptor on the cell, which then allows the glucose to pass into the cell, where it "makes change" in the form of energy. In people without diabetes, insulin is always around to open the door to allow glucose into the cells, and your energy keeps pouring out like so much loose change.

The problem in the most common form of diabetes—Type II (formerly known as non-insulin-dependent or adult-onset) diabetes—lies with the change machine itself. The cells just won't accept insulin (they're said to be insulin-resistant), and glucose, deprived of its entry ticket, is rejected by the cell. The pancreas reacts by making more insulin, but the amount of glucose circulating in the blood stays high and can get even higher with time.

There are ways to get blood sugar levels back in control. Insulin injections are commonly used for Type I (formerly known as insulin-dependent or juvenile-onset) diabetes. For people who have this type, insulin resistance is not the problem; instead, their bodies don't produce any insulin. For people who have Type II diabetes, drugs called oral hypoglycemics are often used to bring down glucose levels. But for both types of diabetes, exercise is a big help, too. It's a natural way to improve insulin action, allowing excess glucose to move out of the blood and into the cells for use and helping to lower the risk of heart disease.

Doing In Diabetes with Exercise

Exercise helps in two very important ways, by burning off body fat and by building up more lean muscle mass, says John Ivy, Ph.D., director of the exercise physiology and metabolism laboratory at the University of Texas in Austin.

Obesity is the number one risk factor for Type II diabetes. In fact, 9 out of 10 people with this form of the disease are overweight. For reasons that are still unclear, carting around excess body fat—especially the spare-tire or abdominal variety—appears to make us more resistant to the effects of insulin. But luckily, "when abdominal fat goes away," says Dr. Ivy, "you see much better insulin use."

The mass of your muscles also plays a potent part in beating diabetes. Specifically, muscle cells are glucose guzzlers. The more muscle you have, the less sugar there will be to build up in your bloodstream. When glucose levels are kept in check this way, you run a much smaller risk of developing diabetic complications down the road.

But now for the big news: Research has shown that for people with mild to moderate Type II diabetes, regular physical activity does more than anyone ever dreamed it could. Exercise (combined with a healthy diet) may eventually allow them to stop taking medication, according to George King, M.D., senior investigator at the Joslin Diabetes Center in Boston. Some experts would even say that a lifetime commitment to exercise can actually cure your diabetes—a feat unheard of by any other means.

So what are you waiting for? Get on the road to natural glucose control. Here are the rules to keep in mind when you're putting your body in gear.

Get the green light. If you have either Type I or Type II diabetes, be sure to get your

---

Smart Moves to Tame Type I

We know that being more active can bring a wealth of wellness to people with Type I diabetes, including a reduced risk of heart disease and mood-lifting effects. But working out when you have Type I (formerly called insulin-dependent or juvenile-onset diabetes) requires some juggling skills. In order to keep blood sugar levels stable, anyone with this kind of diabetes needs to carefully balance physical activity, insulin treatments, and the amount and type of food he eats as well as keep careful tabs on glucose levels, advise diabetes experts.

Exercising too much or when you didn't plan to brings a serious risk of hypoglycemia, or low blood sugar. Hypoglycemia ranges from mild to severe, and the symptoms include anything from sweating and trembling to mental confusion, seizures, and even coma. Exercise also changes your need for medication, but the effects are unpredictable and different for everyone.

To safely benefit from exercise when you have Type I diabetes, you'll need your doctor's guidance. He can help you figure out the best workout and medication schedule to follow and suggest the right exercises for you to do.

doctor's okay before starting on your new exercise program. If you have Type II diabetes and are over age 35, an exercise stress test may be recommended. If you are on oral hypoglycemics or insulin, you'll need to self-monitor your glucose levels and make any necessary adjustments in your food and medications. And if you have the eye-related complication called diabetic retinopathy, neuropathy (nerve damage), or high blood pressure, physical activity could make matters worse.

Fuel yourself first. Eat a healthy snack one to three hours before exercising, says Dr. King. This is especially important for people who are taking insulin or the oral hypoglycemic medications known as sulfonylureas. These drugs use up glucose in the blood, but so does exercise. Extra fuel in the form of carbohydrates (which are turned into glucose in the body) can keep blood sugar levels from dipping too low during your workout. Try a piece of fruit or half a bagel.

Do it routinely. Exercise increases your body's sensitivity to insulin for 24 to 48 hours. That's a good thing, because ultimately you may need less medication to control your blood sugar. To maintain the momentum of better glucose use, you need to exercise at least every other day, says Dr. Ivy.

Try both kinds. Since weight loss and muscle building each help to control Type II diabetes, be sure to try activities that encourage both. You could do cross-training by alternating days of aerobic exercise and resistance training. Or do a combination workout just three days a week. You might follow a set of weight lifting with a brisk walk or bike ride, for instance, suggests Dr. Ivy.

When to See a Doctor

Because half of the people who have diabetes are unaware of it, you should see a doctor if you're experiencing any of the following symptoms.

▶ Frequent urination
▶ Extreme hunger or thirst
▶ Unusual weight loss
▶ Extreme fatigue
▶ Bruises that are slow to heal
▶ Tingling in the hands or feet
▶ Blurred vision

Feet First When Working Out

Foot care is important for any active person, but it's even more so for the active person who happens to have diabetes, says Dr. King. So before you lace up those hightops, take a good look at what you'll be pounding the pavement with—your feet.

After years of diabetes, nerves exposed to high levels of blood sugar can show signs of damage. The feet are especially vulnerable because the nerves running from your spine to your feet are very long, says Dr. King. "The signal has to travel farther, making trouble more likely," he explains.

People with diabetes also have higher rates of heart disease, stroke, and impaired circulation than people without the disease. It's thought that unusually high amounts of blood sugar eventually cause abnormalities in red blood cells, which in turn affect the arteries and blood vessels. Poor circulation and narrowed blood vessels are especially a problem in the feet and legs. Without a free flow of blood, minor foot injuries may turn into infections that just won't go away. In extreme cases of poor healing, people can get gangrene, a condition in which the tissue blackens and dies.

On the other hand, the same exercise that puts the heat on your feet can also help protect them. Since regular exercise steadies blood sugar levels, it may prevent nerve damage from happening in the first place. And if you're already experiencing some numbness, exercise can help keep it from getting any worse. It can prevent or reverse the effects of poor circulation by bringing health-rendering oxygen to the tissues of your legs and feet. And regular exercise will also lower your risk of heart disease and stroke.

It's obvious from this why foot protection is essential if you want to keep reaping the benefits of exercise. Here are five rules of thumb to help keep you in tiptop shape from heel to toe.

Keep them in sight. First and foremost, it's important to give your feet a thorough visual inspection every day. Make a point of concentrating on one whole foot—top, bottom, and sides. Search like a detective for any cuts, cracks, blisters, hotspots, or other injuries that you might not be able to feel. Then scrutinize foot number two. And when you visit your doctor, be sure to have her to take a look, too, says Dr. King.

Opt for alternatives. Running and walking are not the best exercise choices for folks with foot trouble. To sweat without stomping, try low-impact activities like rowing or cycling in-stead. Or, says Dr. Ivy, consider an aquatic workout. Pool-based exercises are a double bargain. You'll get a challenging workout, plus the water pressure on your legs gives circulation an extra boost.

Air them out. After swimming, showering, or bathing, let your feet get completely dry before covering them with socks or shoes, says Dr. King. "This is the biggest mistake people make when it comes to looking out for their feet." Dampness provides a happy home for bacteria, and bacteria growth is what often leads to infection. If you're in a hurry, use a hair dryer set on cool after you towel-dry your feet. Check between the toes to make sure no dampness lingers. Keep things dry as a desert to ensure that no unwanted "tenants" decide to move in and wreck the joint.

Go for the good moisture. Using a rich moisturizer on your feet is a pleasant way to prevent cracks that can form in dry or callused skin, says Dr. King. Choose a cream you like and slather it on. But again, don't be in a hurry to buckle up your boots. Wait a minute or two for the lotion to soak in, and don't slip into your socks until your feet feel dry to the touch.

Find the perfect fit. This one may seem obvious, but if you're already experiencing some numbness in your feet, says Dr. King, finding shoes that fit correctly is no cakewalk. Take your time when shopping, and find a salesperson who knows how to measure your feet. If you need to, be prepared to explain your situation so that the person helping you knows just how important the right size is.

DRIVER FATIGUE

Get in Shape and Stay Safe

Even legends get tired behind the wheel. Indy car superstar Michael Andretti may spend his days jockeying for position at 200 miles per hour, but during those long family trips in the car, he fights fatigue just like the rest of us.

To Andretti, staying alert while driving—at any speed—starts with a commitment to general fitness. He exercises every day, mixing stationary biking and time on a step machine with a little weight lifting. "I'm in better shape now than when I was a rookie in 1984," he says.

Being in good overall shape helps fight fatigue in countless situations. Here are a few tips for fighting road weariness from the man who has won more races and led more laps than any other active Indy car driver.

Prep with a straight eight. The day before a race, Andretti always scores eight hours of sleep. Follow his lead. Before you head north to your beloved fishing cabin, rack up a few extra hours of rest. You won't feel as sleepy after a day on the road.

Stretch before you leave. Andretti always does a series of leg and foot stretches right before he climbs into the cockpit. "I do just enough to relax the muscles. Relaxed muscles translate into fewer injuries, more efficient reaction times, and increased tolerance for stress," he says. It doesn't

matter that you have less horsepower; you have the same muscles. Spend 10 minutes loosening up before you buckle up.

Be ready for anything. Andretti feels that extreme caution will fight drowsiness. "I always try to think of the worst that can happen and then be ready for it." If you can keep a vigilant edge while driving, it will keep you from getting sleepy.

Shrug it off. Shoulder shrugs help Andretti release tension. Anything that keeps your body moving behind the wheel will energize you a bit.

Experts also offer these tips to help you stay peppy out there on Highway 61.

Don't strangle the wheel. The thicker the traffic and the longer the drive, the tighter some of us tend to grip the steering wheel, says Jane Sullivan-Durand, M.D., a behavioral medicine physician in Contoocook, New Hampshire. This death grip requires much more energy and increases fatigue. Loosen your grip and remember to breathe. "Breathing will bring oxygen to your brain, making you more likely to stay awake," Dr. Sullivan-Durand says.

Take a break. If you get tired, take the time to pull over, focus on your breathing, and rest, Dr. Sullivan-Durand advises. Just a few minutes of relaxation can revive you for your long ride. "Don't take chances by pushing through fatigue. That's how accidents happen. It's better to arrive 10 minutes late than not at all," she says.

Swallow less coffee and more air. Gulps of air may be as effective as coffee when you feel a bit sleepy. Roll down the window, take some deep inhalations of air, and then blow them out, suggests Natalie Manor, creator of Karkicks, a fitness-while-you-drive audiotape.

The extra oxygen inside you is a natural way to keep you alert, adds Dan Hamner, M.D., a physiatrist and sports medicine specialist in New York City. He suggests doing the following exercise to get maximum benefit from taking in air: Open your window and take 20 or 30 short breaths, then suck in as much air as you can and blow it all out. This will help rejuvenate you for a half-hour. He warns, however, that this technique should not be used as a substitute for sleep.

Stop, stretch, and meditate. Experts agree that taking periodic breaks is important. "If you are trying to stay awake, pull over and let yourself take a breather," says Kimbra Kimball, a licensed massage therapist and co-owner of Massage Therapeutics in Allentown, Pennsylvania. "A brief meditation break will allow you to relax fully and help you feel that you have rested or slept. Follow your meditation with a full-body stretch, which will bring oxygen into cells and help you feel rejuvenated and refreshed, says Kimball.

To do a full-body stretch, Dr. Sullivan-Durand offers the following technique: Lift your arms over your head and stretch to one side, hold for 10 seconds, then repeat on the other side.

Join the army. Rodger Koppa, Ph.D., transportation researcher at the Texas Transportation Institute and industrial engineering professor at Texas A&M University in College Station, suggests adopting the army's approach.

Beware the Postlunch Slump

It's no surprise that the middle of the night—midnight to 6:00 A.M.—is a peak time for driver fatigue. But there's another drowsy period that may be more unexpected. According to a National Highway Traffic Safety Administration study, the hours between noon and 3:00 P.M. are sleepy times for drivers, too. Who's most likely to get sleepy behind the wheel? Males ages 16 to 29, swing-shift workers, and people who have sleep apnea, a condition in which regular sleep is interrupted by breathing difficulties.

"The armed services have a convoy rule," he says. "They take 15-minute breaks every two or three hours or every 100 to 150 miles, whichever comes first. It's a good rule for all of us."

Drop and give me 20. Kimball fights fatigue by getting out of her car and doing 20 pushups, jogging in place for 30 seconds, jogging around the car, or jogging up and down the shoulder or at a rest stop. "If time is of the essence, a quick up-and-back will help," she says.

The key is to get oxygen pumping through your bloodstream through aerobic exercise, says Dr. Hamner.

Remember your radio. A cross-country trek may go more quickly if you tune in to a lively talk show. Or if Howard Stern or Rush Limbaugh isn't your slice of pie, use audiobooks to fight the boredom that can lead to fatigue, says Dr. Koppa. Try a tape of a Stephen King thriller to keep you alert, he says.

ELBOW PAIN

Relieving Aches Off the Court

A riddle: What does your elbow have in common with a thoroughbred horse farm?

Answer: They're both pretty stable joints.

But seriously, folks... in general, the elbow—the muscles and tendons that connect three bones, the radius, the ulna, and the humerus—is a fairly sturdy structure. But when things do go awry in there, you suddenly become painfully aware of how often you bend and bang your arm. Elbow ache is a nuisance.

Two of the most common elbow afflictions are quite sporty: tennis elbow and its less famous but sometimes equally obnoxious sibling, golfer's elbow.

Tennis Elbow, Anyone?

You don't have to be a Wimbledon qualifier or even a club hacker to get tennis elbow. Anyone who repeatedly rotates their elbows or flexes their wrists while gripping a heavy object is at risk. That includes people who work with tools, do a lot of gardening, or just tote a heavy briefcase. Tennis elbow is the inflammation of the wrist extensor muscles where they attach to the outer knob of the elbow's upper arm bone. It's the most common form of chronic elbow pain.

"Tennis elbow is a catch-all term for pain in the outer elbow joint," says Dale Anderson, M.D., an urgent-care physician in Minneapolis and author of health humor books such as *The Orchestra Conductor's Secret to Health and Long Life.*

Elbow pain can make turning a doorknob or wringing out a towel difficult, says Dr. Anderson. But exercise can help. One Swedish study showed that exercises that stretch the forearm muscles improved range of motion and reduced elbow pain. These exercises may help you ace tennis elbow.

Warm up for the match. Here's a stretch that's a good warmup for elbow activity, according to Michael Ciccotti, M.D., an orthopedic surgeon and director of sports medicine at the Rothman Institute at Thomas Jefferson University in Philadelphia. Try it before you play Beethoven's Fifth or head for the courts with your neighbor.

►Stretch your left arm in front of you, keeping your elbow straight and your palm down.

►Slowly bend your wrist downward until your fingers point toward the floor.

►Place your right hand on top of your left hand and gently press until you feel a tension stretch in your left forearm. Hold for 15 seconds.

►Repeat the same steps with your right arm.

Do the Kalamazoo. Jeffrey Willson, a certified athletic trainer at K Valley Orthopedics in Kalamazoo, Michigan, suggests this basic stretch. He and the medical staff at K Valley Orthopedics recommend doing this exercise four or five times a day, holding the stretch for 10 seconds at a time.

▶Begin with your aching arm hanging at your side.

▶Bend your elbow to 90 degrees so that your palm is facing up (*1a*).

▶Slowly rotate your palm to face the ground.

▶Flex your wrist by bending your fingers toward the underside of your forearm (*1b*).

▶Finally, keeping your wrist flexed, straighten your elbow (*1c*).

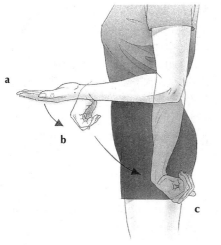

1

Try the tendon glide. This exercise will help alleviate and possibly prevent elbow pain, according to Robert Markison, M.D., a hand surgeon and associate clinical professor of surgery at the University of California, San Francisco. "It is a good stretching exercise for all the tendons in the wrist and elbows," he says, "especially before you plan to do a lot of computer keyboard work."

▶Extend one arm, palm up, in front of you.

▶Use your other hand to pull back on the fingers of the extended hand, bending your hand at the wrist.

▶Let go of your fingers and turn the extended arm over so that the palm faces down.

▶Again pull back on your fingers with the opposite hand, bending at the wrist.

▶Repeat with the other arm, then do the entire sequence once more.

Master massage. To relax the muscles surrounding your elbow, massage your forearm, suggests Meir Schneider, Ph.D., a licensed massage therapist, founder of the Center and School for Self-Healing in San Francisco, and creator of the Meir Schneider Self-Healing Method.

Starting at your wrist, make large, unhurried circles with your fingertips, kneading the muscles up to your elbow. Massage both the front and back of your forearm. Alternatively, you can grasp your forearm with your thumb on the underside of your wrist and your fingers on top and massage both surfaces at the same time. Knead in circles with your thumb and fingertips. The total massage should take about five minutes to complete.

Golfer's Elbow

Golfer's elbow is the flip side of tennis elbow. While tennis elbow is a problem with the muscles that help the wrist straighten and extend, golfer's elbow affects the flexor muscles that help the wrist and elbow bend. This problem isn't limited to Tiger Woods wannabes, according to Dr. Anderson. The golf moniker stems from the fact that golfers who dig up too much of the grass with their swings often strain their flexors.

Try the swan position. To assuage the pain,

try this "fold and hold" exercise recommended by Dr. Anderson. If you position your wrist and forearm correctly, he says, the tender spot should go away or improve by 75 percent.

▶Fold your upper arm close to your body by bending your elbow into a tight angle. Bring the back of your wrist close to your shoulder.

▶Bend and flex your wrist with the fingertips pointing downward. The palm of your hand should face the ground. Your arm should resemble a swan, with your hand as the head.

▶Use your other hand to fine-tune your wrist and forearm into the most pain-free position.

▶Place your other hand on top of the hand that's bent (2).

▶Hold for 90 seconds, then release slowly.

Making Your Elbows Stronger

Here are two more ideas for beefing up your elbows.

Get a grip. To work the elbow muscles, grab a hand gripper or squeeze a tennis ball, says Dr. Ciccotti. Grip and hold for five seconds, then release. Repeat this 15 times. The steady motion of gripping, squeezing, and releasing works the flexor and extensor muscles in the hand and elbow. If you feel sharp pain in your elbow or wrist, however, stop the exercise, he says.

Resist yourself. Hold your left arm in front of you with the palm facing down. Make a fist and bend your left wrist upward. Cup your right hand over your left fist and try to push straight down with your right hand as you resist the downward force with your left fist. Tense the muscles for 10 seconds, then relax. Try five times, then switch hands and repeat.

Doing this exercise two or three times a day can bolster the muscles around your elbows and wrists, says Dr. Ciccotti.

EMPTY NEST SYNDROME

Cognitive Calisthenics and Annual Competitions

Listen carefully. . . . You don't hear anything, do you? That's because the kids are gone. The last of them has trekked off to the University of Minnesota, and suddenly the house is quiet. But it's not a nice quiet, it's a lonely quiet. After all the lively tumult of watching over kids, it's sometimes hard to adjust to the serenity of an empty nest.

"When the kids go away, the parents may wonder, 'What are we going to do with all this time on our hands?'" says Carol Goldberg, Ph.D., a clinical psychologist and president of Getting Ahead Programs, a New York–based corporation that offers workshops on stress management, health, and wellness. "They may also wonder, 'What are we going to do that's meaningful in our lives?'" she says. After all, they've been Mom and Dad for a long time.

Experts agree that exercise is one of the best ways to use the time that you used to spend doing the kids' laundry. There are several reasons why working out is a perfect fit for parents who are suddenly home alone and not happy about it.

First, exercise is a proven mood booster. A regular exercise plan not only elevates the level of feel-good neurotransmitters in your brain, it also gives you energy and a feeling of pride and self-discipline that can help beat the blues. Despondence is often worst for people who feel tired and powerless to control their lives, says Dr. Goldberg. If you're part of an aerobics class or you're running three miles a day, you'll get both an endorphin rush and a feeling that you have at least a modicum of control over your life.

Second, exercise can be a great opportunity to form new friendships. Once the kids have left, lots of people feel as though they've lost their most powerful human bond. It's easy to miss your kids a lot and to have feelings of loneliness and isolation. Exercise can be a convenient way to link up with people.

Join a health club. Take an exercise class. Check out the local canoeing club or the square-dancing classes offered at the Y. You'll make a smooth path to new friendships, says Dr. Goldberg. Make a regular workout date with a friend. When you get into the habit of exercising with others, you'll feel stronger and miss your kids less.

Of course, experts agree that when dealing with the sadness of having an empty nest, the most important thing to exercise is your brain. Try to change the way you think about parenthood and your role in your kids' lives.

Gray-Matter Workouts: Different Dad, Different Mom

If you're gloomy about the kids' departure, keep in mind three encouraging truths:

You're Mom (or Dad) forever. "Your last child's departure from home isn't the end of your relationship with your children," says Dr. Goldberg. "It's the start of a new phase of that relationship." True, Harry Jr. may no longer need you to fix macaroni and cheese just the way he likes it, but he still needs your support, affection, and love. And he always will.

"Children need to separate from parents—that's a healthy part of normal development," says Florence Kaslow, Ph.D., director of the Florida Couples and Family Institute in West Palm Beach and visiting professor of psychology at Duke University School of Medicine in Durham, North Carolina. "But separating doesn't mean severing ties."

You're more than just Dad (or Mom). Sure, parenthood was an important role in your life. But it wasn't the sum total of you. Dr. Goldberg suggests that empty-nesters redefine themselves and help each other reclaim the pre-parenthood spirit of their bond.

"You should try to see your spouse in a different light, not simply as 'Mommy' or 'Daddy' but as an intimate companion," says Dr. Goldberg. "Think back to those days before the kids appeared and before your free time and disposable income vanished.

"You can have a more intense relationship and be more intimate emotionally once you're on your own," she says. "Since you have more privacy, you also have more freedom to be sexually intimate."

The future may well be richer. Once your kids have flown from the nest, your relationship with them may well get better, predicts Dr. Goldberg. She points out that most kids leave the nest in the waning days of adolescence, the Cuban Missile Crisis of parent-child relations. As children mature, they experience the way the world works and begin to appreciate their parents. "Kids will also develop some additional adult interests that you can share," she says.

The Family That Exercises Together Stays Together

Physical exercise—or special events built around it—can be a great way of keeping the family together. Even if there is tension between parents and children, an exercise date lets you be together without having to talk about touchy issues, says Dr. Kaslow. So if the kids live nearby, schedule golf, bowling, tennis, bike rides, or Sunday evening family strolls.

If your offspring are too far away for regular get-togethers, consider initiating one or all of the following exercise events. (Attendance is mandatory, and prizes will be awarded.) Turn them into regular family traditions.

► The Jones Family 4th-of-July-Positively-Cutthroat Badminton Tournament.
► The Smith Family First-Weekend-in-October-Bring-Your-Friends-Intergenerational Touch-Football Challenge
► The Doe Family Annual Hike Up Mount Doe

Since the events are scheduled ahead of time, everyone can fit them in. Let sports and wellness bring you and your grown kids together.

ERECTILE DYSFUNCTION

Try an Uplifting Approach

All right, fellas, we need to talk. Some truth—straight from the shoulder. Yes, erection problems are depressing. For years, you've taken your erections for granted, so there's no question that the first time you can't rise to the occasion is a roundhouse right to the sexual psyche.

You should keep two important facts in mind. First, more than 20 million American men admit to having erection problems on a regular basis, and most experts suspect that recurring difficulties are actually a lot more common than that. The best guess is that at some point in their lives, a huge majority of men aren't the phallic phenoms of fantasy. Second, and even more important, erectile dysfunction is a medical situation, like a bad back or a bum knee. It doesn't mean that there's anything amiss with your manliness.

What does erectile dysfunction mean? Most likely, it means that your circulation isn't all it should be. And maybe that you're under some stress. There's strong evidence that exercise can help keep your erections more frequent and firm.

Erection Directions

An erection is a simple story of fluid dynamics. When a man becomes aroused, blood surges southward, engorging the penis and making it stand tall. If the blood vessels in the penis are somehow compromised, the blood can't flow as strongly or it won't stay put. The result? A less-than-satisfactory lift.

Circulation can be inhibited by lots of things. Smokers have a high incidence of erection problems. Some prescription drugs can cause flaccidity, too. High cholesterol and atherosclerosis (hardening of the arteries) are major contributors to the reduction of blood flow to the penis. Diabetes can also adversely affect your ability to become erect, because this sneaky disease constricts blood vessels all over the body, including the multitudinous tiny ones that supply the penis.

But even if you don't have any of these problems, Father Time can slowly mess with the way your body responds to the call of the wild. Although erectile problems can happen at any age, they're more common as men get older. Over time, our blood vessels—even if they're not significantly narrowed by disease—tend to become less elastic. They don't dilate as easily or as fully as they once did. And since they react more reluctantly to the inflow of blood, you'll have a tougher time achieving engorgement.

The erection-protection secret is to keep those blood vessels as clear and as accommodating as they can be for as long as they can be, explains Dudley Danoff, M.D., senior attending urologic surgeon at Cedars-Sinai Medical Center in Los Angeles and author of *Superpotency*. Keep those vessels vital, and the blood will flow freely. Exercise is essential to erections that endure.

Aerobic Amour

The evidence is overwhelming: Men who exercise regularly have far fewer erectile problems than sedentary men. And thin men seem to have far fewer problems than men who are overweight. Exercise not only keeps blood vessels clear, it also boosts testosterone, which may power sexual arousal. Regular exercise—along with eating a healthy diet—also stacks the odds against those erection wreckers, diabetes and heart disease.

Virtually any aerobic exercise helps keep you in good vascular shape, says Dr. Danoff. Your best bet is a lifetime exercise habit. Thirty minutes a day, three or four times a week, forever. But if it's a little late for that, start moving today. Go for a walk and make this vow: Gentle but steady exercise will become part of your daily routine.

Run yourself rigid. A regular jogging routine—perhaps a couple of miles a day—is a great option to boost circulation and preserve nerve and blood vessel health, says Alan Hirsch, M.D., neurological director at the Smell and Taste Treatment and Research Foundation in Chicago and author of *Scentsational Sex*. It can also keep your blood pressure under control, which protects your veins and arteries from the constant inside-out assault of high blood pressure.

Make a splash. The aerobic benefits of a four-times-a-week swimming program can also offer erection protection, says Dr. Hirsch. Regular workouts maintain heart health so that circulation stays strong all over. The better your heart pumps, the less likely that you'll be limp. Plus, exercise does wonders for your physical stamina, so lovemaking won't leave you breathless.

Beware of biking. This is one aerobic exercise to avoid if you're having erectile problems, says Dr. Hirsch. "The pressure from long-term sitting on a bicycle seat," he says, "may eventually compress the blood vessels supplying the penis, as well as the nerves." While this shouldn't worry weekend pedalers, intensive riders should take note. If you feel persistent numbness after riding, you may need to reduce your time in the saddle by biking less, even if you aren't experiencing erectile dysfunction. And when you're out for a ride, stand on the pedals every 10 minutes or so to encourage blood flow.

When You Need Extra Support . . .

There are still other ways to ensure ample blood flow to your nether regions. Try these tips to preserve your pep.

Use it or lose it. The more erections you have, the less chance that you'll have erectile problems, explains Robert Birch, Ph.D., a psychologist and sex therapist in Columbus, Ohio, and author of *Male Sexual Endurance*. Obviously, intercourse can't cure the complete inability to have an erection, but it can help fortify an erection that's firm enough for penetration but not as firm as you'd like. Every erection gives the blood vessels in your penis a workout. There is evidence that a vigorous sex life may help to prevent erectile problems. Plus, being intimate whenever possible will do wonders for keeping your love relationship rosy.

Love the one you're with. Say you're feeling eager, but your wife's away for the weekend. Or maybe you're single. Well, then, just take matters into your own hands. Masturbation is a perfectly healthy, enjoyable way to give your body a chance to function normally, says Dr. Birch.

Prefer pumpkin. Certain smells have the amazing effect of increasing blood flow to particular body parts, according to Dr. Hirsch. His research at the smell and taste foundation suggests that the combined scent of pumpkin pie and lavender, and the smell of vanilla, have the effect of improving blood flow to the penis. Even if baking doesn't appeal to you, you can get the fragrance in the air with pumpkin-scented, lavender-scented, or vanilla-scented candles or incense. Along with the ambiance of sweet smells and candlelight, you just might get the capillary capitulation that helps sow the seeds of success.

Worry Makes It Worse

Most cases of erectile dysfunction will be found to have some organic or physical cause. Only a small fraction of men with chronic impotence have a strictly psychological problem. But even in erectile dysfunction caused by disease, there's frequently an emotional component involved as well, says Dr. Birch. "A man can make any erection problem much worse by worrying," he says. Try these tips for relieving the anticipatory anxiety that can make sex seem like a command performance.

Tell Your Doctor What's (Not) Up

About half of the time, erectile dysfunction isn't a symptom of a serious underlying medical problem. The other half, however, it can be a sign of diabetes or vascular disease or a side effect of medications, says Alan Hirsch, M.D., neurological director at the Smell and Taste Treatment and Research Foundation in Chicago and author of *Scentsational Sex*. So it's important to tell your doctor if you've been experiencing trouble for more than a week with getting or maintaining an erection. Move past the embarrassment and be open and honest with your doctor. That's the only way he can make sure that you're getting the medical care you need. There are lots of treatments for erectile dysfunction, and your doctor may be able to help solve the problem. But he can't help if he doesn't know about it.

Make intercourse off-limits. For a while at least, you and your partner should make full intercourse verboten. This takes away any fear of failure, because an erection isn't necessary when penetration isn't on the program. Of course, there's more to sex than intercourse, says Dr. Birch. "Have fun exploring other ways to share sensuality."

Get multisensual. Too often, we limit our attention to touch and don't get the extra arousal benefits of engaging the other senses. Focus on all your senses when being intimate.

Use smell, taste, touch, and hearing to take in the whole picture. "Think of how good being intimate feels," says Dr. Birch.

Touch her in the morning. Some men who can't get an erection whenever they want one find themselves ready to go first thing in the morning, says Dr. Birch. Morning erections are really nighttime erection leftovers. If your partner is quickly aroused and you both have the time, you may want to take advantage of these "freebies" now and then and make your mornings together special.

Add hypnosis to your arsenal. A study at the Medical School of Yuzuncu Yil University in Istanbul revealed the potential healing power of hypnosis. For cases of erectile dysfunction that had no discernible medical cause, hypnosis worked 75 percent of the time. This suggests that hypnosis may make a good add-on treatment for sexual dysfunction. If you'd like to try it, be sure to find a licensed physician or psychologist who is certified in clinical hypnosis and who has worked with sexual problems before, recommends Dr. Birch. The American Society of Clinical Hypnosis will send you a list of such experts in your state if you send a self-addressed stamped envelope to 2200 East Devon Avenue, Suite 291, Des Plaines, IL 60018.

EYESTRAIN

Protecting Your Peepers

Imagine little eyeballs in T-shirts and shorts hoisting barbells, pedaling stationary bikes, and sweating through an aerobics class. After a good workout, they leave the gym feeling energized and refreshed.

That image is difficult to accept because when we exercise, we often bypass our eyes. And while your eyes do get a workout every day, you can bet that they rarely wind up a day on the job feeling energized and refreshed. On the contrary, it's more than likely that your eyes get more strained than an out-of-shape runner at the end of a marathon.

"Many of us go to work every day, and we're stuck doing one thing—computers or reading," says Jacob Liberman, O.D., Ph.D., a Colorado optometrist and president of Universal Light Technology, a company that researches and manufactures light and color technology for healing, in Carbondale, Colorado.

Computers are especially rude when it comes to eye assault. Those glaring, flickering screens that hold us in their spell exact a heavy toll. Many people who use computers regularly report itching, burning, dry eyes or blurry vision.

But computers aren't the only eye attackers. Every day, your attention may be held by some attraction or distraction that could ultimately cause eyestrain—such as can't-put-down video games, a suspense-filled novel, or an evening in a smoke-filled tavern. Any of these can ultimately tax your tender eyes.

Even the eye-catching displays in gleaming department store windows can eventually start to affect your eyes. "After a trip to the shopping mall, do you notice how tired your eyes are?" asks Robert-Michael Kaplan, O.D., a behavioral optometrist, consultant in vision care in Gibsons, British Columbia, and author of *The Power behind Your Eyes*. "It's because the mall's visual environment is designed to catch your attention at store after store that you pass." So let's fight back.

Gazing More Easily

Whether it's the visual assault at the mall or the daily grind of computer viewing that assaults your peepers, there are tactics that you can use to give your eyes some rest. Here's what experts recommend.

Book it ahead. If you're the type of reader who can become totally engrossed in the latest John Grisham courtroom novel, there's a good chance that your eyes will stay glued to the page even after they start to ache and show signs of strain. Maybe what you need is a traffic cop inside your books.

Use a bookmark as a stop sign, suggests Dr. Liberman. Write the words, "Put me 10 pages ahead" on the bookmark, and when you come to the marker, move it 10 pages ahead. Each time you reach the bookmark, that's your reminder to stop reading, blink several times, and rest your eyes by looking out the window or across the room. Pause for at least 30 seconds before you resume reading, suggests Dr. Liberman.

Let your eyes rove. Anyone who has ever driven long hours on a blazing sunny day knows how that glare can cost your eyes. To ease eyestrain, keep moving your vision from one focal point to another, advises Dr. Liberman. After you've been looking at the road for a while, glance at the dashboard, look at the license plate on the car ahead, or scan the scenery for a moment. As a matter of fact, if left to their own devices, your eyes will naturally and continuously move on their own. You don't want to sacrifice good defensive driving habits, of course, but try to take in as much as you can with your roving glance.

"Just constantly letting your eyes shift from the dashboard to the distance is extremely important," says Dr. Liberman.

Say no to Nintendo. Aficionados of computer games think nothing of spending a couple of hours fending off aerial bombardments on the gleaming screen. But while you may win the war on aliens, you can lose the game of coddling your eyes.

"When we play video games, we get into a hypnosis-like trance," observes Dr. Kaplan. "Our eye muscles become locked in focus, and as a result, we get blurry vision and tired, burning, watery eyes. Our eyes must move in order for vision to be stimulated."

If the pull of the game is irresistible, put it on

Relief for the Weary-Eyed

Many of us end the workday with glazed-over computer eyes. For this reason, it's important to give your eyes a break from the glare of your screen for at least 1 minute out of every 20, says Robert-Michael Kaplan, O.D., a behavioral optometrist, consultant in vision care in Gibsons, British Columbia, and author of *The Power behind Your Eyes.* He recommends trying some of these eye-moving exercises.

Roll with it. Gently roll your head from left to right and then from front to back. Keep your eyes open and let your gaze follow the direction of your head movements.

Zoom it. Take your focus from your computer screen and zoom it to a point off in the distance in front of you. Wait for five breaths before returning your attention to the screen.

Get cross. If you wear glasses, take them off for this exercise. Focus both eyes on the tip of your nose so that your eyes cross and you can see both sides of your nose. Then quickly zoom your eyes to a point off in the distance in front of you. Wait for three to five breaths before going back to work.

Try a New Hue

To avoid the dreaded computer "zombie" stare, pump some life into your eyes by moving them every 20 minutes, suggests Robert-Michael Kaplan, O.D., a behavioral optometrist, consultant in vision care in Gibsons, British Columbia, and author of *The Power behind Your Eyes.*

But you need something worthy as a distraction. Dress up your monitor, Dr. Kaplan advises. Place brightly colored round stickers on each of the four corners of the frame and another between each set of corners so that your monitor is bordered by three stickers on each side. Every 20 minutes, imagine that your eyes are like a bug that can hop from one sticker to the other. First move from the middle dot on the left side of your screen straight across to the middle dot on the right side. Next move from the middle dot on the top of your screen to the middle dot on the bottom. Finish up with a diagonal move from the bottom right corner up to the top left corner of your screen. As your eyes dart from sticker to sticker, breathe deeply and enjoy the ride.

The secret is to keep your eyes moving. As long as they are moving, they are under less strain and you won't be tempted to glare-stare at your computer screen. To keep this exercise interesting, try randomizing the left-right, up-down, and diagonal eye movements, says Dr. Kaplan.

hold every once in a while and take a break, Dr. Kaplan advises.

Practice some palmistry. The best relief from eyestrain may lie in the palm of your hand. Dr. Kaplan suggests trying this simple exercise.

► Close your eyes and cover them with the cupped palms of your hands.

► Breathe slowly and deeply and try to feel the warmth of your palms penetrating your eyes.

► Take about 20 breaths.

► With each breath, feel yourself growing more relaxed as you let strain and tension flow out of your eyes.

► Slowly remove your hands, open your eyes, blink, and let in the light.

This exercise is particularly effective if you do it near a sunny open window, according to Dr. Kaplan. When you open your eyes, you'll get the contrast of bright sunshine and a cool breeze.

FARSIGHTEDNESS

Celebrate Birthdays with Clear Vision

For a lot of people, 40 flaming candles on a cake do not ignite as much alarm as the need for bifocals at some time around this milestone birthday.

Although most of us expect to get farsighted as we age, the reality still seems to sneak up on us. While reading, we have to move the book farther and farther away to see the words. Finally, our arms aren't long enough. Call in the bifocals or reading glasses!

What's happening is that the focusing mechanisms in our eyes are off-kilter. The cornea (the clear covering of the eye) lacks sufficient curvature, so light enters the eyes out of focus. When you reach your forties and your lenses stiffen and prevent you from seeing close objects clearly, you have presbyobia.

No medication or exercise can reverse presbyobia, according to the American Academy of Ophthalmology. But some behavioral optometrists (optometrists who have conventional training but also see the value of the mind and body working together in a healthy way) disagree. They say that regular eye exercises and a can-do attitude can improve vision or slow its degeneration.

"When you deal with vision from a conventional point of view, you are only touching about 10 percent of what vision is all about. Ninety percent of vision is actually in the brain and the mind," says Robert-Michael Kaplan, O.D., a behavioral optometrist, consultant in vision care from Gibsons, British Columbia, and author of *The Power behind Your Eyes*. "Some people can increase visual fitness naturally by relaxing and stimulating the eyes."

Regularly exercising the eyes is important, agrees Meir Schneider, Ph.D., a licensed massage therapist, founder of the Center and School for Self-Healing in San Francisco, and creator of the Meir Schneider Self-Healing Method. Dr. Schneider credits eye exercises, self-massage of the face, and an open mind for correcting his childhood blindness, which was caused by cataracts, farsightedness, and other serious problems. He experienced the benefits firsthand and is now able to legally drive a car.

"Vision can improve," says Dr. Schneider. "I've worked with hundreds of people who were legally blind and helped them become partially or fully sighted. Vision is composed of sensation, conception, and perception. It is the brain, more than the eyes, that determines how your vision will be."

Behavioral optometrists urge people with farsightedness to practice patience. Vision improves gradually and only when you dedicate yourself to regular exercises designed to relax and stimulate your eyes, says Dr. Schneider.

To prevent or overcome farsightedness, you need to increase circulation to your eyes, put a stop to the habit of tensing your eyes in the presence of strong light, and create a more relaxed, flexible way of looking around the world, says Dr. Schneider. Both he and Dr. Kaplan recommend several eye-strengthening and eye-relaxing exercises designed to improve movement and vision.

Try this wick trick. Here is an exercise you can do with a candle to strengthen your eyes, says Dr. Kaplan.

▶Place a lit candle on a shelf or a dresser so that you will have to look up to see the flame when you're seated.

▶Pull up a chair so you are two or three feet away from the flame.

▶Look up at the flame and take a few deep, relaxing breaths.

▶Close your eyes. Feel tension evaporate from your eyes, neck, shoulders, and the rest of your body. Imagine that you are enjoying a warm bath.

▶For 20 breaths, try to picture a peaceful, stress-free life.

▶With your eyes closed, try to see the candle flame in your mind. Focus on the flame's center, then its edges. Try to distinguish the flame's many colors.

▶Open your eyes slowly and look directly at the flame.

▶Notice objects around the candle as you look into the flame. Blink and breathe.

▶After 20 breaths, cross your eyes for a few breaths: Look at your nose so that you can see both sides simultaneously. When your eyes are crossed, you should see two candles.

Aim skyward. Try this "sunning" or "skying" exercise to give your eyes the flexibility to comfortably accept light without strain, suggests Dr. Schneider. The key is to make a habit of stepping outside often throughout the day for natural light and relaxation. You'll need a total of 10 to 20 minutes a day. Here is how it's done.

▶Find a sunny spot outdoors. Remove your glasses or contact lenses.

▶Lightly close your eyes and raise your face toward the sun.

▶Turn your head slowly 20 times.

▶Keep yourself from squinting by lightly massaging around the eye on the side of your face away from the sun.

▶Turn your head away from the sun and warm your hands by rubbing them together.

▶Gently cup your warmed hands over your eyes and breathe deeply and slowly for 20 breaths.

▶Continue to alternate two minutes of sunning with one to two minutes of "palming" with your warmed hands. From time to time, blink your closed eyes rapidly a few times.

A few safety rules: Avoid very strong sunlight, do not sun through glass, and keep your eyes closed at all times.

Skying is a good substitute for sunning on cloudy days. Go outdoors, face the brightest part of the sky, open your eyes, and turn your head continuously. Follow the directions for sunning, but in the absence of sunlight, it's safe to keep your eyes open.

Massage your visage. Reserve some time to loosen your strained and stressed face with a full-face massage. This exercise is designed to relax your eyes as well as your entire face.

▶Rub your hands together to generate warmth, then use your heated fingertips to gently massage your face.

Harlem "Professor" Studies Eye Exercises

Sitting in the back row of his first-grade class in Harlem, New York, seven-year-old Jim Sharps knew the alphabet by heart. He just couldn't see specific letters on the front blackboard.

"My teachers thought I was dumb until, finally, someone suggested that it might be my eyes," recalls Sharps, who is now more than 45 years older and living in Alameda, California. "An eye doctor told my parents that I was farsighted and that I had astigmatism, and he gave me a real strong prescription for glasses."

But Sharps paid a premium for stronger sight.

"When I wore the glasses, I was nauseated, the sidewalk seemed to come up at me, and I thought I was going to fall down all the time," he recalls. "The eye doctor told my folks that I would get used to it. I did, and my eyes got worse and worse."

Today, Sharps relies on glasses only for reading and night driving. He credits eye exercises and changes in his career, diet, and attitude for improving his eyesight.

Sharps believes that his farsightedness was triggered by shutting out traumatic events he saw in the troubled streets of Harlem.

"I have no memories before age five except for flashbacks of someone telling me, 'You didn't see that,'" Sharps says. "I think that had a traumatic emotional effect on me to the point that I stopped seeing to some degree."

Although optometrists told Sharps that he would need bifocals when he was in his early forties, he decided to go another route. He turned to natural vision teacher Meir Schneider, Ph.D., a licensed massage therapist, founder of the Center and School for Self-Healing in San Francisco, and creator of the Meir Schneider Self-Healing Method. Dr. Schneider gave Sharps a program aimed at teaching him to relax and stimulate his eyes—"palming," "sunning," and lots of massage, especially around the eyes. "Through eye exercises, I learned how our eyes work and how to improve vision," Sharps says.

Palming is among his favorite exercises. While seated in a comfortable chair, he rubs his hands together for 30 seconds to generate warmth and energy. Then he cups his hands and places them over his closed eyes. He palms for 30 minutes each day. "Palming gives my eyes a chance to rest and get stronger," he explains.

After palming, he spends time on an exercise to get his eyes working together and seeing close objects. He extends his right arm and lifts his index finger skyward. He focuses on that finger, bringing it slowly toward him. When the finger looks like two, he stops. His goal is to inch that finger closer to his nose, to the point where double vision would normally occur, and still be able to see only one finger.

These and similar exercises have produced such remarkable changes in his vision that Sharps is now a strong believer in eye exercises.

"In the ghetto, I earned the nickname Professor because I wore glasses," Sharps says. "Now there are some days when I don't wear my glasses at all."

Take a Closer Look

For the farsighted among us, Robert-Michael Kaplan, O.D., a behavioral optometrist, consultant in vision care in Gibsons, British Columbia, and author of *The Power behind Your Eyes,* has created a spin-off version of the traditional eye chart that he calls the Near Eye-C Chart.

Dr. Kaplan created this take-anywhere eye chart to improve your focusing powers at times convenient for you. "The Near Eye-C Chart is a dynamic way to monitor the effectiveness of your therapy," he says.

First, photocopy the chart at right, enlarging it to 155 percent. Place the chart on a shelf or a wall and stand about 12 inches away. As you try to identify each letter, take deep, relaxed breaths. Note how many letters appear clear and determine if the clearness seems to fluctuate. Then think of a happy memory and look at the chart again to see if this triggers a change in the way you perceive the letters. Finally, recall an unhappy time and see what happens to your vision when you look at the eye chart.

vzuerpnrundnvenhndurpdhzenmyxkvzuerpnrundnvenhndurpdhzenmyxkmyxk

frzezfvdxhpnfuqlmprnzfpzpnfrzezfvdxhpnfuqlm

HPNVEZFWHPNVEZFWHPNVEZFWHPNVEZFV

THMARCOADZCNPQEKLSGJ

CBFRAOSZMND2

LSFPONZ3

CDFUR4

VZFY

PR

O

Dr. Kaplan contends that doing eye exercises with this chart will give you improved focusing powers. He also stresses that the eye chart can act as a barometer for the way you view your career, relationships, family, and everything else in your life.

►Massage along your jawbone, moving your fingers from your ears to your chin.

►While massaging, try opening and closing your mouth to help stretch and relax your jaw muscles. If you feel a yawn coming, don't fight it.

►Use your fingertips to massage from the bridge of your nose along your cheekbones and up toward your temples, using small circular motions. Then go back to the top of your nose and use the same motion to massage above, below, and directly on your eyebrows. Spend extra time in the area between the eyebrows,

as this place typically harbors a lot of tension from using your eyes.

►Use long, firm strokes across your forehead, then switch to gentle, circular motions in the temple areas.

►Continue stroking lightly from your temples to your scalp as you imagine you are washing away tension from your eyes.

The total facial massage should take about 10 minutes. Dr. Schneider recommends that you do this exercise daily. Even if your eyesight does not improve, he says that your eyes will feel more relaxed and refreshed.

FATIGUE

Get Moving with Sir Isaac

You could say that Sir Isaac Newton knew a thing or two about getting moving. That, after all, was the subject of his famous First Law: A body in motion will stay in motion unless acted upon by an outside force, like gravity.

That's worth considering if you're feeling fatigued and finding it hard to get moving. The fact is, Newton's First Law applies to you, too.

Experts say that one of the best ways to fight fatigue and ensure that you have the energy you need to keep moving throughout your busy day is to get moving—to exercise. Resist the gravitational pull of your couch, get out there and walk, run, or dodge falling apples, and you'll have more energy to do all the things you need and want to do. Expend some energy exercising, and you'll wind up with more energy, not less.

"With regular exercise, your overall capacity to do anything increases," says Jack Wilmore, Ph.D., professor of health and kinesiology at Texas A&M University in College Station.

There are dozens of ways that exercise can boost your energy level. Aerobic exercise like walking, jogging, and swimming helps strengthen your heart so it becomes more efficient at pumping energizing, oxygen-rich blood through your body. Exercise strengthens your other muscles, too, so they're better able to hold you up and carry you through your day. Regular exercise can ease energy-draining depression. It can also enhance sleep. Studies show that regular exercisers may not only fall asleep faster, they may also enjoy a longer stretch of deep, extra-refreshing snoozing.

Expanding Your Universe of Activity

While inactivity is a common cause of fatigue, other things can leave you down and out, too. Stress, thyroid gland snafus, overwork, dehydration, lack of sleep, anemia, depression, and assorted other illnesses can all cause fatigue. Have you been feeling run-down for more than a day or so? See your doctor, says Dr. Wilmore. If your doctor doesn't find anything amiss, and you've taken care of the other sources of energy drain in your life—by getting more sleep, for instance—exercise could well be what's missing.

You don't have to run a marathon to find more energy. It's easier than you might think. Consider the following modest yet revitalizing activities—posture, breathing, aerobic, flexibility, and strength-training exercises that will fit into the busiest day.

7:00 A.M.: Lift your breastbone. Once you're out of bed and on your feet, lift your breastbone, the V-shaped bone at the top of your rib cage. To get this right, imagine that someone has tied a string to your breastbone and is pulling it forward and up. With that conscious morning lift, your posture will shape up admirably. "Your ears will line up over your shoulders, and your shoulders will line up over your hips," which is where they should be, says Peggy Anglin, P.T. a physical therapist who practices at the Duke University Medical Center in Durham, North Carolina.

If you start out that way, you're less likely to slouch during the day. And your upright posture is a fatigue fighter. Slouching contributes to back and neck strain, which in turn contributes to fatigue, says Anglin. Moreover, slouching can interfere with deep breathing. To breathe deeply and draw in all the vitalizing oxygen you need, you have to stand and sit tall so your chest can expand.

7:01 A.M.: Flex your diaphragm. Once you have yourself in proper alignment, spend a few minutes practicing deep breathing. Your diaphragm—the muscle that lies between your lungs and stomach—should move out when you inhale and in when you exhale. This ensures that you get plenty of air and energy-boosting oxygen with each breath.

To get into the deep, diaphragmatic breathing habit, Anglin suggests this exercise: Hold your palms against the bottom of your rib cage. While inhaling deeply through your nose, fill your lungs with air so that your lower rib cage presses out and against your hands. Breathe out slowly through your mouth, exhaling completely, so your rib cage sinks back inward.

10:00 A.M.: Haunt the hallways. Feeling fa-

When Fatigue Is Fine

Let's say that you get up early, take the Pekingese for a jog, roust the kids and your partner from slumber, make the coffee and the French toast, pack lunches, usher the young scholars to school, earn an honest day's living, endure the commute on the expressway, pick up the milk, help cook the lasagna and vegetables just the way they like them, check the math homework, dry the tears over the math homework, cuddle the spouse, and then, at 9:00 or 10:00, feel fatigued.

"That's perfectly understandable," says Jack Wilmore, Ph.D., professor of health and kinesiology at Texas A&M University in College Station.

It's nothing to worry about—not a symptom of hidden illness or a moral lack. After a long, arduous day like that, you're entitled to be tired. So put your feet up and rest.

tigued around coffee break time? Skip the java and walk the halls. Or climb up and down a few flights of stairs, says Dr. Wilmore.

Even a short walk will get your heart pumping, improve your circulation, and boost your energy level, says Anglin. Whenever you're feeling fatigued after a long, arduous period of immobility, get up and walk around a bit, she suggests. "We're not designed to be still for hours at a time," she says.

Noon: Exercise your right to a constitutional. Finish your food during the first half of your lunch hour so you can spend the second half getting some aerobic exercise. Try walking, which is probably your most convenient exercise option on a workday. (On weekends, of course, you can choose your sport.) To keep your weekday walks from growing repetitive, spend 15 minutes ambling wherever whim takes you, then turn around and backtrack. A half-hour walk will help strengthen your heart, improve circulation, and keep your cells well-supplied with zest-giving oxygen. Shoot for 30 minutes of aerobic exercise at least three days a week, suggests Dr. Wilmore.

You say you don't have the energy to do 30 minutes at a pop right now? Don't worry. If you haven't done lengthy exercise in a while, start with 5-minute sessions, then build. Work your way up by adding another 5 minutes to your routine every week or two. Whatever you do, don't push yourself to the brink of sweat-soaked exhaustion straight off the bat. "Going from no exercise to doing a six-mile uphill hike is guaranteed to leave you feeling both exhausted and ready to give up," explains Dr. Wilmore.

12:30 P.M.: Drink heavily. Carry a bottle of water with you when you exercise and drink as you go. Or drink up before you head out for your workout, says Dr. Wilmore. When you work out, your body loses water through sweat. In fact, you can lose a quart of water over the course of a moderate 45-minute walk if you're not drinking water. That's enough to make you feel truly tired. Remember: Dehydration can cause fatigue. Replace the water you've lost, and you'll feel re-energized.

3:00 P.M.: Stretch. Even when you can't find time to slip away from the office, you can slip in some energizing flexibility exercises, says Anglin. To stretch your back, shoulder, and leg muscles and to give your circulation an energizing boost, try these energy-lending contortions.

The standing back bend: While standing, place your hands on your lower back, with your right palm on your right side and your left on the left. Arch your back slowly, holding your chin down. Count to three. Return to an upright position. Repeat, placing your hands slightly higher, on your mid to lower back. Repeat again with your hands a bit higher still.

The shoulder stretch: You can do this at your desk. Sit up straight and tuck in your chin. Place the palms of your hands against the back of your head (right hand on the right side and left hand on the left). Then push forward with your hands without letting your head move. Count to three, then slowly relax. Repeat.

The foot pedal: While sitting, pump the heels of your feet up and down, keeping your toes on the ground and lifting your knees.

5:30 P.M.: Build your many muscles. You want your muscles to do your bidding all day without getting fatigued, and the best way to get them pumped for daily movement is with resistance exercises. Most gyms and YMCAs have weight-training equipment and offer training classes for beginners, so there's no reason to be shy. To avoid injury, have someone show you how to use the equipment before you jump in, says Dr. Wilmore. Learn the correct form and moves, then pace yourself. Start with light weights and a minimal number of repetitions, then build from there step by step. Weight training twice a week should do the trick, says Dr. Wilmore.

Walk Away from Chronic Fatigue—With Baby Steps

If ordinary fatigue were a brownout—the kind that sets the lights flickering on hot summer afternoons—chronic fatigue syndrome (CFS) would be more like the great blackout, plunging a body into a dark abyss of slumber.

CFS doesn't just dampen your energy, it exhausts it, says Sue Anne Sisto, P.T., Ph.D., a physical therapist and co-director of the CFS Rehabilitation Program at the Kessler Institute for Rehabilitation in West Orange, New Jersey.

To make matters even worse, the exhaustion of CFS is often accompanied by other unpleasant symptoms, such as headaches, sore throat, swollen glands, difficulty concentrating and remembering, and muscle and joint pain.

No one knows exactly what causes CFS—which may affect up to a half-million Americans—but many researchers suspect that it might be caused by a virus or by some immune system snafu.

"It's as though people with CFS get the flu and just don't recover," says Dr. Sisto, who is also co-investigator for CFS exercise studies at the New Jersey Cooperative Chronic Fatigue Syndrome Research Center at the University of Medicine and Dentistry of New Jersey in Newark.

Now, exercise may be the last thing you feel like doing if you have CFS. But it could well be a key to your recovery. In a British study, volunteers with CFS reported significant improvements in their energy levels after just 12 weeks of graded, moderate exercise. The 29 volunteers started out walking for 5 to 15 minutes five days a week. They built from there very gradually, adding 1 or 2 minutes per week until they reached a maximum of 30 minutes daily. People with CFS often suffer relapses or bouts of intensified symptoms when they try more strenuous exercise. But moderate, gradual workouts seem to help ease symptoms, says Dr. Sisto.

"One component of CFS is deconditioning," she explains. "If you have had CFS, usually you've been sedentary for a while, so your muscles have become weaker and your cardiovascular system is not as tolerant of exercise. This contributes to fatigue. With a graded exercise program that leads toward general reconditioning, people improve."

If you have CFS and you want to start exercising, consult a physical therapist who can design a program that's tailored to your abilities and supervise your progress, Dr. Sisto advises. A well-designed program should get you started with stretching exercises to improve your flexibility and resistance training to recondition and strengthen your muscles.

After that gradual introduction, you can begin doing aerobic workouts, building your routine gradually. You should always stay with a particular workload for five to seven days and make sure it's not too intense before you ratchet it up, she adds.

Mixing the Routine

Of course, following an unvarying schedule can be tiring in itself. In fact, the same old routine day after day can be downright fatiguing.

Ditto for the same old exercise routine. If your workouts have gotten humdrum, here's how to reinvigorate them and yourself, suggests Benjamin S. Fialkoff, Ph.D., a clinical psychologist and consultant in sports psychology who practices in Ridgewood, New Jersey, and New York City.

Change your timing. Try exercising on different days of the week or at different times of day. If your lunch hour constitutional has gotten routine, try waking up with the pigeons and walking early in the morning. Or walk at dusk when the shadows get long and interesting.

Change your event. Remember, walking isn't your only option. Consider swimming or water polo or water boxing—or regular boxing or kick boxing, for that matter. There are all sorts of variations on all manner of exercise themes. Sign up for a new exercise class at the local Y or your health club.

Find interesting exercise company. "Find a buddy to exercise with," says Dr. Fialkoff, and choose someone who shares interests similar to yours.

Swim along with *Yellow Submarine*. Put your workout to music—all sorts of music. If you usually listen to Springsteen while you run, try Sibelius. You can even accompany swimming with music if you leave a watertight portable radio blaring nearby in a dry place.

Get ruthless. A little competition can liven things up. Join the local Masters' swim team and enter the butterfly competitions. Or sign up for the annual charity golf tournament at the public course. Whatever your favorite sport, from badminton to bocci, you're likely to find a competition to keep you challenged.

FIBROMYALGIA

The Great Pretender

The Platters spun to the top of the charts in 1955 with "The Great Pretender." Today, it could be the theme song for fibromyalgia, an often misdiagnosed condition whose symptoms mimic those associated with arthritis, multiple sclerosis, chronic fatigue syndrome, and even tennis elbow.

"It is getting trendy when people complain of feeling bad all over to label it fibromyalgia," says Norman Harden, M.D., a neurologist and director of the Center for Pain Studies at the Rehabilitation Institute of Chicago. Sometimes, treatable diseases may be overlooked because fibromyalgia has become a catch-all diagnosis for generalized symptoms of aches and pains.

Fibromyalgia is an irritation of the connective tissue that supports structures throughout the body. Although its symptoms are clear—debilitating muscle aches, stiffness, and fatigue, sometimes accompanied by headaches and a case of the blues—its causes aren't. If your shoulder aches for no apparent reason, for instance, you might have fibromyalgia. "It spans the spectrum of pain from annoying to disabling," says James Stark, M.D., a physiatrist at the Center for Sports Medicine in San Francisco.

Studies suggest that people are born with an inherited tendency toward fibromyalgia, and it may be brought on by physical or emotional trauma, stress, or exposure to environmental toxins, including tobacco smoke. This condition, like all chronic pain, may also cause depression. One thing that doctors are sure about is that regular exercise routines that emphasize muscle stretches, low-impact aerobics, and relaxation can control the pain of fibromyalgia. Here are some exercises tailored to those three areas.

Muscle Stretches

If you have fibromyalgia, begin and end your days with gentle muscle stretches and massages. They boost your body's supply of oxygen and enhance muscle tone and flexibility.

Stand in the doorway. The doorway stretch prepares your muscles for exertion, says Devin Starlanyl, M.D., a West Chesterfield, New Hampshire, physician and author who has fibromyalgia.

► Stand in the middle of a doorway facing forward, with your arms outstretched and your hands resting firmly on each side of the doorjamb at shoulder height.

► Take a step forward and feel the stretch across your chest. Hold the stretch for 30 to 60 seconds.

Stretching Herself and Others

The phrase, "physician, heal thyself," has special meaning for Devin Starlanyl, M.D.

Her life took a detour when she suffered whiplash and a separated shoulder as a result of a car accident. At the time, Dr. Starlanyl was an emergency room physician and the director of a pharmaceutical plant. She was living a high-stress life, often working 120 hours a week.

For weeks after the accident, Dr. Starlanyl couldn't shake the chronic muscle pain that radiated throughout her body. Gradually, she lost the ability to move her body easily or accurately. "Some days, I couldn't even sew on a button," she says. It took her 10 years to finally find a diagnosis for what she had: fibromyalgia and chronic myofascial pain syndrome. "Other doctors simply told me that my pain pattern didn't exist. But I knew that what I had was real," she says.

Now living in West Chesterfield, New Hampshire, Dr. Starlanyl devotes herself full-time to educating people about fibromyalgia.

She has written two valuable books, *Fibromyalgia and Chronic Myofascial Pain Syndrome: A Survival Manual* and *The Fibromyalgia Advocate: Getting the Support You Need to Cope with Fibromyalgia and Myofascial Pain Syndrome*. She also runs counseling and support groups as well as a fibromyalgia Web site.

Her number one strategy is stretching. "I do a lot of different stretches, but not a lot of repetitions," she explains. She recommends stretching under a hot shower in the morning. While standing with your back to the water, clasp your hands behind your head and slowly bend your head forward, letting the neck vertebrae relax.

She also advocates stretch breaks every one to two hours throughout the day. "You can do a variation of the doorway stretch—extend both arms above your head and reach up, trying to touch the top of a door. And an easy stretch for your calves while seated is to raise both feet off the floor while keeping your heels down," she suggests.

▶Return to the starting position and move your hands farther down on the doorjamb. Again, take a step forward and stretch.

This is a good stretching exercise to do once or twice a day, Dr. Starlanyl says.

Massage those aches. Researchers at the University of Miami recommend massages for people with fibromyalgia. In one study, 10 people with fibromyalgia received 30-minute massages twice a week for five weeks. They ex-perienced decreased anxiety and less stress, pain, stiffness, and fatigue.

Play wall ball. Tennis ball acupressure, anyone? You don't need a racquet or a net to do this tennis ball exercise recommended by Dr. Starlanyl.

First of all, identify the tender spots, or "trigger points," where the pain seems to origi-nate. Place a tennis or lacrosse ball on that trigger point and apply pressure by leaning against a wall or lying on the floor with the ball between your body and the firm surface. When you compress the trigger point, liquids in that area are forced out. When the pressure is lifted, the blood and other body fluids rush back in and flush the area, bringing needed oxygen and nutrients to the tissues.

If the trigger point is under one buttock, for instance, here's how to apply pressure with a tennis ball.

▶Lie on your back with your knees bent and your feet flat on the floor. Your knees should not touch.

▶Raise your hips slightly and slide a tennis ball under one of them.

▶Move your hip down and rest on the ball for 30 to 60 seconds. You can expect to feel some pain, but you can control the pressure that you exert. If you roll the ball around, the pain should ease.

Fibromyalgia 101: Do You Have Pain at the Tender Points?

Since fibromyalgia is so difficult to identify, the Arthritis Foundation offers these helpful clues.

▶Fibromyalgia is a syndrome characterized by generalized muscle pain and fatigue. It does not affect the joints.

▶Pain may start in one part of the body and seem to gradually spread all over.

▶People describe fibromyalgia pain as burning, radiating, gnawing, sore, stiff, and achy.

▶People with fibromyalgia experience soreness and tenderness in at least 11 of 18 identified tender points on the body (*right*).

▶Nine out of 10 people with fibromyalgia experience fatigue caused by interrupted sleep.

▶No single lab test or x-ray can diagnose fibromyalgia.

▶The cause is unknown, but doctors say that stress, physical injury, smoking, poor posture, and other factors may play a role.

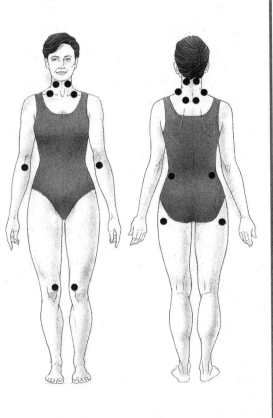

Easy-Does-It Aerobics

Regular low-impact aerobic exercise for at least 20 minutes a day increases blood flow to the muscles and boosts endorphins, the body's natural pain fighters, says Dr. Stark. Here are some low-sweat, non-joint-jarring exercises to combat pain from fibromyalgia.

Spin your wheels. Stationary cycling can reduce tenderness and promote sound sleep, says Dr. Stark. He urges people who have been inactive to gradually ease into cycling or a similar type of program after conferring with their doctors.

Rock away the pain. Never underestimate the aerobic benefits of your favorite living room rocker, says Dr. Stark. Maybe you can gently rock back and forth through your favorite half-hour sitcom. The rocking action activates leg muscles and lessens the effects of immobility, he says.

Stride often. A 20-minute walk, properly executed, can do wonders for fibromyalgia, says Dr. Starlanyl. She offers these walking tips.
► Keep your head held over your shoulders for balance.
► Shift your weight from your heels to the balls of your feet to your toes as you step forward.
► Push off each step from your toes, using your calf muscles.
► Move briskly, but be able to maintain a conversation without being short of breath.

Reach Relaxation

These exercises may help both physically and emotionally. Try these tips.

Guide pain away. Visualize the muscle pain in physical terms such as a burning fire or a terrible monster gnawing on your bones. By mentally extinguishing the flames or taming the monster, you may be able to reduce pain, says Dr. Harden.

Feel great with Alexander. The Alexander Technique focuses on correcting body mechanics and improving posture and movement. This makes it an effective, gentle method of relief for tense muscles, says Dr. Starlanyl. (For more details, see "Alexander Technique" on page 11.)

Hug your dog or cat. A friendly head scratch or hug can make a dog drool in delight and a cat purr with pride. Pets help reduce pain, tension and stress, says Dr. Harden.

Take up tai chi. This ancient Chinese martial art enhances mind-body connection, balance, and harmony, says Dr. Stark. Its slow, fluid, circular motions in graceful sequences strengthen muscles and improve coordination and balance. Think of it as a moving meditation, he adds. (For more, see "Tai Chi" on page 386.)

FINGER STIFFNESS

Count on These Moves to Limber Up Your Digits

Our fingers are amazing tools. Some evolutionary biologists claim that fingers are what makes us human. To be sure, without them we never could have fashioned the first prehistoric tool, not to mention The Mall of America in Minneapolis.

Our fingers face several threats. Arthritis attacks them from the inside, and potential finger foes are everywhere out there, from volleyballs to garden tools. Once our fingers get stiff, we enter a downward spiral. We don't use them as much, which only makes them stiffer.

"The more you use any joint, the more synovial fluid—a lubricant—is produced to keep the joints moving smoothly," says Norman Harden, M.D., a neurologist and director of the Center for Pain Studies at the Rehabilitation Institute of Chicago. The equation is straightforward: Less lubrication equals less freedom of movement.

Further, if you use a stiff finger less and less, the muscles in it get weak. "Not using any body part causes muscle atrophy," says James Stark, M.D., a physiatrist at the Center for Sports Medicine in San Francisco. That's why it's important to work your way through finger stiffness—whether it's arthritis-induced or a result of a blow or repetitive strain—while taking care not to aggravate your symptoms.

How to do it? With two types of exercises: flexibility work and strength training.

Four for Finger Flexibility

Our hands work so well because our fingers and thumbs have great range of motion. It's vital that we not settle for a constricted range after an injury or the onset of arthritis. Here are a few exercises to sustain suppleness and keep your fingers movable in several directions.

Mock Spock. You don't have to be a Trekkie to get a handful of benefits from Mr. Spock's famous split-fingered Vulcan greeting. For this finger flexor, pair up your index and middle fingers and your ring finger and pinkie. Then try to separate the two pairs of fingers as much as you can (1). Hold the pose for 5 to 10 seconds, then relax. Repeat five times.

1

"Maintaining flexibility is crucial for people with arthritis in their fingers," says Robert Markison, M.D., a hand surgeon and associate clinical professor of surgery at the University of California, San Francisco.

Stop traffic. You can make like a traffic cop and alleviate finger stiffness with this exercise. Extend your right hand in front of you like a policeman signaling "Stop." Make sure your thumb is pointing outward. Next, drop all your fingers and press them against your palm. Do this five times and then repeat with the other hand, suggests Teri Bielefeld, P.T., a physical therapist and certified hand therapist at the Zablocki Veterans Affairs Medical Center in Milwaukee. "Besides flexibility, this exercise also works on your coordination," she says.

Try some finger bows. Put your stiff fingers at center stage of this daily workout offered by Bielefeld.

Place your right hand palm down on a flat surface such as a table or a thick phone book. Make sure the middle joints of your fingers are over the edge (2). Using your left hand, gently bend each finger downward, one at a time, over the edge (3). Hold each bend for five seconds and feel the gentle stretch in the joint. Do this five times for each finger and then repeat with your left hand.

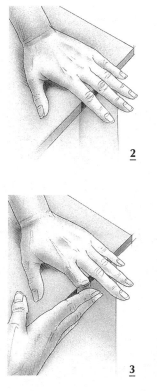

2

3

I'm Okay— And So Are My Fingers

When life feels good, people often flash an "OK" sign. You can turn this sign into an easy exercise to ease stiffness in your fingers, says Teri Bielefeld, P.T., a physical therapist and certified hand therapist at the Zablocki Veterans Affairs Medical Center in Milwaukee.

Start with your right hand. Gently bring your thumb and index finger toward each other until the fingertips touch. Your other three fingers should be raised skyward to complete the gesture.

"Then touch your thumb with the fingertips of each of the other fingers, one at a time, all the way down the line," says Bielefeld. "When you form a perfect "O," you are using the flexors in your hand the best." Do this five times for each finger, then repeat the steps with your left hand.

Find the dime. To keep your fingers flexible, try this treasure search game from Susan Isernhagen, P.T., a physical therapist and director of Isernhagen Work Systems in Duluth, Minnesota.

Fill a quart bowl with uncooked rice and drop in 10 dimes. Stir the contents with your fingers. Keeping your eyes closed, reach into the bowl with your right hand and use your fingers to pluck out the dimes. Put the dimes back in the bowl and try the hunt with your left hand.

"You are stretching and moving all of your fingers as well as increasing your awareness of touch—and having fun," says Isernhagen.

Pumping Finger Iron

Your fingers need strength as well as flexibility. Resistance training is the best way to fortify them. If you've got stiffness, try these handy workouts.

Do the yellow pages walk. Find a flat surface and let your fingers take a healthy stroll. Move them forward in ascending order, from the pinkie to the thumb. Then march them backward, putting a little pressure on them as you do. Alternate hands. This exercise helps build strength, says Mary Ann Towne, P.T., a physical therapist and director of rehabilitation and wellness services of the Cleveland Clinic–Florida in Fort Lauderdale.

Make a healthy splash. While relaxing in a bathtub or jet-powered spa filled with warm water, you can strengthen your fingers with some gentle water play, according to Isernhagen.

"Try doing hand strokes in the water by cupping your fingers and grabbing at the water," she says. "Cup water and bring it toward you, letting it go between your fingers. Then keep your fingers straight and move your arms out to the sides, creating resistance and some strengthening. Do front and back motions, letting your fingers work against the water resistance. The faster you go, the more the resistance there is. Do this for one to three minutes."

Play hand war. You can improve your finger strength with this do-anywhere exercise offered by Bielefeld.

Place your right hand palm down on a table. Then place your left hand crosswise and flat on

Play It Again, Sam

Playing the piano is a great finger exercise that works to both limber up and strengthen your fingers, says Susan Isernhagen, P.T., a physical therapist and director of Isernhagen Work Systems in Duluth, Minnesota.

Try to span an octave or more. "Stretching sideways is something that people rarely think about, but it's a great exercise," Isernhagen says.

Even if you can't play the piano, just concentrate on moving your fingers sideways up and down the keyboard. You don't have to make a tune. And if you don't have a piano, pretend.

4

top, covering all the knuckles. Gently press down with your left hand and try to lift your right index finger as high as you can (*4*). Hold for a count of five. Repeat this lift one finger at a time with each of the remaining fingers on your right hand for a total of five repetitions. Then repeat the entire process with your right hand on top of your left hand.

FLATULENCE

Fighting Poison Gas

Hitler tried to defeat his chronic flatulence by taking charcoal pills and nonlethal amounts of strychnine and belladonna. Stalin gave up the prevention battle altogether and tried to disguise his backdoor burps by rattling an ice-filled carafe on his desk. If the most powerful tyrants in history couldn't control their involuntary emissions, what chance do simple folks have against the enemy known as passing wind?

Don't despair. We may never vanquish flatulence entirely, but in exercise, there is hope. The key is gas control.

"Gas is a result of fermentation of unabsorbed material in the colon," says William Ruderman, M.D., a physician at Gastroenterology Associates of Central Florida in Orlando. "Milk products are among the biggest causes of gas," he says. Other dietary octane boosters are beans, cabbage, broccoli, and cauliflower, so steer clear of these if you're afflicted with flatus. Once your problem diminishes, slowly reintroduce those foods one by one. You may find that you can tolerate some of them and that some of the others are okay in occasional small servings, says Dr. Ruderman. Once you're on a quiet diet, try these exercises to help mute the toots.

Give yourself a morning massage. People who eat big meals or gas-generating foods late at

---

Written on the Wind

Let's clear the air about flatulence. It is a natural trait of being human. Consider these flatulence facts, based on a survey done in the early 1990s.

- On average, we pass gas 9 to 14 times a day.
- Dating couples wait, on average, 93 days before expelling gas in front of one another.
- Once flatulence escapes, 52 percent of people will blame others or the family dog.
- About 71 percent of swimmers admit that they release gas in a lake, ocean, river, or pool.

---

night can expect to have a gas-filled stomach when they wake the next day. Here's an easy five-minute exercise to release unwanted gas before your feet hit the floor, says Dr. Ruderman.

- Lie flat on your back in bed.
- Place the open palm of your right hand on the lower right side of your belly just above your pelvic bone. This is the seven o'clock position.

▶Press firmly and begin kneading that spot with your fingers.

▶Move your right hand clockwise to the 11 o'clock spot, just below your ribs on the right side.

▶Repeat the firm pressure and finger kneading.

▶Repeat these actions at the one o'clock and five o'clock spots, moving across and down the left side of your belly.

"You can get rid of a lot of gas by massaging your belly in a clockwise circle," suggests Dan Hamner, M.D., a physiatrist and sports medicine specialist in New York City.

The circular motion tends to clear gas from the intestines, adds Dr. Ruderman.

Go aerobic. Yup, on top of all its other health benefits, raising your heart rate a bit is a gas fighter, too.

"Aerobic exercise is helpful because it tends to clear gas from the intestinal tract and move things along," says Dr. Ruderman. He especially recommends the two Rs.

Rip off a run. "Nothing lets flatulence escape more quickly than a fast jog," says Dr. Ruderman. "There are almost no constipated runners."

Row, row, row your bloat. Rowing is particularly effective at forcing the gas in your stomach to move faster through the colon, says Dr. Ruderman.

If you have equipment and access to a lake, you can actually take to the water. A scull is best, but a rowboat will do. So will a kayak or canoe.

For the landlubber, Budd Coates, an exercise physiologist and fitness director in Emmaus, Pennsylvania, offers these rowing machine tips for stroking your way toward intestinal silence. It's a tricky motion, so you might want the help of a fitness instructor to get you started off with the right form.

The *Fartiste* Wows France

If you think you can't stop the gas, think again. Joseph Pujol, a true master of posterior escapades, flatulated for francs during the late 1800s, bringing fame to what some may consider a bodily act of shame.

Known around Paris as *Le Petomane* ("The Fartiste"), this father of nine often deliberately and skillfully passed gas for cash before large audiences at the famous Moulin Rouge and other theaters in France. He grew so popular that a street in Paris bears his name.

For each performance, Pujol stepped on stage in an elegant silk suit and white linen shirt. After a brief monologue, he turned his back to the audience, leaned forward, and placed his hands on his knees. Relying on his unique ability to contract and flex his anal and abdominal muscles, Pujol then released odorless flatulence that captured the ranges of tenor, baritone, and bass. He would duplicate the delicate gas release of a bride on her honeymoon or deliver a thunderous shot that sounded like cannon fire.

The world's only man known to earn a living passing gas, Pujol retired during World War I. He and his family ran a bakery in Marseilles until he died in 1945 at age 88.

▶When you're strapped into the rowing machine, bend your knees to pull the seat up to the starting position and keep your heels as flat as possible against the foot plates.

▶Be sure to hold the handles at the ends as you pull on the handle of the machine and push off the foot stretchers, straightening your legs.

▶Continue to pull evenly, pulling your arms straight in toward your chest.

▶Make a small clockwise circle with the handle, circling toward your body—but don't drop the bar to your lap—then lean forward, bend your knees, and return to the starting position in one smooth motion.

Rock away your gas. When you feel painful gas buildup, try heading for the rocking chair. The vigorous rocking motion helps to hasten the release of unwanted gas from your body, says Dr. Hamner.

And here's how to exercise some short-term restraint when you're in a situation where you *don't* want to release gas.

Don't blast your boss. The next time you're trapped in an elevator with your boss and feel gas ready to rumble, try relaxing, suggests Dr. Ruderman. Your instinct may be to hold back the blast, but that strategy could—ahem—backfire. "Getting tense and worrying about it may make it more likely to happen," he warns.

If you try to keep your sphincter muscle tense during the elevator ride, you'll probably lose the struggle because the muscle will tire, he says. So when you spot the head honcho, relax. Take a deep breath and exhale. Squeeze the muscles in your rectum for a few seconds and relax again.

FOOT PAIN

How to Be a Sole Survivor

If you've got achy feet, you'll never squawk alone. Three out of every four Americans share your problem in their pedal extremities. It's no wonder. Our feet are intricate systems, featuring 26 bones and 33 joints each, and the average person takes 8,000 to 10,000 steps a day. That's a lot of stress on our poor little doggies.

Foot pain is also an unfair affliction that affects a lot more women than men. According to the American Podiatric Medical Association, women have about four times as many foot problems. Wearing high heels is often the culprit.

Some foot pain is simple to diagnose, particularly if it's caused by ill-fitting shoes. Arthritis and circulation problems may also be culprits. If blood flow is compromised in any way, our feet, at the far reaches of the circulatory system, feel it most acutely. Exercise won't necessarily help with corns and bunions, but for many other foot faults, especially arthritis and circulation aches, motion is a great pain-fighting weapon.

To take arms against throbbing feet, experts suggest three steps.

1. Do low-impact exercise regularly—about 30 minutes three or four times a week. Try bicycling or swimming.
2. Do stretching exercises.
3. Do strengthening exercises.

When to See a Doctor

If foot pain persists for more than two to three days, seek professional care to rule out any serious problems such as infection or circulatory problems, advises Leonard A. Levy, D.P.M., professor of podiatric medicine and past president of the California College of Podiatric Medicine in San Francisco. People who may be at special risk—the elderly, people with diabetes, and those with circulatory problems, for instance—should seek medical attention if the pain persists for more than several hours.

Healing with Heel Stretches

Heel pain is a common complaint among foot patients. Frequently, the source of the pain is a condition called plantar fasciitis, named after the plantar fascia, a flat ligament band on the underside of the foot. The plantar fascia, which acts like a bowstring to maintain the arch of the foot, can become inflamed when it's repeatedly placed under tension. The irritation can be caused by being overweight, by standing for long periods, or excessive running or walking.

The pain may progress from a dull, intermittent heel ache to a sharp, persistent pain. Classically, it's worst in the morning with the first few steps you take, says Steven Lawrence, M.D., an orthopedic surgeon at the Lancaster Orthopedic Group in Pennsylvania.

Simple stretching exercises relieve heel pain in many cases. "The key to treating plantar fasciitis is stretching," says Thomas Meade, M.D., an orthopedic surgeon and medical director of the Allentown Sports Medicine Clinic and Human Performance Center in Pennsylvania. He suggests these heel-health exercises.

Take the first step. Here is a do-anywhere stretch to combat heel pain. You can do it while brushing your teeth, washing dishes, or standing and watching television, Dr. Meade notes. To give yourself those options, use a stepstool or nail together some 2 × 4's to make a portable step. You should feel a stretch in the calf and plantar fascia, but Dr. Meade warns that you shouldn't do this stretch if you feel pain.

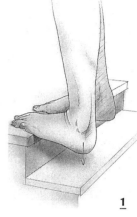

1

►Stand with your legs straight on the bottom step of a flight of stairs, holding on to the railing. Or use your portable step and a table for support.

►Stand so that the ball of the afflicted foot is on the edge of the step. Put the other foot flat on the step.

►Slowly drop your heel down until you feel a stretch in the back of your calf muscle (*1*).

►Hold the stretch for at least 20 seconds. Repeat 8 to 12 times and then switch to the other foot.

If the Shoe Doesn't Fit

The right-fitting slipper brought Cinderella and the prince together. A properly fitted shoe reduces the risk of bunions, corns, calluses, hammertoes, and other foot maladies, according to the American Orthopaedic Foot and Ankle Society (AOFAS).

Here are some shoe-buying tips from the AOFAS that will help you have pain-free feet.

►Judge the shoe by the way it fits your foot and not by the size marked inside the shoe. One brand's size 6 is another brand's size 7.

►Have your feet measured regularly, because foot sizes change as you grow older.

►Shop for shoes at the end of the day when your feet are largest.

►Make sure the ball of your foot fits comfortably into the widest part of the shoe.

►Stand during the fitting process. Check that there is 3/8 to 1/2 inch of space for your longest toe at the end of each shoe.

►Don't buy tight shoes and expect them to stretch to fit later.

Use a therapeutic towel. This exercise is designed to stretch inflamed plantar fascia tissue and relieve pressure on the nerves in the foot, says Dr. Meade. It can help relieve pain and in general tune up your feet to keep future discomfort away. "Let pain be your guide," he adds. "This should feel tight but not painful."

▶ Lie on your back with your legs straight.

▶ Loop a towel around the ball of your sore foot and gently but firmly pull the towel toward you. You'll feel tension in the bottom of your foot.

▶ Hold the stretch for 20 seconds. Repeat 10 times and then do the same with your other foot.

Do the golf ball massage. It's too rainy to play 18 holes? Or maybe your foot hurts too much? Try this sporting stretch suggested by Dr. Lawrence. It's great for people with plantar fasciitis, arch strain, and foot cramps. Simply place your foot on top of a golf ball on a flat surface, then roll the ball forward and backward across the underside of your foot for about two minutes. The ball provides a soothing, relaxing massage.

Strengthening from Heel to Toe

You probably never think about the importance of strong feet. But no matter how much you use your tootsies, it always helps to keep them toned. Experts say that these exercises will banish pain and keep it away.

Take the towel challenge. Does your spouse simply drop the bath towel on the floor after showering? Turn this irritating habit into an exercise opportunity, suggests Dr. Meade. The goal here is to try to pick up the towel with your feet, or more precisely, with your toes.

Start at the near end of the towel and curl it toward you, using only your toes to scrunch it forward. Then try to pick the towel up with your toes and deposit it in the hamper or laundry basket.

For foot-fortifying variety, see if you can pick up 20 marbles or pencils, one at a time, with your toes and deposit them into a bowl. This motion strengthens the small muscles in the foot, which helps fight toe cramps, says Dr. Meade.

Try some toe tugs. For relief from toe cramps, doctors suggest looping a thick rubber band around all of the toes on one foot. Spread your toes apart, feeling the resistance from the rubber band. Try to hold this position for 5 to 10 seconds. Repeat 10 times before you switch and attempt the toe stretch with the opposite foot.

For variety, try a toe-of-war by placing the rubber band around each of your big toes and pulling them away from each other. Be sure to keep your heels in place. Hold for 5 to 10 seconds and repeat 10 times.

Get on your tippies. Even if you can't do a pirouette like Mikhail Baryshnikov, this balletlike exercise will help fight toe cramps, says Leonard A. Levy, D.P.M., professor of podiatric medicine and past president of the California College of Podiatric Medicine in San Francisco.

You can stand or sit for this exercise, but put only as much weight on your toes as you can comfortably handle. First, remove your shoes, then do the following steps, one foot at a time. In a one-two-three sequence, rise to the ball of your foot (2), rise up farther and point your toes (3), then curl your toes under (4). Try to hold each position for 5 to 10 seconds; repeat 10 times.

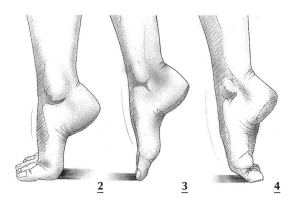

Rubbing the Right Way

A five-minute foot massage can sometimes relieve discomfort as well as make you feel relaxed and rejuvenated, says Dr. Levy. He recommends following these steps.

▶Sit comfortably in a chair or on the floor.

▶Bring your right foot toward you, resting your right ankle on your left knee.

▶Starting at the heel, press and knead the bottom of the foot with your fingers. Move up the middle of your foot, then veer over to your big toe and work across your foot to the little toe.

▶Massage the pad under each toe.

▶Gently squeeze each toe with your thumb and forefinger, moving each toe from side to side and gently stretching each toe out.

▶Wrap your hands around your foot at the arch so that your thumbs are on top and your fingers on the bottom. Knead your instep with your thumbs.

▶Place the outer edge of your palm under the tips of your toes and gently pull back. As you flex your foot, hold for a few seconds, then let go and relax.

▶Gently press the Achilles tendon (on the back of your heel) between your thumb and forefinger. Run your fingers up the back of your heel, ending above the ankle.

▶Switch and repeat with your left foot.

FORGETFULNESS

Dance to Enhance Your Memory

After a few weeks of fox-trotting and lindy-hopping around the dance studio, the students in Dr. Seham's senior dance class started to notice something interesting. They felt stronger, more flexible, and surer on their feet. But that wasn't the most interesting part. They also felt more fit mentally.

"They said they could focus better," says Jenny Seham, Ph.D., a clinical psychologist and dance instructor in New York City. Intrigued by what her students told her, she followed up with a formal study. The results? Her research suggests that dance can enhance the mind and memory.

Research by other scientists who looked at the benefits of other kinds of exercise also shows mental benefits. Whether you're dancing the watusi or taking a brisk walk, growing evidence suggests that exercising can help shape up your memory as well as your muscles.

While your arms are flapping and your feet are tapping, your brain is getting a workout, too. Exercise appears to boost the production and functioning of neurotransmitters, chemical messengers that ferry information from one part of your brain to another and to the body. Studies with rats suggest that aerobic exercise improves blood flow in the brain by stimulating the growth of extra blood vessels. And preliminary results from a University of Illinois study show that older adults who start getting regular aerobic exercise also begin to do significantly better on memory and other cognitive tests, says Arthur Kramer, Ph.D., the University of Illinois psychology professor who is heading the study.

Should You Have Danced All Night?

There have been no scientific studies proving that dance is better for brain health than any other form of aerobic exercise. But consider this logical leap: There is increasing evidence that doing regular mental exercise—reading, playing cards, playing music, doing crosswords—keeps your memory sharp. It stands to reason that aerobic exercise with a cognitive requirement—remembering the patterns of the flamenco as opposed to just running mindlessly on a treadmill—would be the best bet for maximizing mental functions like memory.

Dr. Seham suspects that any complex aerobic exercise, such as dancing through a series of elaborate steps, is good for your heart, lungs, blood pressure, *and* your powers of recall.

If you work your body and challenge your brain at the same time, you may well get more from your memory, says David Masur, Ph.D., director of neuropsychology at Montefiore Hospital and associate professor of neurology at Albert Einstein College of Medicine in the Bronx.

So, are you ready to try this fox trot for one dancer and sharpen your memory? Here goes.

Step out. "Put on the kind of music that makes you want to move," recommends Dr. Seham. She suggests "Lady Be Good" or "Cheek to Cheek" if you want to try the fox trot. Then try this dance that she uses in her own classes.

▶Beginning with the left foot, walk eight steps forward and eight steps back, then four steps forward and four steps back.

▶Slide once to the left (step left, then slide your right foot to meet the left), and once to the right (step right, then slide your left foot to meet the right), then repeat this left-right maneuver three more times for a total of four sets.

▶Starting with your right foot, alternate tapping your toe in front of you and then behind you for a total of eight taps. Repeat with the left foot.

▶Slowly "grapevine" to the right (step to the side with your right foot, then cross behind your right leg with your left foot, then repeat both steps), then march four steps in a full circle to end up facing front.

▶Repeat the grapevine step to the left.

▶Do the train step (step forward with your right foot, step in place with your left foot, then step back with your right foot and step in place with your left foot) and repeat once.

▶Do one box step (step forward with your right foot, cross over with your left foot, then step back with your right foot and step back with your left foot so that it is in line with the right) and put a little bounce into your finishing step.

Pick up the pace. Higher-intensity aerobic exercise may lead to greater improvement in

And Let Me Introduce You to . . . This Memory Strategy

If you're 50 or older, odds are that you're well-acquainted with Mr. Whatshisface, the mysterious and ubiquitous party guest whose name leaves no trace in your memory.

To remember names or other slippery information, try this mental exercise suggested by Fergus Craik, Ph.D., professor of psychology at the University of Toronto.

After you're introduced to the woman in the bright red A-line dress, repeat her name. When your host says, "And this is Shirley McGivens," make a point of saying "Hello, *Shirley*, it's nice to meet you."

"Repeating the name forces you to pay attention and register it," says Dr. Craik.

Repeat the name to yourself a few minutes later. As you're talking to her, think to yourself, "This is Shirley McGivens." Later, as your conversation is winding down and the two of you are exchanging "Nice to meet you," remind yourself, "This is Shirley McGivens." Later, look at her from across the room and say to yourself, "That's Shirley McGivens."

"It's an amazingly effective technique," says Dr. Craik.

blood supply to the brain than lower intensity workouts. Once you've learned the steps above, exert yourself a bit more. "Lift your knees higher or dance more vigorously," suggests Dr. Seham. Or switch to a tune with a quicker tempo—say, a snappy salsa.

Get complex. Give yourself more elaborate steps to remember and more practice at putting your memory through its paces by changing steps. You've got that dance down pat? Then mix things up, says Dr. Seham. "If you're used to starting with your right foot all the time, trying starting with your left foot. Challenge your brain," she says. Another option: Do some of the steps in double time. Instead of taking two beats for a right-left combo, do it in one beat.

Learn to tango. Challenge your memory and musculature by broadening your dance repertoire. Sign up for dance lessons. Call your local community college and ask about adult education dance classes or ring up a local dance academy (check the yellow pages or ask dancing friends). "A good teacher will help stimulate and motivate you," Dr. Seham says.

GLAUCOMA

Steps to Slow the Blurring

It wasn't old age or bad knees that sidelined the Minnesota Twins' Kirby Puckett from baseball. Something sneakier silenced his bat midway through the 1996 season: glaucoma.

"We use the Kirby Puckett story with our patients as an example of the importance of being conscientious about glaucoma and getting tested regularly," says Don Teig, O.D., an optometrist and co-director of the Institute for Sports Vision in Ridgefield, Connecticut.

Indeed, glaucoma is a sneaky eye thief, robbing two to three million Americans of their vision. In fact, glaucoma is the chief cause of blindness in the United States. The chances of getting glaucoma increase if you fit any of these categories: You are over age 40, are African-American, have a family history of the disease, have diabetes, are on steroid medications, or have had an eye injury.

Glaucoma affects a clear liquid called the aqueous humor, which flows in and out of the eye and nourishes the lens, iris, and cornea. In healthy eyes, the fluid drains out of the eye through a group of drainage canals around the iris. But if these drainage pipes clog, pressure builds. Fibers in the optic nerve are destroyed and eventually, blindness can occur.

The American Academy of Ophthalmology recommends that you report any blurred vision, severe eye pain, headaches, haloes around lights, or nausea or vomiting to your ophthalmologist. The academy urges that during their regular annual eye exams, people have their doctors check for any changes in their central and peripheral vision. These are clues that glaucoma may be present.

"A lot of people have their eye pressure checked at health fairs, but that only gives false security," says Paul Planer, O.D., an optometrist in Atlanta and president of the International Academy of Sports Vision in Harrisburg, Pennsylvania. "It is also important to have a visual fields test, which determines nerve performance inside the eye."

Conventional treatments for this disease include eyedrops, pills, laser treatment, and eye surgery. If your eye doctor detects early signs of glaucoma, of course you will want to follow his recommendations. If you need laser treatment or eye surgery, you will need to see an ophthalmologist.

Sight-Saving Skills

In addition to traditional treatment, glaucoma can also be controlled through daily mental

Rubbing Away at a Problem

During bus rides, Lula Vee doesn't mind being stared at. Whenever she begins finger massaging her entire face, other passengers are sure to notice, but their attention doesn't stop her.

Quietly and methodically, she massages in a circular motion. First, she does her eye orbits, the circular bone structure that surrounds the eyes. Then she addresses her forehead, cheekbones, nose, ears, neck, chin, jaw, and the top of her head. She softly recites Bible verses or affirmations referring to the power of light while taking deep, relaxed breaths. The massage might take 10 to 45 minutes, depending on the length of the bus ride.

"Sometimes, I get so relaxed that I fall asleep," says the 62-year-old former English teacher who is now a shiatsu and massage practitioner in Virginia Beach, Virginia.

The relaxation exercises became part of Vee's routine when she was in her late forties. That's when an ophthalmologist diagnosed glaucoma in her left eye. The sight in her right eye was already poor, and when she became blind in her left eye, she felt that she had to work at strengthening her right eye if she wanted to see anything.

At the time, she was taking prednisone, a steroid, to deal with a lymph system disease called sarcoidosis. "The doctors discovered that I was allergic to prednisone, and that's what caused my glaucoma," she says. "They said that my eye would have to be removed eventually because the pain would become unbearable. I took that as a threat and refused to take any more medicine."

Vee turned to natural healers, a macrobiotic diet, and eye exercises instead. Ten years later, her left eye is pain-free. Although she still can't see out of that eye, the pressure has dropped more than 20 points.

Every day, Vee devotes time to exercises that help her relax and may benefit blood circulation to her head and eyes. While standing, she drops her chin to her chest and slowly rolls her head from side to side. After a few repetitions, she is able to relax the muscles in her neck.

She offers another exercise that she likes to do each day. While standing, she lowers her chin to her chest. Slowly, she bends over, leading with her head. She moves her back downward one vertebra at a time, first curving her upper back, then the middle, and finally, her lower back. Then she slowly unrolls in the opposite direction until she is standing erect. She often spends 45 minutes to an hour each day doing this exercise.

"I also do palming (rubbing my hands together to warm them, then cupping them over my closed eyes), yoga, and long, rotating swings from my waist every day," she says. "I have strengthened my eyes so much through these exercises that my right eye has improved and I am now able to drive a car in the daytime again.

"There are few doctors who are open to the possibility of reversing poor vision," says Vee. "I believe we can use the eye as a metaphor and look everywhere for answers for healing."

and physical exercises, say some behavioral optometrists (optometrists who have conventional training but also see the value of the mind and body working together in a healthy way). Specifically, here are some exercises that may help your eyes, according to some behavioral optometrists.

Daydream a healthy feeling. Robert-Michael Kaplan, O.D., a behavioral optometrist, consultant in vision care in Gibsons, British Columbia, and author of *The Power behind Your Eyes*, recognizes the mind's powerful role in treating disease. Here's an exercise that he recommends for people with glaucoma.

Find a comfortable, quiet place to sit down. Close your eyes and feel how easily the watery fluids in your eyes drain from the front to the back of your eyes. Let all of life's pressures drain away as you breathe and relax. Visualize your optic nerve regenerating from the healthy foods you eat and imagine the eye pressure lowering each day. Then pretend that you are in the eye doctor's office and see the surprised look on his face when he tells you that your eye pressure has dropped. Smile at your success.

Play blue-green. Different colors of the light spectrum stimulate different emotions. This mental exercise requires an open, willing mind, says Dr. Kaplan.

With your eyes closed, try to see the calming color blue-green, which is a relaxing trigger used to visualize the healing of tissue and improved eye function. Once you get blue-green in your head, visualize that you are breathing that cool hue into your eyes and to the specific part that you wish to heal. Feel wellness returning to your eyes.

Follow the finger. This three-step exercise

Give Some TLC to Your Eyes

"Begin visualizing your eyes becoming healthier no matter how bad you may think your eye condition is," urges Robert-Michael Kaplan, O.D., a behavioral optometrist, consultant in vision care in Gibsons, British Columbia, and author of *The Power behind Your Eyes*. "It is important to imagine wellness in the parts of your eyes afflicted by disease. These parts are crying out for attention and love."

breaks the habit of straining to see and balances central with peripheral vision, says Meir Schneider, Ph.D., a licensed massage therapist, founder of the Center and School for Self-Healing in San Francisco, and creator of the Meir Schneider Self-Healing Method. If you're doing it outdoors, it's helpful to combine it with "sunning" or "skying." (Lightly close your eyes and raise your face to the sun. Turn your head from side to side 20 times. Massage around the eye that's away from the sun as you turn your head. Then stop and face away from the sun. Rub your hands together for warmth, then "palm" your eyes by cupping your warmed hands over them for 20 breaths. Alternate two minutes of sunning with two minutes of palming.)

▶Stand with your legs about two feet apart.

▶Hold your finger at eye level about two feet away from your face and focus on it.

▶Keep looking at your finger as you move it as far to each side as you can, turning your head to keep your finger in front of your nose. As you look at your finger, see your peripheral surroundings move in the opposite direction.

▶For the next step, sway your torso far enough in each direction so that the heel of the opposite foot lifts off the ground. Remember to see your surroundings in the periphery while looking at your finger.

▶Step three adds more movement. Always looking at your finger and keeping it in front of your nose, stretch tall as you pivot to the left, swoop down into a forward bend in the middle of your arc, and then stretch tall again as you pivot to the right. You may even want to throw your head back and arch your back a little to the left and right, like a drum major. Again, remember to see your peripheral surroundings moving in the opposite direction as you look at your finger.

HANGOVER

How to Survive the Moaning After

A wee bit too much wine at your cousin's wedding, and the next morning—ooooohhhhh!—your punishment arrives: headache, nausea, thirst, muscle aches, and a big dose of despair.

The simple fact is, you've poisoned yourself. Alcohol causes your blood vessels to contract (headache), causes the buildup of poisons such as aldehydes and lactic acid in your cells (body aches), and dehydrates your entire system (thirst). All the hair-of-the-dog folk wisdom notwithstanding, there's only one real way to fix what you've done—get the alcohol out of your body as fast as possible.

Your body knows this. If you've really overindulged, vomiting may purge some of the sauterne from your system. But by the morning after, even a good heave won't help with the alcohol that has already passed into your bloodstream. Exercise, however, may help speed recovery—if you're willing to breathe hard and sweat it out, says Melvin Williams, Ph.D., professor of exercise physiology at Old Dominion University in Norfolk, Virginia.

But there's one small problem. There's an anvil chorus in your head. Lifting your head off the pillow barely seems possible. Still, the fact remains that if you pump up your circulation, respiration, and perspiration, the alcohol will leave your body more quickly and you'll feel better faster. Here, in ascending degree of difficulty, are a few easy exercises that can help exorcise the bourbon and get you back in the game.

Hydro-exercise. Not swimming in H_2O, but drinking it. Doctors agree that the most useful hangover antidote is w-a-t-e-r. "Alcohol makes you dehydrated, and the membranes of the brain scream for water," says Dan Hamner, M.D., a physiatrist and sports medicine specialist in New York City. Quaff some water before you go to sleep; Dr. Hamner recommends at least 24 ounces of water before hitting the sheets.

Then, the next morning, have more *agua*. It not only helps with the rehydration of your body but may also help you urinate. Anything that leaves your body takes at least a trace of the demon rum with it, says Dr. Hamner.

Respiro-exercises. Let your lungs help blow the alcohol out of your body, advises Dr. Williams. When you awaken—even before you brush your teeth—try taking deep breaths and blasting the alcohol-soaked air from your lungs. "You may get rid of the alcohol a little bit faster because you're ventilating more," he says.

Crawl to a sauna. More perspiration means a faster farewell for the toxins. "Alcohol is total poison for every cell in the body. A steam room should help for a hangover," Dr. Hamner says. Even a half-hour in a steamed-up bathroom might help a bit.

Beg for a massage. If you're lucky enough to have a loved one nearby, a massage may help by getting your blood circulating, says Kimbra Kimball, a licensed massage therapist and co-owner of Massage Therapeutics in Allentown, Pennsylvania. Make whatever deal you have to, but get your partner to minister to you softly. Here's the massage step-by-step.

Lie on your stomach with your arms at your sides. Your partner should:

1. Gently squeeze the underside of your heels, pushing the skin toward the end of the heel.
2. Move gradually up your legs with quick, gentle hand pumps that push your skin and muscle upward from the calves and thighs.
3. Keep heading north, massaging your buttocks, back, and shoulders.

Then flip over sunny-side up and have your partner give your stomach a gentle, circular massage in a clockwise direction. "By going

---

Running under the Influence

You may not feel like hard exercise if your head is about to explode, but sometimes a real physical challenge is just what you need to help with recovery.

Melvin Williams, Ph.D., professor of exercise physiology at Old Dominion University in Norfolk, Virginia, remembers one festive night in Russia. Drinks flowed and disco music pounded as his tour group celebrated their final night in Kiev.

"Normally, I just drink beer, but this was a special occasion," says Dr. Williams. "I didn't know how potent vodka can be."

He found out the next morning when a 7:00 A.M. wake-up call pried open his puffy, bloodshot eyes. Somehow, this marathoner willed himself out of bed, dressed, and ran his daily eight miles before boarding the tour bus to the airport.

"I had a pounding headache for the first five or six minutes into the run, but by the end, I felt basically good," says Dr. Williams. He says that the heavy breathing while running possibly helped him excrete the poison in his bloodstream caused by heavy drinking.

As he ran, the rest of the group was trying to fight their hangovers by sleeping in. "When I stepped onto the bus, I was feeling much better, but everyone else looked miserable," he recalls. "The run worked for me."

Whether or not you're a runner, any kind of heart-pounding, blood-pumping exercise can help carry unwanted alcohol from your body, says Dan Hamner, M.D., a physiatrist and sports medicine specialist in New York City. "Try shadow boxing in front of a mirror for several minutes," he suggests. "That should get your heart rate up. You want to get that bad stuff out of you as fast as possible."

clockwise, you are not resisting your normal body functions. And you are pushing everything in the right way for elimination," explains Kimball.

Stretch those muscles. When we drink too much, lactic acid and other poisons such as aldehydes build up in our muscles. Why? Because the liver gets overloaded trying to deal with all the toxins and can't keep up.

"Lactic acid is what makes you ache. Under a microscope, this molecule is abrasive to the body," explains Dr. Hamner. Try this soothing stretch to combat achiness from a hangover. You can do these movements even before you get out of bed in the morning.

►Lie flat on your back and stretch your hands and feet out. Reach for all four corners of the bed.

►Point your toes and stretch out your fingers.

►Slowly bring your left knee to your chest. Hold that pose for a few seconds and then slide your leg down.

►Repeat with your right leg.

Hay Fever

Mind over Pollen

Your mind is a powerful gizmo that's capable of astounding feats—among them turning allergies on and off. Case studies of people with multiple personalities show that the allergies of one personality may not exist in an alternate personality of the same person. For these watery, itchy maladies, the mind is truly the master. But before you send in your application for a personality change, you might want to consider some less chaotic measures.

Hay fever takes hold when your immune system attacks pollen and molds, which it treats like bacterial and viral terrorists. It's really a simple case of mistaken identity. Unfortunately, your body is the battleground, and when it tries to drive out the invaders, you're likely to experience itching, watery eyes, a runny nose, a burning throat, and a whole lot of sneezing.

So you're sick of being one of the 26 million Americans that hay fever makes miserable? Here are some ways to get relief.

Use Your Head

Your mind can be a strong ally when it comes to beating hay fever. Some researchers who investigate mind-body connections think that we may unconsciously see allergens as harmful aliens. By changing this faulty view, you can prevent your body from sending out a SWAT team to kill a bunch of harmless pollen, says William Mundy, M.D., clinical professor of medicine at the University of Missouri School of Medicine in Kansas City and author of *Curing Allergy with Visual Imagery*.

Here are two mind-body techniques that you can use to try to control your allergies. One catch: Be persistent and patient. "For some people, a technique works immediately," Dr. Mundy says. "Others may have to practice as often as possible for several days before seeing results."

Let good conquer evil. Practicing a visualization technique called anchoring can be your gateway to relief. By combining a negative image with a positive one, you can wipe out allergy symptoms 85 percent of the time, says Dr. Mundy.

▶To try it, place your left hand on your left knee.

▶Look across the room, through an imaginary protective glass wall, and see yourself having an allergic reaction to whatever is a trigger for you—let's say weed pollen. Really get into the experience. "Many patients often experience allergy symptoms just by imagining an attack," says Dr. Mundy. But the technique can be just as effective if you don't experience symptoms while imagining, he says.

►Now come back from imagining the nasty weed pollen. Keeping your hand on your left knee, place your right hand on your right knee.

►Look again through the glass wall. Imagine a place where allergies haven't bothered you a bit, and the air is pleasant to breathe. "Most patients who suffer from hay fever use the beach or a snow-covered landscape as their pleasant place," says Dr. Mundy.

►Imagine your allergy-free scene in great detail. If it's the beach, hear the waves crashing and breathe in the clean, salty, ocean air. If it's a snow-covered field, inhale the clear, sharp, tingling air and listen to the vast silence.

►Then recall the first image of yourself having an allergy attack. Move that image closer to your easy-breathing self until the two images become one.

►Focus on the healthy image as you lift the imaginary glass wall that was separating you from your allergy-plagued self, says Dr. Mundy.

►As you continue breathing perfectly, take both hands off your knees at the same time. "When you have a negative and a positive image anchored and both anchors are released at the same time, the mind-body will nearly always choose the positive," Dr. Mundy says.

Organizing an Exercise Coup

Working out is probably the last thing you feel like doing when hay fever takes over your life. But you can still go about your normal exercise routine, says Charles Jaffe, M.D., Ph.D., an immunologist and allergist with the Allergy and Immunology Medical Group in San Diego. Just follow these tips.

Wear a mask. A good face mask that covers both your nose and mouth may let you exercise outdoors again. Choose one that forms a tight seal to your skin and prevents airborne allergens from entering your airways. Dr. Jaffe recommends the Contour Face Mask, which is made of breathable cloth and hypoallergenic neoprene and comes with a replaceable filter. It's available by mail order from Allergy Control Products, 96 Danbury Road, Ridgefield, CT 06877. Most paper masks, because they lack a seal, do not provide adequate protection, he says.

Scrub it off. Take a shower after you've been outside exercising when pollen counts are high, suggests Timothy McCall, M.D., an internist in private practice in Boston. If you wash off the pollen, you won't transfer it from your skin and clothes to your furniture, pillow, or bedding.

Time it right. Check your local weather forecast for the pollen counts and avoid exercising outdoors when they are high, suggests Dr. Jaffe. Morning is generally the worst time to exercise, unless it's rainy or cool, which keeps pollen out of the air. By midafternoon, counts usually subside enough that you can head outside.

When you feel that you can control your allergies, test yourself, says Dr. Mundy. If you're allergic to ragweed, for example, expose yourself for a brief time outdoors without your usual medication and see if you have a less severe reaction or none at all, he says. If you don't notice a change, keep practicing and look forward to good results, or give the next technique a try.

Try a mental allergy shot. To show hay fever the door, try desensitization, a technique that is similar to allergy shots, says Judith Green, Ph.D., professor of psychology at Aims Community College in Greeley, Colorado, and co-author of *The Dynamics of Health and Wellness*.

The idea of desensitization is to mentally subject yourself to a tiny bit of the element that makes you allergic so that your body can adapt to it with no harmful reaction. If you first imagine an environment containing a small amount of what triggers your allergy, your body can develop the capacity to cope with it, says Dr. Green. For best results, practice the following technique daily until you can beat your allergy, she says.

The first step is to relax, says Dr. Green. Relaxation prepares the body to listen to the mind during the visualization phase.

► Sit in a comfortable place and take easy, deep breaths that start way down in the bottom of your lungs, letting your stomach gently move out.

► Let your arms and hands feel heavy and your whole body go limp.

► Scan your body for any tense muscles and concentrate on relaxing any that you find. When your hands feel warm, it's an indicator that you're relaxed, says Dr. Green.

► Once relaxed, picture yourself in a forest that has a tiny bit of your allergen. Imagine the scene in great detail. If mold triggers your allergies, for example, you might see yourself standing among moldy leaves. While imagining this scene, continue to stay relaxed, breathing evenly.

► Imagine breathing in tiny particles of mold and breathing them out again, and say to your body, "Mold is just mold, harmless bits of nature."

► Continue to breathe perfectly, imagining that you are breathing mold particles in and out. Try to recall this image as often as possible until you are comfortable with it.

► Over time, gradually add more and more mold to your imaginary scene until you can breathe perfectly while walking through a forest full of mold, says Dr. Green. Visualize dead trees covered with mold and moldy leaves carpeting the ground. You may even see yourself getting some mold on your hands. But through it all, you stay relaxed and breathe perfectly.

"By imagining little doses of your allergen while relaxed, you can become desensitized to it," says Dr. Green, although it may take several days or perhaps two to three weeks before you reach that point.

When you can handle the moldy forest in imagery without experiencing any allergy symptoms, you can test yourself by going into an environment that triggers an allergic reaction. While there, focus on normal breathing with no allergic reaction. The central idea in desensitization through visualization, says Dr. Green, is to associate complete freedom from the symptoms of hay fever with the allergen. This is why it is essential to not have any symptoms when doing the visualization.

"I think of an allergy as a bad habit that can be changed," says Dr. Green. "My advice is, be patient and persistent."

Acupressure to the Rescue

Acupressure, which is performed by pressing on specific points in the body, is a great way to prevent or reduce the severity of allergy attacks, says Mary Muryn, a certified teacher of polarity (energy healing) and reflexology (a type of acupressure that focuses on the hands and feet) in Westport, Connecticut. Give these techniques a try.

Press the web. For instant relief, press the webbing midway between your thumb and index finger of one hand with the fingers of the other hand. Gradually apply pressure, angling toward the bone that connects with the index finger, located about an inch below the knuckle, suggests Glen S. Rothfeld, M.D., assistant clinical professor in the department of family medicine and community health at Tufts University School of Medicine in Boston and founder and medical director of Spectrum Medical Arts in Arlington, Massachusetts, in his book *Natural Medicine for Allergies*. Then repeat this technique on the other hand. This maneuver can stop an imminent allergy attack, according to Dr. Rothfeld.

Squelch that sneeze. You *can* stop a sneeze, and that's particularly important when you're in a meeting, a church service, or any other place where you'd rather not let loose. The next time you feel a sneeze coming on, press the skin between your upper lip and your nose until you can feel the pressure against the gum, says Michael Reed Gach, Ph.D., director of the Acupressure Institute in Berkeley, California.

Pamper your toes. A nice little toe massage can bring fantastic hay fever relief, says Muryn. "Take the fingers of both hands and massage the tips of every single toe," she advises. By massaging the tips of your toes, you can relieve pressure and pain in the head, the center of most hay fever symptoms, she says. Perform this technique two or three times a day, or more often if your symptoms hang on.

HEADACHES

Relief Can Be Free and Freeing

In ancient Greece, headache sufferers didn't reach for aspirin. Instead, they flopped torpedo fish on their foreheads. These eel-like fish delivered an electric shock that was supposedly strong enough to drive away the pain.

Fortunately, in modern times, we can cast a broader net for headache solutions. We can dodge headaches—or at least soften their punch—through exercise and even a good laugh, doctors say.

Headaches come in many varieties. They can be big or little, frequent or infrequent, disabling or minor. If you're all too familiar with headaches, you may be aware of certain triggers that herald the onset of a painful episode or at least contribute to the misery. Some women get menstrual headaches—migraine-like pain that invades before, during, or immediately after menstruation—nearly every month. People with sinus conditions are well-acquainted with the headaches that congestion can cause. The famous morning-after hangover headache is a predictable aftereffect of alcohol consumption. Allergy headaches, usually accompanied by watery eyes and nasal congestion, can be prompted by any kind of allergic reaction, and even eyestrain can trigger headaches in some people.

In sheer numbers, the tension headache rules with its dull, steady ache and muscle tenderness. In sheer power, the mighty migraine reigns with its arsenal of pulsating head throbs and the visual and speech disturbances and nausea that sometimes accompany it. But there are other kinds as well—the agonizing cluster headache that pierces one temple with searing pain or the rebound headache that is literally the result of a rebound from taking too much headache medicine. Headaches of various kinds can also be triggered by computer eyestrain, potent perfumes, and even ice cream.

Headaches are like wrong telephone numbers. Nearly everybody gets them, and some people get them again and again. In fact, more than 45 million Americans are likely to get repeated headaches. Among that number are the 16 to 18 million people who get migraines each year.

While no one has found a cure for headaches, physical and mental exercises can provide free—and freeing—relief. Doctors recommend a range of exercises almost as varied as headaches' many incarnations. Taking deep breaths, engaging in enjoyable sex, and letting loose a series of laughs are among the options. Even a completely invisible exercise—willing the index finger to a warmer temperature—has helped some people conquer headache pain.

"The body is truly a forgiving and wonderful thing," says Joseph Primavera, Ph.D., a psychologist at the headache clinic at Thomas Jefferson University Hospital in Philadelphia. "If you pay attention to it, you can really get in touch with the body's amazing ability to heal."

Free Ease

Best of all, exercise remedies for headache won't cost you a dime. And that's saying a lot, considering that Americans invest a bundle trying to quiet their throbbing temples. According to the National Headache Foundation, about four billion dollars a year goes toward over-the-counter pain relievers. Six out of every 10 people with headaches first reach for some nonprescription medication such as ibuprofen, according to a survey by Mediamark Research.

That's an investment that may have diminishing returns, some doctors point out. "There is a danger in overusing medicine. You can get rebound headaches—headaches caused by too many painkillers," says Lawrence Robbins, M.D., founder of the Robbins Headache Clinic in Northbrook, Illinois, and co-author of *Headache Help*. Dr. Robbins says that he became a neurologist, in part, to find ways beyond visits to the medicine cabinet to ease his tension and migraine headaches. He takes an occasional painkiller for relief but prefers medicine-free remedies such as imagery, meditation, and deep-breathing exercises.

Dr. Primavera agrees that much can be done before you grab the medicine bottle. "What has happened is that today, people are not used to using their own bodies to be healthy," he says.

"We get bombarded by advertising that says, 'Don't have time for a headache? Take a pill.' I say, why not breathe slowly and deeply instead? When you take control, you just feel better."

Many people find that they can turn to their internal pharmacy for relief, as these doctors suggest. From that pharmacy we can muster endorphins, the internally produced, soothing chemicals that are harbored in our brains and are on call for the headache battle.

"Endorphins provide a kind of pain relief and euphoria," says Dale L. Anderson, M.D., an urgent-care physician in Minneapolis and author of health humor books such as *The Orchestra Conductor's Secret to Health and Long Life*. "Once the endorphins are raised, muscles become relaxed and tensions are eased."

For the vast majority of us, the biggest challenges are dealing with tension headaches, migraines, or both. Some of the following physical and mental exercises might rescue you from these chief offenders and perhaps relieve other "trigger" types of headaches as well. Here are some low-sweat techniques to try.

Taming the Tension Headache

When you find yourself broiling in rush-hour traffic or drowning in a stack of office work, you may feel your neck, shoulder, and back muscles tightening. Then pain pops into your head as another tension headache is born.

More than 90 percent of all headaches are tension headaches. The symptoms are inescapable: pain, usually on both sides of the head; a dull, steady ache; muscle soreness; and a run-down feeling. Here are some ways out of that nagging discomfort.

Find a rush-hour relaxer. For many

people, the worst head-wracker crops up on the highway. "A lot of tension headaches are associated with neck tightness, sometimes triggered by scrunching up inside a car," says Dr. Primavera. "The head is the shape of a small bowling ball. If you push it out in front of your body, it gets very heavy and pulls on those neck and shoulder muscles."

To alleviate that tension, first loosen your grip on the steering wheel, suggests Dr. Primavera. Adjust your seat and posture so that you're sitting up as straight as possible. Then adjust your rearview mirror so that you can use it easily when you're in this position.

Learn the art of self-hypnosis. While self-hypnosis certainly isn't recommended for drivers, it can be an instant reliever on the home front, according to Dr. Primavera. Just sit in a comfortable chair, close your eyes, and imagine a safe, cozy place. Relax all over, until you feel limp and loose from your toes to your head. Then begin breathing evenly and deeply, focusing your mind on the rhythm of your breathing and the sense of feeling sleepy and deeply relaxed. Repeat a soothing message to yourself. "A message as simple as 'make the pain go away' or 'just relax' as a headache begins can be very effective," says Dr. Primavera.

Hitting Headaches Head-On

Ellen Blau went to sleep with pulsating pain in her temples and a pressure-clogged head every day for 30 years. Daily tension headaches and regular migraines rendered her immobile.

Blau, who is now the support group coordinator for the National Headache Foundation (NHF), visited more than a dozen doctors in her search for relief. She depended on various prescription medications such as amitriptyline (Elavil) just to keep going.

"I woke up one day seven years ago and said, 'This is it,'" recalls Blau, who is in her late forties.

She was determined to shake off the cycles of pain. When an article on headaches in a national weekly magazine caught her attention, she highlighted the names of all the doctors quoted in the article and then fired off letters to each of them.

Seymour Diamond, M.D., immediately responded with an invitation to his Diamond Headache Clinic in Chicago.

"I owe my life to him," she says, adding that she has replaced a powerful antidepressant with milder medications, vigorous walks, and a new take-charge attitude. As a result, she no longer faces daily headache misery. Although she might contend with two or three migraines and several tension headaches per month, it's a far cry from the old days.

"I took my life back," says Blau. "A lot of people expect to take a magic pill to cure them, but you need to take an active role in your recovery from chronic pain. I don't wake up anymore worrying that I'm going to get a headache."

For Blau, taking an active role means daily exercise. She wakes up every morning around 6:30, drinks one cup of coffee, puts

Try the ha-ha-hee-hee-ho-ho remedy. Some people can actually laugh away a headache. A 15- to 30-second laughter prescription can stimulate endorphins, which can eliminate a pain-induced frown and provide a euphoric effect. "Not just a little twitter, but a belly-holding, gut-busting, guffaw," Dr. Anderson says. Here's how he describes it.

▶Stand in front of a mirror and give yourself a big, toothy smile.

▶Begin laughing—accelerating from a mild ha, ha, ha to all-out guffaws. Compare it to a car engine trying to start on a cold day, going from a few coughy sputters to a full-powered purr.

▶Keep it going until you're laughing at full throttle, holding nothing back.

Act like a kid. "When you feel the pain of a stress headache coming on, ask yourself how an 8- or 10-year-old would deal with this situation," says Joel Goodman, Ed.D., director of The Humor Project in Saratoga Springs, New York. "Sometimes, a childlike perspective can be a mature adult coping mechanism."

He credits his daughter with this tip. A few years ago, the Goodman family escaped to a resort for some relaxation. But within an hour of their arrival, his wife could not locate her purse. "Her panic brought on a headache," he says.

on her sneakers, and marches outside for a vigorous, arm-swinging, 45-minute walk on the flower-lined roads of Bingham Farms, a suburb of Detroit.

"I take decent strides and walk a 14-minute mile," she says. "My arms are moving very quickly, I'm sweating, my heart rate is up, and I literally pound the pavement to work out things that are stressing me."

When she is working indoors at the computer, Blau fights headaches head-on by keeping an egg timer nearby. Every 30 minutes, it dings, alerting her to step away, walk around the room, and do neck stretches.

"I place my left hand to my right ear, cock my head to the left, and stretch to release tension. Then I do the same to my left side," she explains. "If I feel tension in my head, I put my fingertips on my temples and firmly massage them in small circles."

As NHF support group coordinator, Blau fields daily calls from people all across the country who are eager to shed their headaches. She matches patients with local hospitals and doctors and helps create support groups in cities throughout the United States, delivers other doctor-approved pain-relieving tips, and also provides personal encouragement.

A favorite headache prevention tip that she offers to women is to lighten their load by shedding overstuffed purses. "A lot of women tend to carry purses weighing 5 to 10 pounds or more on their shoulders," Blau observes. "With all that weight hanging on one side of your neck, the muscles contract—a common trigger for tension headaches." (For a referral to an NHF member physician near you, contact the NHF at 428 West St. James Place, 2nd Floor, Chicago, IL 60614.)

Act like a Turtle

This easy exercise—recommended by Roger Cady, M.D., director of the Headache Care Center in Springfield, Missouri—can help uncoil the tight muscles in your neck and shoulders and relieve a tension headache.

▶While sitting or standing, inhale deeply.

▶Tuck your chin in to your chest.

▶Exhale fully, and as you do, raise your head and stick your chin out as far as you can. The curve of your neck should be exaggerated.

Do this chinning exercise three to five times, advises Dr. Cady. You may feel like a turtle, but you should also feel better.

Then their eight-year-old daughter made a suggestion. "She said, 'Hey, Mom, just think, now you have to get a new driver's license. Maybe the new photo will look better than the old one,' " recalls Dr. Goodman.

Stress—and the risk of headache—was swept away in laughter. "We still had a missing wallet, but we were no longer missing our sense of humor," Dr. Goodman recalls.

Mastering the Migraine

The migraine, the most menacing of all headaches, often attacks the entire body. About 24 hours before its arrival, people who are migraine-prone may feel depressed or euphoric, sensitive to light and noise, and prone to excessive yawning.

A migraine can last anywhere from 4 hours to a marathon 72 hours. Many people feel nauseated, have achy, tender neck and scalp muscles, and feel an incessant, throbbing drumbeat on one side of the head. Some may experience an aura, which can cause a temporary loss of speech or visual signals such as flashing lights or a dark blind spot in the center of their vision.

Migraines are sometimes triggered by irregular sleep cycles or by skipping meals, caution National Headache Foundation officials. Other triggers include loud noise, bright lights, and certain foods, such as chocolate, pickled herring, caffeine, or bananas.

During a migraine, blood vessels leading to the brain expand. As the blood pumps through, a throbbing pain begins. Serotonin, an anti-inflammatory chemical in the brain that is designed to help prevent headaches, is in short supply. But even so, you can fight back with easy exercise if you know the way. Here are some techniques.

Fire up your finger. You can outmuscle a migraine merely by using your index finger if you master the practice of finger-warming. But this is a mind-over-body exercise that takes some concentration and practice.

For headache-free people, the temperature of the index finger is usually around 85°F. But for some reason, people with migraines usually have finger temperatures in the 70° range—a full 15 degrees or so below "normal." If you can raise your finger temperature to 96°, you can "burn off" a migraine, according to Roger Cady, M.D., director of the Headache Care Center in Springfield, Missouri.

"About 50 percent of people can reduce the frequency of migraines by 50 percent using this type of temperature biofeedback," says Dr. Cady.

If you want to practice, find a quiet place at

home where you can either sit or lie down. Make yourself comfortable, but don't cross your arms or legs, says Dr. Cady. Place an oral thermometer on the fleshiest part of your fingertip and secure it with tape. It should read below 95°F when you start. Then:

▶Close your eyes.

▶Breathe in slowly, stretching your abdomen so that you suck in breath through your nose. Inhale to a count of four.

▶Hold your breath for another four seconds.

▶Exhale through your mouth as you count to eight.

"When you breathe in, say to yourself, 'My hand . . . ,' and as you exhale, say, '. . . is warm,' " says Dr. Cady. While you're doing this exercise, you should make sure that you clear your mind of interfering thoughts or worries and imagine that the sun is beaming heat into your finger. Continue this exercise for 20 to 30 minutes twice a day.

"The goal is to get the temperature to 96°F at will," Dr. Cady says. "If they have open minds, most people can start to do this." If you do this twice a day to fend off migraines, within four or five days, you'll begin to see results, he says.

Give yourself a pinch. A pinch in the right spot on the body can banish an approaching migraine, says Michael Blate, an acupressure expert and founder of The G-J Institute in Davie, Florida. "Suffering is the doorbell signaling that it is time to use acupressure," says Blate, author of more than 20 books on natural healing.

For a person who is in good health, acupressure, or the art of applying pressure to ease pain, works fast and effectively, says Blate. But it might not be for everyone. If you're pregnant, using a pacemaker, or taking daily medications for cancer or diabetes, you should check with your physician first, Blate advises.

The easiest method of relieving migraine with acupressure is to press on a spot that's literally right under your nose. Here's what Blate recommends for relief.

▶Locate the philtrum—the area of skin between the bottom of your nose and your upper lip.

▶Press with the knuckle of your index finger until you feel a small area of discomfort. Massage that spot deeply until the point becomes increasingly tender.

▶Stop when you achieve a sense of warmth, perspiration, or clamminess (called an acupressure reaction) across your cheeks or forehead or elsewhere in your body.

The whole process should take only between 30 and 45 seconds, according to Blate. "What happens is that you are changing the energy flow in the body to the head," he says. Blate adds that it can not only help headaches but might give you some side benefits as well. "This point is also useful for nighttime foot aches and certain toothaches," he says.

Say yes to sex. Your partner is in an amorous mood, but you feel a migraine coming on. This is no time to say no, urges Dr. Cady.

Satisfying lovemaking is an exercise that Dr. Cady recommends to stop a migraine dead in its tracks—and it might also conquer a tension headache. Enjoyable sex reduces stress and inspires a sense of well-being, he says. It also may elevate your levels of serotonin.

The Headache-Free Lifestyle

You can take steps to reduce headache intrusions into your life, say doctors. "Part of good health is that willingness to nurture yourself," says Dr. Cady. "You have to learn to invest in yourself." That may sound vague and optimistic if you're someone who's prone to headaches, but doctors point out that there are many specific things you can do to "nurture yourself." Here are some of their recommendations.

Don't worry, be happy. Don't overlook the power of smiling as a mental exercise to rid yourself of headaches.

"I don't ever recall a patient coming in with a headache who said, 'Every time I'm terribly happy, I get a headache,'" says Dr. Anderson.

The first step to combat headaches is to boost your mental attitude by improving what Dr. Anderson calls your "smileage." Simply, smile more. Think happy thoughts. Along the same lines, do something that makes you happy, he suggests. For some people, that might mean buying a new blouse or shirt. For others, it's sitting by a sunny window listening to music.

"Anything you do that puts you in a happy place helps your endorphins," says Dr. Anderson. "They also relax muscles," he adds.

Learn to conduct yourself. Ever want to conduct an orchestra? Well, you can—and it might do your head a world of good. To drum out a headache, Dr. Anderson recommends an exercise he calls J'ARM (jog with arms). First, turn on the radio or put on a favorite CD. Maybe you prefer an energetic Bach classic or a contemporary tune like the Macarena.

"Every time we exercise our arms, our brains say thank you, because by moving our arms, we improve the blood circulation to the brain," Dr. Anderson says. He adds that the life expectancy of orchestra conductors was reported in 1980 to be about five years longer than that of the average person.

For your first class in J'ARM school, find a makeshift baton. Simply use whatever is handy—a spoon, a chopstick, or even a handkerchief. As the music starts, raise both arms high and then move them vigorously up and down in exaggerated, enthusiastic conductor motions, keeping time with the music. Sing along. If possible, watch yourself in a mirror during your performance.

If music is not available, you can do this exercise by "playing" a favorite tune in your head and still get the same benefits, says Dr. Anderson. Or just hum to yourself while you conduct your favorite hummable music.

Do some arm twisting. If you're stuck in the office—where wild, J'ARM-style conducting might be frowned upon—you can wrestle with headaches in a more underplayed way. When you've got a spare moment, stand up and let your arms hang relaxed at your sides. Turn your palms inward so they're pressed loosely against your thighs. Inhale deeply, then exhale, and as you do, rotate your arms so that your palms face forward and your thumbs point out. With the next inhalation, again turn your palms toward your thighs. Repeat a few more times, always with a relaxed, effortless motion, says Dr. Cady.

By doing this exercise three to five times daily, you'll begin to dissolve tension in your shoulders, neck, arms, and upper back.

Shrug off pain. If you're weighed down by worry, your back muscles may tighten and your shoulders may slump, triggering a tension headache, says Dr. Cady. To counteract this tension, try a sequence of tensing and relaxing your shoulders.

Stand with your arms hanging at your sides and your shoulders relaxed. Inhale, lifting your shoulders as high as you can. Then exhale, re-

Pressure Places That Get the Ouch Out

The body has a number of acupressure points that can bring quick relief from tension headaches. Here are two that are recommended by Michael Blate, an acupressure expert and founder of G-J Institute in Davie, Florida.

► With your right hand, squeeze your left thumb as close as possible to the forefinger, forcing a mound of skin to rise between them.

► Place the tip of your right forefinger on that mound and the thumb underneath. Press deeply from above and below, relaxing your left hand. Keep pressing until your right thumb and forefinger discover the tender "ouch" point (called the trigger spot) in the webbing between your thumb and forefinger (*right, a*).

► Massage that trigger spot deeply until the skin becomes tender and you achieve an acupressure reaction, indicated by a feeling of warmth as well as some clamminess or perspiration.

► Continue squeezing while you count slowly to 15.

► Repeat the steps on your opposite hand. For tension headaches caused by neck and shoulder pain, try this exercise for applying acupressure to the forearm.

► Extend your left hand so that your palm faces the floor.

► Using your right hand, bend your fingers back until you see a crease appear on the top of the left wrist. Release your hand, letting it relax.

► From the position of the crease, measure upward two thumbs-widths toward your elbow. Place the middle finger of your right hand in the middle and on top of your left arm and find the space between the two bones— the radius and ulna (*right, b*).

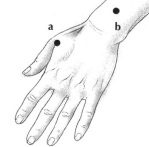

► Press down with your middle finger, massaging deeply as you count slowly to 15.

► Stop when you feel the warmth or clamminess that signals the acupressure reaction.

► Repeat, using your left hand on your right arm.

laxing your shoulders again. On the next breath, lift your shoulders only half as high and then exhale. On your last lift, you should barely raise your shoulders. Repeat this sequence three to five times a day.

Take a walk. Walking is a great headache preventive and offers quick relief for tension pain, say doctors. You just need good walking shoes and a willingness to make walking a habit.

If you're a first-timer, begin with a brisk, 15-minute walk with the ultimate goal, over time, of covering four miles in an hour, suggests Dr. Cady. As you walk, enjoy the scenery and clear your head of worry. Inside, your body is boosting its endorphin supply, which sets off a chain reaction. Endorphins stimulate circulation, provide more oxygen for our tissues, and flush out body toxins more quickly.

Use your snooze power. Your whole body—head included—needs a good long rest stop every day. To help fight headaches, the key is a regular sleep pattern amounting to six to nine hours per night, according to Dr. Cady. About an hour before you climb into bed, start to get into the mood for snoozing, he advises. Try to relax your mind and body with a warm bath, an enjoyable book, or deep, regular breaths.

Meet my headache. This is really a mental exercise, and it may seem strange when you first try it, but Dr. Robbins vouches for it: Try giving your tension headache a first name, like George. And once you've named it, firmly insist that George behave.

What you are doing, in essence, is using your mental powers to psych out a headache. "If people can name their headache and wake up saying, 'I won't let George ruin my day,' they distract themselves and can possibly displace the pain," says Dr. Robbins.

Prepare for a bland entrance. If you know that the second you walk in your front door, you will have a dozen urgent phone messages waiting for you or be besieged by family demands, sit in your car for a few minutes and unwind, advises Dr. Primavera. Take deep, relaxing breaths or listen to a relaxation tape on your car cassette player. Or walk around the block before you enter the house. And when you do go in, remember to use good stress management by setting realistic goals and prioritizing the duties you need to tackle first.

HEARTBURN

Firefighting 101

Can you guess the heartburn capital of the United States? You know, heartburn, that searing sensation in your chest and that unappealing sour taste in the back of your throat. Well, it's not New York City with its frantic pace, strip steaks, and triple cherry cheesecake desserts. Nope, it's Birmingham, Alabama, where 72 percent of people report having heartburn. Perhaps it's all that rich, down-home Southern cooking.

Heartburn occurs when stomach acid finds its way up into your esophagus. There's a little round muscle called the lower esophageal sphincter that's the gatekeeper between your throat and your stomach. It's supposed to allow food down and prevent anything from coming back up. (Think of those spikes in parking garages that will puncture your tires if you exit the wrong way.) But if acid surges past the little muscle back into the esophagus, you get heartburn.

For most people, heartburn is a nuisance, not a serious health threat. For a small but growing population, however, heartburn can lead to cancer of the esophagus. Here's how.

Sometimes, in an attempt to protect the esophagus from repeated burning, your body starts replacing normal esophageal cells with ones that resemble cells in the stomach lining. This condition, called Barrett's esophagus, can lead to cancer, says M. Michael Wolfe, M.D., chief of gastroenterology at Boston University School of Medicine and Boston Medical Center and co-author of *Heartburn: Extinguishing the Fire Inside*.

Barrett's is most common in people who have had chronic heartburn (more than twice a week) for five years or more. "The problem with Barrett's is that it doesn't necessarily have any symptoms that would alert a person that they have it," says Dr. Wolfe. That makes early detection difficult. "But if you've had heartburn for a long period of time, that's very worrisome," he says.

If you're experiencing chronic heartburn, difficulty swallowing, weight loss, weakness, or paleness or if you've had heartburn for five years or more and the nature of the heartburn pain suddenly changes (gets worse or better), you should see your doctor.

Although people of all shapes and sizes can experience heartburn, it's more common among people who are overweight, especially folks who have their extra pounds around the middle, where they can put pressure on the sphincter muscle. Losing weight around your belly is a great way to stop heartburn.

Losing Weight the Low-Burn Way

There's one small problem, though. Exercise, which is usually a crucial part of weight loss, often makes heartburn worse. Why? If you start running, stomach acid sloshes around down there, and it's more likely to splash back up into your esophagus. Before you know it, "feel the burn," that exercise mantra of the 1980s, has a whole new meaning.

Any gentle weight workout that keeps you vertical and doesn't jostle your stomach contents can be an important weight-loss complement to aerobic exercise, says Stanley Lorber, M.D., former chairman of the gastroenterology department at Temple University School of Medicine in Philadelphia and former team doctor for the Philadelphia 76ers basketball team. The more muscle you have, the faster your metabolism. The more quickly your system burns calories, the more easily you'll lose weight. Here are two great exercise options for losing weight without getting burned in the process.

Pedal painlessly. Riding around your block or pedaling in front of the TV is a great way to lose weight without lighting the heartburn fire. Since your tummy doesn't move, your stomach acid stays where it belongs, says Dr. Wolfe.

If you're out of shape, you should build up your biking distance and speed slowly. For the first day, just ride easily for five minutes, says Iona Passik, a certified personal trainer, certified master Spinning trainer, and certified group fitness instructor at Chelsea Piers, a fitness center in New York City. (A Spinner is a specialized stationary bicycle that simulates outdoor biking and is used as indoor training for people who bike outdoors.) Each day, add another minute or two, and if you're working out on a stationary bike, increase the resistance a little bit. "Increasing the level of difficulty over time helps you build strength and muscle," Passik says. Work up to three 20-minute sessions a week.

Give yourself a lift. Shape up without burning up by using 5- to 10-pound hand weights, says Dr. Lorber. Doing exercises with light weights should help turn fat into muscle without causing heartburn, he says. For starters, try this lateral lift.

► Sit upright in a straight chair with a dumbbell in each hand, your arms at your sides with your elbows slightly bent, and your palms facing your upper thighs (*1a*).

► Lean forward slightly at the waist, keeping your shoulders back and your back straight.

► Slowly raise your arms up and out to each side until the dumbbells reach shoulder level. Your arms should be straight and perpendicular to the rest of your body, with your palms facing the floor (*1b*).

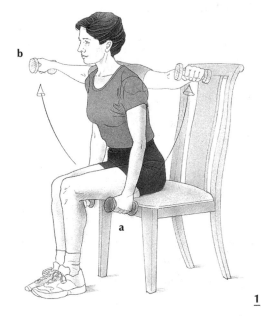

1

▶Hold the position for a second, then slowly lower your arms to the starting position.

▶Do a few to start, then gradually work up to two sets of 8 to 12 repetitions.

Turning Down the Burning

You say that you're slim, trim, and still cursed with heartburn? Try these other heartburn halters recommended by experts.

Breathe deeply. This yoga strategy will help douse the fire. Deep breathing makes the diaphragm massage the stomach, and that can relieve heartburn by speeding up digestion, says Richard Miller, Ph.D., a clinical psychologist and co-founder of the International Association of Yoga Therapists in Mill Valley, California.

Most of us just don't breathe deeply enough, says Dr. Miller. When you're taking deep, even breaths, your belly should rise and fall steadily and your shoulders should hardly move at all. If you want to test for this kind of breathing, just put your hand on your belly and watch whether it moves up, down, and out. You should check your breathing a number of times during the day to make sure your stomach is rising and falling if you want to stop heartburn.

Loosen up. Relief may be as easy as loosening your belt a notch. Wearing pants that are too tight can create the abdominal pressure that can aggravate your heartburn, says Dr. Wolfe. When you exercise, for instance, you might want to trade in your tight-fitting spandex pants for a pair of looser-fitting sweats.

Chew gum while you run. If you have a passion for running and heartburn has you sidelined, try chewing sugar-free gum while

---

Anti-Heartburn Etiquette Exercises

Some experts believe that heartburn is caused less by what we eat than by how we eat it. Consider this advice from Joyann Kroser, M.D., a gastroenterologist at Presbyterian Medical Center and assistant professor of medicine at the University of Pennsylvania, both in Philadelphia, and Stanley Lorber, M.D., former chairman of the gastroenterology department at Temple University School of Medicine in Philadelphia and former team physician for the Philadelphia 76ers basketball team.

Eat smaller portions. The more food you put in your stomach, the more acid it secretes for digestion. And more acid could mean more heartburn.

Eat slowly. If you eat more slowly, you are likely to eat less. To reduce the gulp factor, get a pair of chopsticks. Unless you're an accomplished user, they'll slow you down. Plus, your new skill will impress your friends the next time you go for Chinese food.

Chew completely. When your stomach receives partially chewed food, it has to secrete more acid to break it up for digestion. If you become a master masticator, you cut the acid level.

you run, says Dr. Wolfe. Chewing gum stimulates saliva, which contains bicarbonate, a natural antacid.

Be an early bird. Run in the morning before you've eaten anything, says Dr. Wolfe. "I am a runner who suffers from heartburn, and I almost always run in the morning," he says. "But if I have to run after a meal, I take a swig of antacid, such as Maalox, beforehand." Try not to exercise for two to three hours after eating, he says.

Be upright after meals. Wait at least three hours after eating before you lie down. In fact, consider some of these easy activities after a meal, says Dr. Lorber. They may help you relax and minimize stomach acid.

►Develop a green thumb: Stand at a picnic table to repot some plants.

►Gather your pals for a friendly game of pinochle or Pictionary.

►Join a choir that practices in the evening.

►Take a pleasant after-dinner stroll.

HEART DISEASE

Giving Your Pump a Perfect Day

When Dr. Bruno Cortis listens to his patients' hearts, he doesn't always use a stethoscope. Sometimes he simply listens—while they talk about their families, their work, and their frustrations and satisfactions.

"It's important to know patients as people, to ask about their lives and feelings," says Dr. Cortis, a Chicago cardiologist and author of *Heart and Soul: A Psychological and Spiritual Guide to Preventing and Healing Heart Disease.*

Studies bear this out, suggesting that feelings, particularly feelings of anger, loneliness, and stress, have a significant effect on the health of your heart. While high-fat diets, smoking habits, and sedentary ways are key culprits in the onset of heart disease—the number one killer of American men and women—emotional distress appears to do its share of damage, too.

The upshot? Keeping your heart healthy requires not only going easy on the corn chips and exercising regularly but also learning to manage stress, deal with anger, and connect with people, says Dr. Cortis.

Does that sound like a lot to handle? You can do it all in a single day. Here's just one example of a perfect day of heart disease prevention from several heart experts. It includes the perfect combination of good eats, healthy workouts,

and some key cognitive exercises that will help you keep calm, cool, and connected.

This day embraces some active exercise breaks. If you are obese, smoke, have high blood pressure or high cholesterol, have a family history of heart disease or diabetes, or haven't exercised in a long time, you must be especially careful and see your physician before starting any exercise program. But once you do that, you're ready to go.

7:00 A.M.: Get moving. After a refreshing night's sleep, awaken early enough for a sit-down breakfast and a brisk walk.

7:20 A.M.: Eat a heart-smart breakfast. Once you're dressed and ready, fix yourself a perfect breakfast: a bowl of your favorite whole-grain cereal with nonfat milk and ripe strawberries. Among other reasons, this is a perfect breakfast because it's low in fat, particularly saturated fat. The kind of fat that's found in abundance in whole milk, butter, bacon, and other foods that come from animals, saturated fat is a prime contributor to heart disease, says Gerald Fletcher, M.D., professor of medicine at the Mayo Clinic in Jacksonville, Florida.

Saturated fat raises levels of a certain type of cholesterol, called low-density lipoprotein (LDL), in your bloodstream. And if there's too

much LDL in your blood, it can start accumulating inside your blood vessels, laying the foundation for a waxy substance called plaque. A significant buildup of plaque in the arteries leading to your heart can limit blood flow to that most important pump, thus setting the stage for a heart attack. To keep your arteries squeaky clean, the American Heart Association (AHA) suggests that you get no more than 30 percent of your day's calories from fat and no more than 10 percent from saturated fat.

Not only is your perfect breakfast low in fat, it's also high in fiber. Indigestible plant stuff that's found in abundance in fruits, vegetables, whole grains, and beans, fiber fills you up before you eat too many calories. Consequently, it helps you control your weight. That's important because overweight increases your risk of heart disease, partly because excess body fat boosts your LDL levels. The kind of fiber found in strawberries, whole oats, and oat bran (called soluble fiber) can also lower your LDL levels.

7:20–7:45 A.M.: Show support. While eating your perfect breakfast, get cozy with your spouse. Whisper something nice in your partner's ear. Talk about your day ahead and listen attentively, empathizing while your spouse talks about what's coming up. That's good for your heart, too.

Research has found that people who are in supportive relationships are less likely to develop heart disease than isolated types. "In part, it's because social support provides a buffer against stress," explains Redford Williams, M.D., director of the behavioral medicine research center at Duke University Medical Center in Durham, North Carolina, and co-author of *LifeSkills*. Stress is bad for your heart in any number of ways. Stress hormones damage the linings of your arteries, and damaged arteries are more likely to accumulate plaque. Stress hormones also raise your blood pressure, and high blood pressure damages artery linings, too. As if that weren't enough, stress hormones also trigger reactions that make your blood more likely to clot. That's truly troubling because most heart attacks are caused by blood clots that lodge in plaque-narrowed arteries.

7:45 A.M.: Stretch and stroll. Share one last cuddle before your partner leaves for work, then put on your walking shoes and head out. Before you start, warm up and stretch. Warming up gets you ready for exercise by gradually increasing your heart rate, and stretching leaves your muscles more flexible and therefore less vulnerable to injury, advises Glenn S. Rothfeld, M.D., assistant clinical professor in the department of family medicine and community health at Tufts University School of Medicine in Boston and founder and medical director of Spectrum Medical Arts in Arlington, Massachusetts, in his book *Natural Medicine for Heart Disease*. To warm up, spend 5 to 10 minutes strolling. Then try these injury-preventing stretches.

For your shoulders and upper back: Raise your right arm straight above your head, then bend your elbow and reach as far down your back as you can. Starting with your left arm down by your side, bend your elbow and try to touch the fingers of your right hand. Hold for 5 to 10 seconds and repeat with your arms reversed.

For your calves: Leaning forward on the balls of your feet, lift your heels, then very gently raise and lower them 20 times. Then stand slightly more than arm's length from a wall and lean toward it, reaching out with both arms. Keep your palms parallel to the wall so that you end up supporting your weight on

both hands. Keep your legs straight and your heels on the ground so you feel tension in your calves. Hold the stretch for 10 seconds.

7:50–8:10 A.M.: Hit your stride. Now for that walk. Make it brisk. Any kind of moderate aerobic exercise like brisk walking, doing the tango, cycling, or swimming is perfect for your heart. For starters, it helps lower high blood pressure and boost levels of high-density lipoprotein (HDL), the good cholesterol in your bloodstream. HDL is truly good—it actually cleans fatty plaque from your arteries.

Not only does regular aerobic exercise give HDL a leg up, it also helps you lose excess weight and keep it off. This is good for your heart because obesity raises the odds that you'll run afoul of heart disease.

Exercise does still more. Research suggests that it may elevate levels of special vitamins, called antioxidants, in your bloodstream, and antioxidants can help prevent some of the changes that give plaque a toehold in your arteries. On top of all this, exercise is a powerful balm for stress. It may even condition your body to produce smaller quantities of stress hormones in high-stress situations, says James Blumenthal, Ph.D., professor of medical psychology at Duke University and founder of Behavioral Fitness in Chapel Hill, North Carolina.

So keep up the pace, but don't get so focused on moving fast that everything else is just a blur, says Dr. Cortis. While you walk, enjoy the scene. Feel the breath of air on your face. Listen to the sounds of the neighborhood. To take the edge off busy, potentially stressful days, tuck in as many pleasurable experiences as possible, Dr. Cortis advises.

All told, shoot for 40 minutes of aerobic exercise at least three days a week, suggests Dr.

Fletcher. This can be divided up over the course of the day. Your total for a week will be at least two hours.

7:55 A.M.: Clock your ticker. Now that you're making tracks, check your heart rate. To get maximum heart benefits, you need to exercise intensely enough that your heart beats a certain number of times per minute, says Stephen Wood, M.D., a family practitioner in Salt Lake City and co-author of *Conquering High Blood Pressure*.

To calculate your target heart rate, simply subtract your age from 220, then multiply by 0.7. You can do this at home with a calculator. If you're 60 years old, for instance, the math is $220 - 60 = 160$; $160 \times 0.7 = 112$. Your optimal heart rate is 112 beats per minute.

To check your progress, take your pulse while walking. Hold two fingers to your wrist, count the beats for 10 seconds, and multiply the number by six. If the total is below your target and you feel that you can do a little more, then go for it. Step up the pace a bit or pump your arms more vigorously, says Dr. Wood.

8:05 A.M.: Cool down. Finish the last five minutes of your walk at a leisurely pace. You'll avoid stiff muscles tomorrow. You'll also reduce the odds that you'll have an abrupt and dizzying drop in blood pressure, which can happen if you suddenly stop moving after exerting yourself. Finally, repeat your prewalk stretching routine. Back home, change into your office attire and head for work.

8:10 A.M.: Assuage the rage On the way to your job, choose a route that steers you clear of school buses and other traffic-jam magnets. If you minimize annoying traffic problems, you'll do your heart a lot of good. Researchers used to think that people with so-called Type A be-

The Anatomy of Heart Disease

Heart disease starts quietly when a certain type of cholesterol, known as low-density lipoprotein (LDL), starts accumulating in the arteries leading to your heart, which can happen when you have too much LDL floating around in your bloodstream. Nothing boosts blood levels of LDL like foods high in fat, especially those like butter and cheese steaks, which are loaded with saturated fat. That's why a high-fat diet is bad for your heart.

Studies suggest that LDL is most likely to start accumulating in the linings of arteries that have been damaged. And all sorts of things seem to damage them, such as high levels of LDL in your blood, stress hormones, high blood pressure, smoking, out-of-control diabetes, and a sedentary lifestyle, explains Gerald Fletcher, M.D., professor of medicine at the Mayo Clinic in Jacksonville, Florida. Consequently, inactivity, smoking, high blood pressure, and the like are bad for your heart, too.

Once inside an artery, LDL lays the foundation for a waxy, no-good substance called plaque. Over time, a significant amount of plaque can build up inside the arteries leading to your heart, narrowing the openings inside and limiting the flow of oxygen-rich blood to your ticker. Insufficient blood flow can cause angina, or chest pain, a signal that your heart isn't getting enough oxygen.

Some people have angina only when they're exerting themselves or are under stress, since the heart needs extra blood in both situations, explains Dr. Fletcher. If plaque buildup limits blood flow to the part of the heart that regulates your heartbeat, this can cause arrhythmia, or abnormal heartbeat. (You can also have an irregular heart rhythm without having gunked-up arteries, since lots of caffeine and spicy foods can throw your beat out of sync.)

If blood flow to your heart is severely limited, the result is a heart attack. A few different scenarios can lead to an attack. In the most common one, the plaque in an artery leading to your heart ruptures, triggering formation of a blood clot that blocks the flow of blood, explains Dr. Fletcher.

The most common symptoms of heart attack are uncomfortable pressure, fullness, squeezing, or pain in the center of your chest that lasts more than a few minutes. You might also feel pain spreading to your shoulders, neck, or arms or experience chest discomfort accompanied by lightheadedness, fainting, sweating, nausea, or shortness of breath. Women tend to have somewhat atypical heart attack symptoms—the sudden onset of shortness of breath, fatigue, weakness, or nausea.

If you have these symptoms, go straight to the emergency room of the nearest hospital, says Dr. Fletcher; getting immediate help can save your life if you're having a heart attack.

havior patterns—ambitious, pushy, hurried, red-faced, hostile types—were especially prone to heart disease. But more recent research points a finger at hostility alone. "Studies show that the greater a person's hostility, the higher his risk of heart attack," says Dr. Cortis.

Rage sends blood pressure skyrocketing, makes your heart pound, and triggers the release of damaging stress hormones. Intense rage, like intense stress, can even cause heart rhythm abnormalities and artery spasms that trigger heart attacks. To stay healthy, you need to avoid situations that make you blow your cork and learn to deal with irritation when it's unavoidable, says Dr. Williams. Since rage tends to become a habit, the more you blow up, the more likely it is that you'll blow again. Worse still, people who are chronically enraged show even higher jumps in blood pressure, heart rate, and stress hormones when angered than do those on a more even keel.

8:30 A.M.: Take a hike. Park as far as possible from your office. That way, you can get in a little extra walking on your way to your desk. Not a lot, admittedly, but every extra bit of exercise helps.

A study at the Jean Mayer USDA Human Nutrition Research Center on Aging at Tufts University found that people who move around a lot in the course of a day have significantly less body fat than those who tend to stay put. This is important because excess fat not only raises levels of LDL in your blood, it also lowers levels of helpful HDL. Extra pounds also elevate your blood pressure. As if that weren't enough, they also boost your risk of diabetes, which in turn boosts your risk of heart disease. So the more activity you fit into a day, the better. "Exercise shouldn't start and end in the gym," says Dr. Cortis.

8:35 A.M.: Temper your temper. Try a heart-friendly way of dealing with an irritating situation. Let's say you're heading toward your office when you run into Houndstooth from the public affairs department. "Company talent show's coming up—you got an act?" he asks, jabbing you in the ribs with a bony elbow. Immediately, you're hot under the collar. Your act in last year's show—shooting Mr. Jetsam out of a cannon—was a flop. And Houndstooth has been needling you about it ever since. You're miffed. Before you do anything else, Dr. Williams suggests, ask yourself four questions.

1. "Is this important?" If the answer is no, you need to talk or distract yourself out of being angry (see below for pointers on doing this). If you answer, "Yes, this is important to me" (as in this case), move to the next question.

2. "Is my feeling appropriate?" Consider only the facts when you answer this one. Don't make any assumptions about Houndstooth's motivations. After considering the facts, you answer, "Yes, my irritation is appropriate, because Houndstooth is clearly needling me." If you had answered no, you would need to talk or distract yourself out of being angry. But you said yes, so move to the next question.

3. "Can I do anything to change the situation?" Sometimes you can't do anything about what has you hot under the collar. In those instances, you have to distract yourself or just talk yourself into a better mood, says Dr. Williams. In the situation with Houndstooth, though, you can say, "I feel embarrassed and upset when you remind me about the talent show, Houndstooth. I'd appreciate it if you didn't mention it." Once you've decided that you can do something, ask this final question.

4. "Is this worth doing, given the consequences?" Consider how your response will affect both you and the other guy. In this case, there's no real harm done, and some good done, if you speak up and let Houndstooth know how he's coming across. So do it. Sometimes, however, the consequences warrant doing nothing.

If you decide after answering the first question that you need to distract or talk yourself out of feeling angry, there are a number of ways to do that. Distracting thoughts might be enough, such as imagining yourself in a grass hut in Bali, for instance. Music is also helpful, whether it's Frank Sinatra or Frank Zappa. You can also calm down by breathing deeply. Breathe slowly and inhale deeply, so that your stomach pouches out with each inhalation, then exhale so completely that your belly flattens with each breath. Keep breathing deeply until you've calmed down completely.

Noon: Round up some friends. On the way to get the perfect lunch, pick up your buddies in advertising. Social connections of all stripes are good for your heart, explains Dr. Williams. In fact, a 29-year study of roughly 7,000 Californians found that those with few social ties were more likely to die from coronary artery disease and other major illnesses. "Connections buffer the effects of stress," Dr. Williams explains. Researchers have found that simply having a supportive friend around reduces the heart rate and blood pressure hikes that people experience when subjected to distressing circumstances.

12:00–12:45 P.M.: Go for your goals. Fix yourself a terrific lunch: At the salad bar, toss together lots of dark leafy greens, carrots, tomatoes, and broccoli florets; garnish with slivers of turkey; and dress with a couple of teaspoons of olive oil and a drizzle of balsamic vinegar. Grab a crusty whole-wheat roll and a ripe peach for dessert. Do this, and you'll be well on your way to meeting some of the AHA's dietary goals: at least five vitamin- and mineral-rich fruits and vegetables and six or more whole-grain breads, cereals, or starchy vegetables daily.

Throughout lunch, chat with your buddies. Make plans to see the Charlie Chaplin film festival together next Sunday. Laughter is also an antidote to stress, Dr. Williams notes.

12:50 P.M.: Make a date with your doc. Call your doctor and make an appointment to have your blood pressure and cholesterol levels checked. Be sure to get a test that will tell you your total cholesterol, LDL, and HDL levels, says Dr. Cortis. A healthy total cholesterol level is lower than 200 milligrams per deciliter (mg/dl) of blood, a desirable HDL level is more than 35 mg/dl, and a desirable LDL reading is less than 130 mg/dl. A healthy blood pressure reading is no higher than 140/90. (See "High Blood Pressure" on page 179 for more.)

Have both your cholesterol and blood pressure checked periodically, says Dr. Cortis. If your readings are high, your doctor should help you come up with a comprehensive plan to lower your risk of heart disease and give you an idea of how much exercise is good for you right now.

1:00–5:00 P.M.: Address the stress. At that afternoon meeting, deal with conflicts in a way that keeps stress to a minimum. Follow Dr. Williams's advice on conflict-limiting listening and speaking habits. When the marketing director takes over the meeting to explain her battle plan, there are ways that you can make your own points without confrontation. Don't interrupt. Make eye contact. Uncross your arms and lean toward her while she talks. Use body language that says, "Hey, I'm listening." This will make her less defensive.

After the marketing director has finished speaking, let her know that you've heard her. Repeat what she's said by paraphrasing it. Try starting with, "What I hear you saying is . . ." or "It sounds like you want . . ." followed by words that capture the gist of what you heard.

Be clear and respectful. If you disagree with what the director just said, be diplomatic. Don't attack. Rather, explain your thoughts and feelings, using what Dr. Williams calls "I statements," for example, "I'm confused because I don't understand the connection between Bullwinkle and our client's new hand lotion . . ." (rather than "This Bullwinkle idea is dumb.").

Then listen while she explains.

3:00–3:20 P.M.: Meditate on your break. During coffee break, skip the coffee and meditate instead. When you meditate, your blood pressure drops and your pulse and breathing slow. "Meditating helps quiet your mind and leaves you feeling more peaceful," says Dr. Cortis, who suggests that you practice the technique twice a day for 20 minutes at a time.

To meditate, sit in a comfortable chair with your hands in your lap and your eyes closed. Focus on your breathing—as you inhale and exhale, pay attention to the feeling of the air going in and out of your nostrils. Repeat a calming word to yourself each time you exhale. Try *love*, *peace*, *God*, or anything else with spe-

---

What's Good for the Heart Is Good for the Brain

Take good care of your heart and you'll be taking good care of your brain, too.

That's because the same things that can lead to heart attacks—smoking, fatty foods, inactivity, overweight, and high blood pressure—can also set the stage for strokes.

Like most heart attacks, most strokes happen when blood clots form in arteries that have been narrowed by fatty goo called plaque. Fatty foods, inactivity, smoking, and the rest all contribute to the formation of this plaque.

If a blood clot forms in an artery leading to your heart and blocks blood flow, the result is a heart attack. If a clot forms in a blood vessel leading to your brain and shuts off its blood supply, the result is a stroke, explains Gerald Fletcher, M.D., professor of medicine at the

Mayo Clinic in Jacksonville, Florida. Strokes are so similar to heart attacks, in fact, that Dr. Fletcher calls them brain attacks.

Sometimes, strokes are preceded by mini-strokes, which occur when blood flow to the brain is temporarily blocked. The effects of mini-strokes are usually short-lived and reversible. For any kind of stroke, the symptoms are sudden weakness or numbness in the face, arm, or leg on one side of your body; sudden dimness or loss of vision, particularly in one eye; difficulty talking or understanding speech; sudden, severe headaches with no apparent cause; and inexplicable dizziness, unsteadiness, or sudden falls. If you have any of these symptoms, head for an emergency room immediately, says Dr. Fletcher.

cial significance. Distracting thoughts such as "Did I turn off the stove this morning?" will inevitably crop up. Don't try to repress them; that will only make them dig in their heels. Instead, acknowledge the thoughts, then refocus on your breathing and your important word. (*Note:* On really busy days, when you don't have even 20 minutes to spare, try to allow yourself at least a few minutes for meditation.

5:30 P.M.: Get a lift. At the end of the workday, get out of the office and onto the treadmill at the fitness center. Warm up by treading at a leisurely pace for 10 minutes, then stretch and try some weight lifting.

Even if you're in perfect health, you should have an instructor show you how to lift weights if you've never done it before. Lifting helps build muscles, which are big-time calorie burners. Muscle burns far more calories per pound—even at rest—than fat does. "So supplementing aerobic exercise with resistance exercise will help you control your weight," promises Dr. Fletcher. Try to include weight training in your schedule twice a week, he advises.

When you're working out in a gym, you can build both muscles and social ties at the same time. "Part of the benefits of exercise come from the social support you get if you're exercising with other people," says Dr. Williams.

7:00 P.M.: At home, make contact. Greet your family warmly. Talk with your partner—try to spend 20 minutes every night talking to one another without interruption, says Dr. Williams. Talking about what matters most to you fosters intimacy. After you've each had a chance to talk, help each other make the perfect dinner.

7:40 P.M.: Have a winner of a dinner. Time for your evening meal. A nice piece of broiled, skinless chicken, a baked potato with a little low-fat yogurt as a topping, and plain vegetables, followed up with a fresh fruit sorbet or frozen yogurt, should keep you going through the night. Remember, according to the AHA, it is best to keep your intake of saturated fat as low as possible. You can achieve this by eating plenty of fruits, vegetables, breads, and cereals, along with low-fat dairy products.

8:30–10:00 P.M.: Share an evening. Go dancing with your partner. Make each other feel special.

After observing scores of couples in a special "love lab"—a small apartment outfitted with cameras, microphones, and cooperative men and women—researchers at the University of Washington concluded that partners who stay together have a ratio of at least five positive encounters to every negative one. Sample positive encounters include touching, smiling, talking, laughing, and sharing enthusiasms. Sample negative ones include criticism, defensiveness, and withdrawal. To emphasize the positive, Dr. Williams suggests that you and your partner make dates together at least once a week.

10:30–11:00 P.M.: Head for the bedroom. Get really cozy. Sex is good for your heart for all sorts of reasons. For starters, it's a form of aerobic exercise. Second, it fosters stress-relieving closeness and involves touch, which is good for your heart. Not only does it lower the levels of stress hormones in your bloodstream, it boosts levels of feel-good hormones, which make life more pleasant and joyous. To maximize satisfaction, take turns asking one another to try new and exciting things, suggests Dr. Williams. Go with whatever suggestions the two of you are comfortable with, such as changes in lighting, pace, attire, and music.

HICCUPS

Block Those Nerves and Get Quiet

The hiccup is that true rarity in human physiology, a body function with no apparent purpose. Even sneezing and burping do your body some good, ridding it of unwanted germs and air. But hiccuping accomplishes nothing.

The medical term for hiccups is *singultus*, from the Latin "singult," which means a sob or speech broken by sobs. It's a pretty good description of what hiccups sound like.

"It's an abnormal, useless reflex," says Irwin Ziment, M.D., chief of medicine at the University of California, Los Angeles, UCLA School of Medicine Olive View Medical Center in Sylmar.

Hiccups are often the wages of overeating, swallowing too much air, and drinking carbonated beverages—all of which distend your stomach and irritate your diaphragm, the dome-shaped muscle and membrane between your stomach and lungs. This irritation seems to stimulate the nerves that run between your diaphragm and some sort of "hiccup center" in your brain. Activated, these nerves instruct your diaphragm to contract quickly. When it does, your lungs expand equally quickly into the space that your diaphragm has just vacated. This quick expansion forces you to inhale suddenly, which creates that goofy "hic" sound.

When Hiccups Spell Trouble

Long-lasting hiccups are no joke. They can actually be a symptom of some serious health problems, says Greg Grillone, M.D., assistant professor of otolaryngology, head and neck surgery, at Boston University School of Medicine.

Hiccups that last for more than two days may be a sign of injury to the brain caused by strokes or tumors. They can also be symptoms of growths in the stomach, lungs, or diaphragm. So if your hiccups linger more than 48 hours, see your doctor for a checkup, suggests Dr. Grillone.

It doesn't take much to stop the flow of "Now hiccup!" commands flowing between brain and diaphragm. You probably already know a remedy or two—sipping ice water, eating granulated sugar, and biting a lemon, for example—that appear to be effective hiccup enders. Chewing on a whole clove may help as well, since cloves have local anesthetic properties that

Try Stunt-Drinking

The next time the hiccups strike, consider this remedy from the far side: Fill a glass with water and sip it from the far side of the glass.

Research shows that far-sided sipping really can help quash hiccups. It somehow seems to short-circuit the flow of impulses between the "hiccup center" in your brain and your diaphragm, the sheet of muscle below your lungs that is the source of hiccups.

could interrupt the nerve reflex, says Dr. Ziment. Here are some other simple ways to trick the hic.

Stretch your tongue. Simply grasp your tongue and tug gently. Research shows that this trick can stop hiccups in their tracks. "It can interrupt the flow of messages from brain to diaphragm," explains Dr. Ziment.

Puff out your stomach. Change your breathing pattern, suggests Greg Grillone, M.D., assistant professor of otolaryngology, head and neck surgery, at Boston University School of Medicine. Inhale slowly and deeply, expanding your diaphragm so that your stomach pouches out. Then exhale slowly and completely, contracting your diaphragm so that your stomach flattens.

Roll into a ball. While either lying on the floor or sitting, pull your knees to your chest. This position seems to interrupt the impulses speeding between your diaphragm and hiccup center as well, says Dr. Ziment.

Be a bag breather. When you breathe into a paper bag, you end up inhaling a fair amount of carbon dioxide along with your customary oxygen. "High CO_2 in your blood will trigger deep breathing," says Dr. Ziment, and switching from your normal breathing pattern to a deep-breathing pattern—as you inevitably do when you breathe into a bag—may stop hiccups. Just don't overdo it. "Try it sitting down, and stop if you start feeling lightheaded or notice tingling in your lips or fingers," he says.

Get alarmed. When someone scares you, you hold your breath momentarily, says Dr. Ziment. That makes you hold your diaphragm rigid and may short-circuit hiccups.

Do the uvula lift. The uvula is that fleshy little structure that dangles over the cavelike opening to your throat and moves up and down when you say "Aaaahhhh." Try lifting it gently with a spoon and holding it up momentarily. You guessed it—this, too, seems to interrupt the flow of neural impulses that trigger the hiccups. The connecting nerves run through your neck.

Even if you're like most people and end up gagging instead of lifting, the attempt should have a beneficial effect because even gagging can have an inhibitory effect on hiccups.

HIGH BLOOD PRESSURE

Low-Intensity, Low-Stress Exercise to the Rescue

Take 14 sedentary Italians with high blood pressure, send them to the gym three times a week for 12 weeks, and what do you get?

Eleven Italians with normal blood pressure and, well, three holdouts who still have blood pressure that's high, but not nearly as high as before.

That's what researchers at the Instituti Clinici di Perfezionamento in Milan, Italy, found when they convinced 14 *patate di sofa*—all with high blood pressure—to exercise regularly for a few months.

Surprised at the rapid results? The experts weren't. Increasingly, researchers exploring the benefits of exercise are coming to the same conclusion: Regular exercise can help take mild high blood pressure down a few pegs—in a matter of months.

"Aerobic exercise can be as effective as drugs in treating mild high blood pressure," says Stephen Wood, M.D., a private practitioner in Salt Lake City and co-author of *Conquering High Blood Pressure*. "And it doesn't take that long."

Blood pressure is more or less what it sounds like—the pressure that your blood exerts against the walls of your blood vessels. But it's actually the result of two forces, the force that your heart creates each time it beats and pumps blood through those vessels, and the force that your blood vessel walls exert as they push back against the blood passing by. If your heart pumps too hard or your blood vessels stiffen up too much to "give" a bit when blood passes through, or both, the result can be high blood pressure.

While high blood pressure usually doesn't cause any obvious symptoms, it can cause plenty of harm. Over time, it can damage the arteries leading to your heart, setting the scene for heart disease and heart attacks. It can also cause something called cardiac hypertrophy, an enlargement of your heart that leaves it extra-vulnerable to heart attack. High blood pressure can lead to strokes and may, over the years, slowly shave points off your IQ. It can also result in serious eye and kidney damage.

Why So High?

No one knows exactly what causes most cases of high blood pressure, says Harry Gavras, M.D., professor of medicine and chief of the hypertension and arteriosclerosis section at Boston University Medical Center. But the experts do know that certain things stack the deck in favor of developing it.

For reasons that aren't entirely clear, African-Americans are more apt to develop it than Caucasians. And as we get older, we're all more likely to get high blood pressure, in part because our arteries become less elastic and can't give quite as much.

At any age, smokers run a higher risk of developing high blood pressure, and anyone who's highly stressed is also at risk. That's because stress hormones like adrenaline make blood vessels tighten up and increase the heart rate, explains Dr. Wood. And finally, some people see their blood pressure rise when they eat a lot of salt, although why this happens still isn't clear.

Among the other factors that you can control are exercise and weight. Those of us who are inactive or overweight run a higher risk of developing high blood pressure.

Reversing the Climb

Of course, the first thing you need to do is find out if you have high blood pressure, which is easy to do with one trip to the doctor. But whether or not your pressure is high, there are things that you can do to keep it within a reasonable range and lower it if it's heading skyward. Among the tactics: Quit smoking, cut

Lift Weights, Lower Your Pressure

Doctors used to think that weight lifting was out of the question if you had high blood pressure. All that face-reddening strain, they reasoned, would push your internal pressure sky high.

Research suggests, however, that lifting can be perfectly safe, even beneficial, as long as you do it right, says Stephen Wood, M.D., a family practitioner in Salt Lake City and co-author of *Conquering High Blood Pressure*.

There's a lot to be said for weight lifting. It builds muscle fast, and muscle is a key ally in fighting overweight, which is a major culprit in high blood pressure. Muscle helps you lose excess weight and keep it off because it burns a lot of calories. "Even when you're at rest, " says Dr. Wood, "a pound of muscle burns far more calories than a pound of fat."

What's the right way? Follow these three rules, Dr. Wood suggests.

1. Get your doctor's okay first.

2. Have an exercise physiologist (who can be found at your local fitness center or exercise spa) teach you how to lift and breathe properly. "A lot of people use improper breathing techniques—they hold their breath and bear down when they're lifting—and that's very dangerous," says Dr. Wood. "People have had strokes lifting that way."

3. Stick to the lighter weights. The strain of lifting heavy weights can trigger a potentially dangerous blood pressure spike, says Dr. Wood, so pass those by. You don't need the big ones, anyway. "You can get the same benefit with the lighter weights simply by doing extra repetitions," he says. Start with 10 to 20 repetitions and gradually build up. It's best to check with a trainer at your gym to design a personal program, says Dr. Wood.

back on salt, start exercising regularly, lose excess pounds, learn to cope with stress, and take the blood pressure medication your doctor prescribes, says Dr. Wood.

Exercise is helpful for a number of reasons. It helps you lose weight and manage stress, and it triggers certain physiological changes in your body that help reduce blood pressure. Regular aerobic exercise conditions and strengthens your heart in the same way that it strengthens your other muscles. Since a stronger, well-conditioned heart can pump more blood with every beat, regular exercise actually slows your resting heart rate. That's helpful because a slower heart rate translates into lower blood pressure. Second, regular aerobic exercise triggers changes that help make your blood vessels more flexible, so they give more each time your heart beats and pumps blood.

The key is getting the right kind of exercise—not too intense and not too competitive—and working your way into the exercise habit gradually, says Dr. Wood. Before you get started, get your doctor's okay if you already have high blood pressure, heart disease runs in your family, you have high cholesterol levels or diabetes, or you are sedentary, Dr. Wood cautions.

Once you have the green light, shoot for 40 minutes of aerobic exercise, such as walking, cycling, swimming, dancing, and the like, three times a week, Dr. Wood says. If you're just starting to exercise, though, start gradually. The first week, spend just 5 minutes exercising each time out—at any pace that's comfortable. Make it 10 minutes daily the following week, then 15, and work your way up to 40 minutes. Here are some guidelines.

Aim low. If you have high blood pressure, lower-intensity aerobic exercise—walking, swimming, cycling, or dancing just hard enough that you can't sing but can talk—is your safest bet, says Dr. Wood.

But all exercise boosts your blood pressure somewhat because your heart has to pump harder to get extra blood to your hard-working muscles. When your heart works harder, your blood pressure rises. Lower-intensity exercise is safer than high-intensity exercise simply because it doesn't trigger as great an increase in pressure.

Among lower-intensity options, Dr. Wood prefers walking. "It's ideal because you don't need a lot of equipment or a gym membership to do it, and if you swing your arms, you use almost all of your muscles," he says. If you'd prefer an alternative, read on.

Get all wet. Swimming gives most of your muscles a workout, too, notes Dr. Wood. Other watery variations, such as water aerobics, synchronized swimming, and water polo, are equally good choices.

Hop, skip, and jump to the music. Lower-intensity, low-impact aerobic dance also does the trick, says Dr. Wood. It works almost all of your muscles, and you can do it with a buddy. "If you have an exercise buddy, you're more likely to keep exercising," he adds.

Turn competitive sports into cooperative ones. Competitive sports are yet another good pick, unless you're the type who's overwhelmed by the competitive urge. "If you curse, kick the ground, and smash your clubs when you miss a shot in golf, that's no longer beneficial exercise—it's too stressful," says Dr. Wood.

But even if you're highly competitive, you can adapt the rules so they're more cooperative in spirit, he suggests.

Total team tennis: Out on the court, never

mind keeping score for a while, just as long as you can keep a volley going. Have each player run for shots, but not to the point of exhaustion. Think of yourselves as members of the same team, with a shared goal—to give each other a good workout.

Assisted one-on-one basketball. Play the usual game, with one twist on the rules: You get two points for assisting the other guy with a shot and one point for making a basket yourself.

Compassionate golf. With this one, you and your fellow players give one another an empathetic below-par advantage: You subtract points for "heartbreak shots," shots that should have been easy but went terribly astray. If the ball is six inches from the hole and you whack it through the woods to Grandma's house, you get a break.

Other options include subtracting points for shots that trace the letter "S," or the double helix, or whatever shape you and your buddies decide gets a "compassion break" on this particular day. It'll add an element of excitement to the game. Remember: To get your requisite aerobic exercise, walk from hole to hole.

Change your attitude. Another option is to play the game the way you always have, but change your hypercompetitive attitude, says Dr. Wood.

Try this advice from Judith Beck, Ph.D., director of the Beck Institute for Cognitive Therapy and Research in Bala Cynwyd, Pennsylvania, and clinical assistant professor of psychology in psychiatry at the University of Pennsylvania School of Medicine in Philadelphia. Before you get out on the links or under the hoops, give yourself a talking to. Instead of telling yourself, "I've got to do my best, or this will be a disaster," or "I have to win, or it'll be terrible" (which is what you've no doubt been telling yourself, even if you didn't realize it), change the message. Remind yourself that the point is to get some exercise and have some fun

Blood Pressure Readings: Safety in Numbers

High blood pressure rarely causes symptoms, which is why you have to have your blood pressure checked regularly.

The American Heart Association suggests having your blood pressure tested at least once every two years. When you have it checked, you'll get a two-number reading. The first, and larger, number represents systolic blood pressure, the pressure inside your blood vessels each time your heart contracts, or beats. The second, smaller number, the diastolic pressure, is the pressure inside

your vessels when your heart rests between beats.

A healthy systolic pressure is around 120 or less, and a good diastolic reading is about 80 or so, but any reading below 140/90 is considered normal.

A single high reading, though, doesn't mean that you have high blood pressure, since your pressure fluctuates over the course of the day. If you get a high reading, your doctor should check it again on two separate occasions to see if that first elevated reading was a fluke.

and that how well you do is not a reflection of who you are.

Quit comparing. Also in the category of attitude adjustment, try to catch yourself when you're in the midst of comparing your performance with someone else's. Remind yourself that you don't have to be the best, and you don't have to top your own best score. "Instead, try to be satisfied with a range of performance," says Dr. Beck. Some days you're hot, others, not. Some days it's sunny and warm, others, it's raining locusts. Accept that.

Distract yourself. Fuming because that near-birdie just turned into a bogie? Distract yourself. "Start a conversation with one of the other players or look around and enjoy the scenery, " suggests Dr. Beck.

Give yourself time. If you've been highly competitive all of your precocious life, it will take time to shift your sights from the winner's cup. Don't try to set the world record in the change-of-competitive-heart category. "Each time you play, try to reduce your competitiveness a little more, " says Dr. Beck. "Over time, you'll make inroads."

High Cholesterol

Accentuating the Positive

We've been brainwashed, over and over again, that when it comes to cholesterol concerns, LDL is the bad guy. And it really is: Low-density lipoprotein lives in the bloodstream like a circling stretch limo, offering a free ride to cholesterol. The more LDL you have, the more cholesterol there is cruising your highways. Like traffic on a Los Angeles freeway at rush hour, the crush of cholesterol builds until it comes to a dead stop—right on the walls of your arteries. The classic outcome: Traffic jams. In arterial terms, that means high risk of heart attack and stroke.

To lower our levels of that lousy LDL (and avoid heart-harming outcomes), we forgo the meat and fat, fill up on veggies and fruits, and keep a standing breakfast date with the Quaker Oats guy. And that's all to the good.

But there's another side to cholesterol control, one that you may not be as familiar with. Enter HDL—the superhero of the heart—the healthy high-density lipoprotein whose motto is "the more the merrier."

Bumping up your levels of HDL gives cholesterol a run for its money. Instead of cholesterol being able to hitch a luxury ride on some slow-rolling LDL, HDL grabs it by the collar and shows it the door by way of the liver. So the more HDL you have hanging around, the less dangerous cholesterol there is free-floating in your veins and arteries.

Unfortunately, while a special diet can bring down levels of LDL, going low-fat has no effect on raising your HDL count. Read on to discover the secret of growing your own healthy HDL heart patrol.

The Exercise Connection

In the 1970s, researchers started seeing a connection between low HDL and high risk of heart disease. Not long afterward, someone noticed that exercise can help raise HDL levels. Now we know for a fact that exercise is one of the only ways to specifically increase healthy, heart-helping HDL. And, boy, does it do the job.

"There's no pill that even comes close to what exercise can do," says David Nieman, Dr.P.H., professor of health and exercise science at Appalachian State University in Boone, North Carolina, and author of *The Exercise-Health Connection.*

Exercise—specifically, aerobic exercise—recruits muscle tissue on a search-and destroy mission. Muscle cells need to munch on fat for fuel during moderate- to high-intensity exercise. They will gladly grab that fat in the form of very low density lipoprotein (VLDL), another type of cholesterol carrier that carts around triglycerides, another form of fat in the blood. When muscle burns triglycerides for energy, the reaction literally spawns baby HDL particles, says Dr. Nieman. This means that your local gym doubles as an HDL nursery.

Let Your Body Do the Talking: Using the RPE Scale

You don't need fancy electronic heart rate monitors or lightning-quick, in-your-head calculations to figure out your most healthful pace when working out, says Lee Lipsenthal, M.D., medical director of the Preventive Medicine Research Institute in Sausalito, California. Using a simple scale called the Rate of Perceived Exertion (RPE), you can get directly in touch with how hard your body is working.

The scale ranges from no activity (6) to exhausting activity (20). The target to aim for, says Dr. Lipsenthal, is between 11 and 13, which is "fairly light to somewhat hard" activity. That level of exertion approximately equals a pulse rate near 70 percent of your maximum—the place to be to garner the most HDL-raising effects from your workout.

You're probably wondering how to know when an exercise feels "somewhat hard." That's the secret behind the RPE scale's effectiveness. "What you perceive—what you feel internally—as hard or easy is actually very accurate in terms of heart rate," says Dr. Lipsenthal. In other words, you really can tell how hard your heart is working.

What's more, this scale ensures safety but won't let you off the heart-pumping hook as you get in better and better shape. If you're new to exercise, you might get to level 11 by walking at a slow pace for a half-hour. That's where your body feels that the exertion is "somewhat hard." Over time, you'll find yourself becoming more fit and your heart growing stronger, and you'll need to walk faster and longer, or maybe even break into a run, to bring yourself up to that same level of moderate exertion.

Here's what the whole RPE scale looks like. To use it, simply estimate how much you are exerting yourself and match it to the chart. Remember, Dr. Lipsenthal recommends that you aim for the 11-to-13, or "fairly light to somewhat hard," range.

6: No activity
7: Very, very light
8
9: Very light
10
11: Fairly light
12
13: Somewhat hard
14
15: Hard
16
17: Very hard
18
19: Very, very hard
20: Exhausting

There's one catch to all this grow-your-own magic, though. To see results, you need to do more exercise than you might think, says Dr. Nieman. "If you can put in close to an hour a day of some good, brisk activity at least five days a week—that's where the benefits are very measurable."

A study at the Hospital del Mar in Barcelona, Spain, proved that moderately intense exercise builds the most heart-healthy HDL. In a group

Stress: The New Cholesterol Culprit

You can add another health problem to the list called "blame it on stress." Now it seems that even cholesterol levels are affected by our mental states, says Lee Lipsenthal, M.D., medical director at the Preventive Medicine Research Institute in Sausalito, California.

Research shows that people who visit emergency rooms for injuries of any kind will have elevated levels of a dangerous type of cholesterol called very low density lipoprotein (VLDL) and artery-clogging triglycerides. And it seems to be the emergency situation that causes this spike in VLDL levels. Researchers found that the amounts of these cholesterol-raising substances had subsided when people who went through that emergency room experience were tested again six weeks later.

Stress brings on double waves of stress hormones called catecholamines and gluco-corticoids. Catecholamines come on at the start of a stressful experience, and if the stress continues for a while, your blood becomes crowded with glucocorticoids. When these hormones build up, the liver is prodded to crank out that troublemaking team of VLDL cholesterol and triglycerides. And if they're not burned by muscles as energy, they in turn may produce the sticky, road-blocking plaque that can lead to heart disease.

Acute, momentary stress—such as trips to the emergency room—happens to all of us, at least some of the time. Nonstop stress, though, is something we'd be wise to do without. "Each of us has a limit to the amount of stress we can bear," says Dr. Lipsenthal. "When it's chronic, we wind up living too close to our stress threshold."

What's the answer? Either reduce the stress in your life or raise your threshold for handling it, advises Dr. Lipsenthal, who helps people accomplish both goals. "It can be as simple as asking yourself what is more important—your job or your life?" he says. Even better is learning ways to lift your threshold for stress.

For this, the tools favored by Dr. Lipsenthal include meditation, yoga, and tai chi. (For specifics, see "Meditation" on page 249, "Tai Chi" on page 386, and "Yoga" on page 426.) These methods teach you that the troubling stuff that happens from day to day isn't terribly important; it's really the big picture that matters. And that, says Dr. Lipsenthal, is the key to declawing the stress monster. "For stress to affect you, it relies on a certain limited perspective. When you alter that perspective, stress loses its power to harm," he says.

of 537 men ages 20 to 80, activities that used up more than seven calories per minute were the ones associated with higher levels of HDL and a better overall good-to-bad cholesterol ratio.

Swimming, running, and bicycling are three common effective exercises. Doing these on a regular basis, and enthusiastically, will improve your HDL count. But if you'd like to try something new when you go for the burn, sample the five calorie crunchers below, says Lee Lipsenthal, M.D., medical director of the Preventive Medicine Research Institute in Sausalito, California. A person weighing around 140 pounds can burn seven calories a minute or better with any one of these activities.

Savor some strokes. No, we're not talking golf strokes here. Stroke the ocean with a vigorous bout of rowing. If you don't have the requisite rowboat and placid body of water to go with it, try using a rowing machine at your local health club. They're almost always available, and they'll give you a lot of bang for your gym-break buck. You'll burn about seven calories per minute with a rowing workout. Before you start rowing, though, be sure to check with a trainer, who will show you the proper technique. Using improper rowing technique is a fast way to damage your lower back, says Dr. Lipsenthal.

Make a racquet. Racquetball is a fast-moving sport that's guaranteed to get you huffing and puffing. Find a partner who plays at your level and have a ball. You'll burn 12 calories a minute on the court.

Jump around—and around. This is definitely not kid stuff. Rope jumping is the way boxers traditionally get pumped up. Now you can do it, too. The equipment is cheap—all you need is a supportive pair of sports shoes and a quality rope from your local fitness shop (good ropes cost around $10). When you skip briskly, you burn nearly nine calories a minute.

Dance cholesterol away. Doing the lambada or even the twist is a workout that's perfect for a Saturday night. Using up 10 calories a minute makes even a vigorous Virginia reel a good bet for cholesterol control.

Take a day hike—and don't forget the pack. Backpacking is an outdoorsy endeavor that gives you fresh air along with high calorie consumption. Fill a pack with just 11 pounds of lunch fixin's, beverages, and other picnic necessities, and take an hour's walk to your favorite spot. You'll fire up 8 calories a minute on the way and 7.5 on the way back (with an empty pack).

HIP PAIN

Get Yourself an Elvis Pelvis

The biggest fear among Elvis impersonators may be pelvic bones that howl like a hound dog while getting all shook up on stage.

It's tough to remind anybody of The King if you're hobbled in the hips. But Elvis impersonator Michael Bartle of San Francisco doesn't worry anymore. Why? Because before every performance, he stretches and strengthens his hips with a series of water exercises. "Since I started water workouts before my shows, I am the most limber I've ever been. And I am more springy on stage," he says.

The origin of pain in the hip area is often complex. Its most common cause is osteoarthritis, the wearing down of the cartilage cushion between the ball and socket of the joint. But hip pain is sometimes caused by strained (or pulled) muscles. It can also stem from malfunctioning tendons in nearby parts of the body—the lower back, the buttocks, and the upper legs.

If you have severe hip pain, you should consult your doctor before embarking on an exercise program. But for mild cases of hip ache and stiffness, experts recommend an easy-does-it, two-pronged workout strategy that combines stretches for hip flexibility with low-impact strengthening exercises for optimal hip power.

Stretch Those Hips

Stretching—elongating the muscles—helps relax muscles and improve flexibility, says Michael Ciccotti, M.D., orthopedic surgeon and director of sports medicine at the Rothman Institute at Thomas Jefferson University in Philadelphia. Try these.

Make a marriage proposal. The position for this hip flexor stretch, recommended by Thomas Meade, M.D., orthopedic surgeon and medical director of the Allentown Sports Medicine and Human Performance Center in Pennsylvania, looks as if you're about to "pop the question" to your sweetheart.

►Kneel on your left knee. Bend your right knee and keep your right foot flat on the floor in front of your body.

►Put your hands on your hips or rest your right hand on your right knee and let the other hand hang by your side (1).

►Tighten your abdominal muscles and lean slightly forward without arching your back. You should feel a stretch in the front of your left thigh. Hold for 20 seconds.

►Repeat three times and then switch legs.

Push away. For variety, try this stretch recommended by Dr. Meade to relax and improve flexibility in your hip rotator muscles.

►Hold this stretch for 20 seconds, then repeat three times.

►Switch legs and repeat.

►Lie on your back with your knees bent and your feet flat on the floor. To keep your neck from arching, place a pillow under your head.

►Place your left ankle on your right knee.

►Keep your lower back flat on the floor as you use your left hand to gently push your left knee away from you (2). You should feel the stretch in the buttocks area.

Rock bottom. To relieve muscle tightness in the buttocks, Meir Schneider, Ph.D., licensed massage therapist, founder of the Center and School for Self-Healing in San Francisco, and creator of the Meir Schneider Self-Healing Method, offers this hip-rocking stretch from his book, *The Handbook of Self-Healing*.

Sit cross-legged on the floor or on an armless chair or stool with your feet flat on the floor. Shift your weight onto your left buttock, then slowly rotate your upper body in a circle over the left

Bo Knows Hips—And Leg Lifts

Former football and baseball player Bo Jackson is proof that it's possible to maintain a healthy body after hip replacement surgery. Jackson was 29 when his hip was replaced in 1992. "Afterward, it was very important for me to strengthen the muscle groups surrounding my hips," he says.

Now president of HealthSouth Corporation's Sports Medicine Council, based in Birmingham, Alabama, and an actor, Jackson shares his favorite daily hip-strengthening exercises.

►Lie on your back.

►Keeping your right leg straight, bend your left knee and place your foot flat on the floor.

►Raise your right leg six inches off the floor and hold for six seconds. Repeat 10 times.

►Next, lie on your left side with your left leg slightly bent. Lift your right leg without turning your knee toward the ceiling and hold for six seconds.

►Repeat 10 times, then repeat the entire sequence on the other side.

buttock (most of the movement will be in the lower back area). First, lean forward as far as you can. From that forward position, lean to the left as far as you can. Then rotate backward and finally to the right. Remember to keep your weight on your left side. Try 20 full rotations, then relax and repeat the steps while putting your weight on your right buttock, suggests Dr. Schneider.

Press and please. Sit on a rug or mat so that the soles of your feet touch and your knees are out to each side, says Dr. Schneider. Be careful to keep your weight evenly distributed on your buttocks. Place one hand on top of each knee and gently press down, first on your right knee and then the left. Then press both hands down on your knees at the same time. This motion opens the hip joints and stretches your inner thigh muscles, says Dr. Schneider.

Take your hips for a dip. To fight hip joint stiffness and improve flexibility, water exercises may be ideal, says Jane Katz, Ed.D., professor of health and physical education at John Jay College of Criminal Justice at the City University of New York, world Masters champion swimmer, member of the 1964 U.S. Olympic performance synchronized swimming team, and author of *The New W.E.T. Workout.* Water offers a low-impact workout, takes the weight off your aching hip, and strengthens the muscles around the hip. Here is one of Dr. Katz's favorites.

▶Stand in waist- to chest-deep water with your back resting against the pool wall for support.

▶Standing on your left leg, cross your right leg in front of your body, grasping your ankle with your left hand. Use your right arm to hold on to the pool edge.

▶Slowly bend your left leg. This will stretch the muscles at the hip and the back of your right thigh (3).

3

▶Slowly straighten up and change legs. Repeat this cycle five times.

A Joint Effort

Add these muscle-toning exercises to your hip flexibility program.

Kick a game-winning field goal. To strengthen the flexor muscles in front of your hip as well as the thigh, hamstring, and buttock muscles, try a little slo-mo air football, says Dr. Ciccotti.

▶Stand next to a countertop or sturdy railing and hold on with your left hand.

▶While keeping your supporting left leg straight and your foot flat on the floor, raise your right leg as high as you can in front of your body (4a).

▶Hold that pose for no longer than a second before letting your leg fall and slowly swing behind your body as far as possible without straining (4b).

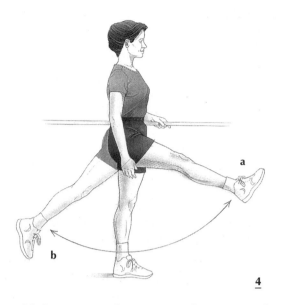

<u>4</u>

►Hold the position for no more than a second, then swing back to the starting position, as if you were kicking a football in slow motion.

►Try 10 swings before switching to the left leg.

By trying to raise your leg higher on each end of the swing, you can increase your range of motion, says Dr. Ciccotti.

Squat down. This do-anywhere exercise helps strengthen the muscles in the fronts of your hips and thighs and your buttocks, says Dr. Meade.

Stand with both feet flat on the floor and hip-width apart. Cross your arms over your chest and face straight ahead, then slowly squat until your thighs are parallel to the floor. Try to keep your torso straight and your knees directly over your toes. Hold the squat for a couple of seconds and then stand up. Try 10 squats twice a day.

Caution: Don't Be a Hip Hero

Most people can build limber and strong hips through daily stretches and low-sweat leg lifts, says Michael Ciccotti, M.D., orthopedic surgeon and director of sports medicine at the Rothman Institute at Thomas Jefferson University in Philadelphia. But it's important that you don't overdo it. Dr. Ciccotti recommends that you seek immediate medical attention if your hip pain:

►Intensifies during or after exercise
►Radiates from the buttock area down to the leg
►Persists and is not relieved with rest, ice, or over-the-counter pain relievers
►Keeps you up at night
►Keeps you from bearing weight on the hip

Try water wading. To strengthen arthritic hips without jarring pain from pavement, Dr. Katz recommends walking chest-high in H_2O.

Start off slowly by marching in place, raising your knees high as if you were leading the high school band onto the football field at half-time. Slowly pick up speed throughout your workout. Do this for one minute and follow it with a minute of rest, then do another minute of stepping followed by rest. Over time, work up to three minutes of continuous exercise.

IMAGERY

Finding the Healer Within

WHAT IT IS

The bowler, cursed by spares, envisions the 10 tooth-white pins crumbling in the thunder of a strike. The hockey goalie sees his glove snatching the icy, spinning black puck hurtling toward the net. Just before he steps into the batter's box, the slugger pictures the ball soaring high over the centerfield wall.

These athletes *will* themselves into action, using their minds in concert with their muscle in a mental technique known as imagery. They've trained their imaginations to picture their strengths and their successes ahead of time. This ability to "image" good results subdues their worry and anxiety, say doctors.

There's no reason for our thoughts to hobble us. What works for the Michael Jordans of the sports world can also work for you. Our inner creative voices can also ease our nerves so that our bodies can function at their best, says Martin Rossman, M.D., clinical associate in the department of medicine at the University of California Medical Center in San Francisco and co-director of the Academy for Guided Imagery in Mill Valley, California.

The mind is not all-powerful. You can't just think yourself cured or calm. But research has shown that mastering imagery can help with both physical and psychological problems.

Feeling chills? Mentally put yourself on the beach, calmly basking surfside on a summer afternoon. Feeling sure of failure in an upcoming exam? Envision yourself acing it. You can use your mind as a wellness tool.

"Imagery is a window to the soul and the inner mind," says Dr. Rossman. But imagery is more than words or pictures in the mind, he adds. It can include all five of your senses. Invoking hearing, smell, and taste sensations can guide you to therapeutic serenity. Using imagery, you can hear the sounds of a picnic, smell the barbecue, and taste the cool lemonade as you mentally return to a favorite family get-together.

Imagery works best when it is customized for each individual, says Belleruth Naparstek, a lecturer and expert in guided imagery in Cleveland Heights, Ohio, who developed an audiotape series on guided imagery. An image that works for an eager golfer may not work for an accountant, but either the athlete or the office worker can move toward a relaxed, healing state by unleashing the power to imagine things vividly.

"Everybody has an imagination," says Naparstek. "What imagery does is take us beyond clock time to transcend the limitations of how we see ourselves so that we can access the healing power within us."

WHAT IT DOES

Imagery is a pain-free health tool that has been shown to help many medical problems. This minding-the-body technique has been used as a pain reliever as well as an emotional healer.

The power of controlled thought can influence your body's temperature, heart rate, blood pressure, oxygen consumption, gastrointestinal activity, sexual arousal, muscle relaxation, and immune system, say doctors.

In battling cancer, imagery has been shown to stimulate the immune system and increase the aggressiveness of fighter cells that attack cancer cells, says Dr. Rossman. "If a pill could be found that stimulated the immune system like the imagination, it would be absolutely mandatory to give it to immuno-compromised patients."

There's no question that the ability to marshal our imaginations can, in many cases, cut stress, speed recovery, and keep us feeling tip-top. It can reduce the levels of stress hormones like adrenaline and cortisol and can bump up the natural tranquilizers like endorphins. That's why imagery is especially useful against illnesses that are caused or complicated by stress.

In a controlled study involving patients undergoing colorectal surgery at the Cleveland Clinic Foundation, it was found that patients who listened to Naparstek's guided imagery audiotapes before and after surgery had less anxiety before the operation, needed less medication after it, and were released from the hospital sooner.

Dr. Rossman offers this persuasive example of imagery's power: If you tell someone to salivate, it may or may not happen. But if you ask them to close their eyes, take deep, regular breaths, and imagine sucking on a juicy, sour lemon, for most people, a physiologic response kicks in and they start salivating.

REAPING THE BENEFITS

Let's look at some tips from the experts to guide you on a successful imagery journey.

Practice, practice, practice. Not everyone is a master at imagery after the first attempt, but most will get better with practice, says Naparstek.

Suggest, don't tell. For best results with newcomers to imagery, Naparstek and Dr. Rossman recommend that you work with trained therapists who offer suggestions and do not deliver commands of specific images. Vivid images of a Thanksgiving family celebration may conjure up pleasant memories for a therapist, but those images aren't right for you if they trigger nightmarish memories of an unresolved family feud.

Put it together with the practical. Imagery is a valuable member of the total medical team, but you should not arbitrarily stop taking prescribed medications or forgo other recommended treatments without consulting your physician first, says Dr. Rossman. "You may still need cardiovascular work done, blood pressure tested, or a stress evaluation done, for example," he says.

Chat with your pain. If something pains you—physically or mentally—experts suggest that you create an image of it and have a chat with it in your mind. Ask it why it is here.

What does it want? What can you do to make it go away? This technique will make the pain less threatening and may help you find solutions, says Naparstek.

Fight negativity with cartoons. Imagery can help you deal with an unreasonable boss or co-worker. Instead of imagining all of the unanswerable questions or criticisms coming from your boss or co-worker, change the image, says David Lee, a mind-body expert from Saco, Maine, who helped develop an accredited program in guided imagery for health professionals. One way is to see that person as your favorite cartoon character or dressed in a chicken suit. "It actually worked for one of my clients who was very anxious about a job interview," says Lee. "It's hard to be intimidated by someone in a chicken suit." By changing the emotion from anger or fear to humor, you're more apt to laugh off an uncomfortable encounter with a nasty boss or relax during an important interview.

"This helps you de-trigger an emotional response by stepping out of a painful memory so that you will not feel it physiologically," Lee says. "What's implied is that you are taking power away from an authority and giving yourself more control in your life."

Sleep yourself toward healing. What happens if you become so relaxed that you fall asleep during an imagery session? Don't worry, says Naparstek. Let yourself snooze your way toward healing. "You don't have to stay awake for imagery to work. In fact, it may work even better for some Type A personalities who have trouble relaxing when they are awake," she adds.

Make imagery a daily habit. Imagery offers you the chance to take a free 15-minute mini-vacation anywhere in your mind, and it allows your body a chance to relax, says Dr. Rossman. So make it a regular part of your lifestyle.

TRIAL RUN

The trick to mental imagery is concentration and using all your senses. It's not enough to just use your mind's eye. You've got to use your mind's nose and your mind's ears as well, say experts. Mental imagery works best when you close your eyes and allow yourself to drift to a pleasant place that makes you feel relaxed. It's a healthy escape from life's distractions.

Ready to give it a try? Here are some steps described by Naparstek that should throw some light on the process.

► First of all, ask yourself where you felt the most safe or at peace in your life. Was it at the beach on the 4th of July when you were a kid? When you curled up on a winter's day with a favorite book and a cup of hot chocolate? Whatever the memory, make sure it's a good one.

► Sit in a comfortable chair and close your eyes. Breathe deeply, all the way down into your belly. Then breathe all the way out as fully and completely as you can, suggests Naparstek.

► Now take yourself back to a time when you felt safe . Key into your feelings at that time in your life. If it's a beach scene when you were a child, summon the way your hand felt as it was held securely by your parent, hear the sound of the waves whooshing back into the sea, smell the brine and sunscreen, feel the texture of the warm sand, and taste the salt on your lips. If it's a winter scene, try to taste the hot chocolate, hear the crackle of ice against the window or the wind blowing snowdrifts, feel the warmth of the logs burning in the fireplace, and listen to the music you could hear playing far away. Engage all of your senses as you let your spirit drift back to that time.

▶As you re-experience whatever time and place made you feel safe and serene, let yourself be carried back. Stay there as long as you like.

▶When you are ready to resume life's daily tasks, open your eyes slowly. Allow yourself a minute or so to readjust. Feel your breath fill your lungs and be aware of your feet on the floor, your buttocks on the chair, and your hands in your lap. You should feel relaxed and refreshed.

Taking even 10 minutes a day and shutting out the demands of the world with the power of your healing imagination will subdue the stress that inhibits your body's ability to heal itself.

INCONTINENCE
Shoring Up Your Leaky Plumbing

Sometimes, it helps just to know that you're not alone. And if you have chronic plumbing problems, you're definitely not. Sometimes, incontinence is caused by anxiety and stress, but most often, it can be traced to a loss of tone in the pelvic muscles that support the bladder. Those weakening muscles aren't able to withstand pressure from the abdominal muscles above, so any ab pressure, like laughing or coughing, can cause unwanted trickles of urine.

For women, giving birth can weaken important pelvic muscles. And often, the drop in estrogen levels that comes with menopause can lead to a slackening in your tissues and muscles. Plain old age can be the cause, too.

There are several different kinds of incontinence, so tell your doctor in detail about what you're experiencing. Depending on the cause, you might be able to avoid the problem if you limit your intake of caffeine and fluids in the evening. Your doctor can also suggest strategies and treatments that will help to significantly reduce or even cure your incontinence.

Get Control with Kegels

The best stress incontinence inhibitor is a type of exercise known as Kegels, according to Dim-itri Kessaris, M.D., co-medical director of URO*Star–The Urology Center in Great Neck, New York. Arnold Kegel (his name rhymes with bagel), M.D., came up with an exercise that actually builds up the interior pelvic muscles, specifically the pubococcygeal (PC) muscles that surround the urethra, the tube from the bladder to the outside world.

Kegel exercises have been proven quite effective. One study in Denmark showed less urinary leakage in 60 percent of the women who were trained in Kegels. When they perform Kegels on a regular basis, women report improvement in almost 80 percent of cases, says Dr. Kessaris. If you do not notice an improvement after six weeks of faithful exercise, you should see your urologist for a full exam. You can begin giving your PC muscles a workout at home by following these guidelines.

First, find them. When you're urinating, try to stop the flow in midstream. The muscles you activate are the PC muscles.

Get a feel for things. If you have trouble isolating the correct muscles, try inserting one or two fingers into your vagina. When you activate the right muscles, you'll feel a little pressure.

Build the tension. Kegel exercises involve core steps that are easy to practice. First, gradually tighten the PC muscles partway. This will build muscle strength and help you prepare for the full tightening, says Dr. Kessaris. (When you insert your fingers, you will be able to gauge how much tension is needed to tighten the muscles partway. Eventually, you won't need to use your fingers as a guide.) Then hold for five seconds. Squeeze the muscles as tightly as you can. Hold for another five seconds.

Let go. Release the Kegel squeeze in three more stages. Loosen a bit and hold. Loosen some more, then hold again. Finally, relax completely. Here again, using your fingers will help you to get a feel for the tension needed for the gradual loosening of your PC muscles.

Be persistent. Once you learn to do Kegels effectively, more is always better, says Dr. Kessaris. Increasing your daily repetitions is the key to working out your muscles. Build up to at least three sessions of 25 Kegels a day. Once you do, frequency of incontinence episodes should diminish after four to six weeks of steady Kegel exercise, says Dr. Kessaris. Incontinence symptoms can return if you stop doing the exercises, so to see a steady and marked improvement, you must do them faithfully, just like a weight lifter.

Finagle some Kegels. These exercises can be done anywhere, anytime—even while driving or cooking, Dr. Kessaris points out. Try them while making love; your partner won't mind a bit.

Biofeedback: Information, Please

If you have trouble isolating your PC muscles, try biofeedback. You can get maximum benefit by learning from a certified biofeedback therapist, says Dr. Kessaris. In fact, some women

Fighting the Urge

If you find yourself unable to get to the bathroom fast enough, or if you urinate very frequently day and night, you probably have what's known as urge incontinence.

"It's sometimes referred to as 'garage door syndrome,'" says Marc Lowe, M.D., chief of urology at Group Health Cooperative Central Division of Puget Sound in Seattle. The closer you get to a bathroom, the more anxious you get and the worse the urge to urinate becomes.

Often, urgency troubles can be resolved with a procedure called bladder retraining. Your urologist can recommend a properly trained nurse or physical therapist to coach you. But here are two strategies that might help prevent accidents in the meantime.

Wait a minute. The next time you feel the need to go, "stop whatever you're doing, take a deep breath, and squeeze your PC muscles," says Dr. Lowe. If you can assure yourself that you have control, it may alleviate panic in less convenient situations.

Look for a distraction. If you need to urinate but a bathroom is a long way off, try to distract yourself until you get there. Think of anything else. Start a conversation, or focus on driving, walking, or riding. Deliberately move your mind to an unrelated topic.

can do more harm than good by performing Kegels without guidance, he says. A certified biofeedback therapist can teach you how to strengthen the right muscles. A small, highly sensitive electrode, which is placed either in the vagina or the rectum, can record tiny electrical signals from the muscles. The muscle activity is amplified by the biofeedback instrument, and you receive a signal from tones and lights on a monitor. By watching the screen and recognizing when you're targeting the right muscles, you can learn what it feels like when you're working the important PC muscles.

"Biofeedback is a way of gathering information about your own body," says Wendy Wright-Kilker, P.T., a physical therapist and certified biofeedback therapist at Boulder Medical Center in Colorado.

Once you learn biofeedback, you can do it on your own without the machines. But until you do, here are some tips from Wright-Kilker to help you get the results you want.

Find an expert. Biofeedback training is only as good as the therapist working the equipment. To find a qualified biofeedback therapist, get a referral from your doctor or call the urology department of a local hospital.

Take an anatomy lesson. Learning a bit about what the pelvic muscles look like and where they are is a big help in biofeedback. "Visual aids are especially helpful," says Wright-Kilker. Visit your local library and check out an atlas of human anatomy.

Think about it. Imagery combined with biofeedback can be doubly effective. "I tell my patients to imagine the muscles squeezing shut when they contract them and see them dropping and opening when they relax," says Wright-Kilker. Get a visual image in your mind of what the muscles do when you work them properly, then recall that image later, when you need to work these muscles without the feedback.

INHIBITED SEXUAL DESIRE

Tune Up Your Body and Mind

Something's missing." "I'm just not myself anymore." "The feeling's gone." When people try to describe low sexual desire, it's often in these vague terms. There's no real standard when it comes to having the urge to merge, says Joel Block, Ph.D., a sex therapist and psychologist in private practice in Huntington, New York, and author of *Secrets of Better Sex*. Some lusty folks like to nuzzle every day, while others have little or even no sex drive.

For some people, the physical changes that come with age subdue the sex drive. Sometimes, physical problems, such as high blood pressure, or drugs, especially antidepressants, can clobber your carnality, so you should check with your doctor to find out whether a health condition or your medication could be cooling you down. But for most people, a drop in desire is connected to the stresses and strains of obligation.

"People work on average 10 hours more a week than they did a generation ago," says Barbara Bartlik, M.D., a researcher in the human sexuality program at the New York Hospital–Cornell Medical Center in New York City. The less-than-erotic reality of work and kids can diminish your sex drive.

Of course, if you're happy with a lessened libido, that's fine, says Dr. Block. An inhibited sex drive is only a problem if you'd like to be more sexual, either to match your partner's lust level or because you miss those old reeling feelings.

Exercise is a great way to feel more frisky. People who are active often have stronger sex drives than sedentary people, says Dr. Bartlik.

Exercise is basically a psychological aphrodisiac. Weight loss and improved muscle tone—often the benefits of a regular exercise regimen—give people a better body image. A study at Brown University in Rhode Island showed that overweight women had improved sexual functioning after losing weight. Lots of research suggests that if we feel more attractive, we're more inclined to cuddle.

The Mind Matters, Too

Of course, sex is about more than muscles. Plenty of people who are in good shape are just too stressed or depressed to delight in the deed, explains Dr. Block. Sex is mental magic, too.

"If you want to feel sexy, you have to think sexy," he says. Like a key in a lock, mental stimulation can open the door to physical arousal. So try oiling up your body and your mental machinery, summoning a little sweat, and getting mischievous in your mind.

The recipe for sexuality has two ingredients. First, get more physically active throughout your day, and second, get sensual. Yes, it's tough to find time to exercise. And, yes, it's tough to think transcendently when you're feeling exhausted. But you can salvage a sex life from the storm of everyday obligation. Try some of these tips—some physical, some mental—to rekindle your own fire.

Take a morning minute. It's the dawn of another day. Before you climb out of bed, though, take a moment and just stretch. Lie there and luxuriate. Stretch your feet, your legs, and your arms. Twist your torso gently. Imagine you're a cat, sinuous and lithe. Focus on the feel of the sheets against your skin. Savor the wonder of you. That's right, the wonder of you. It's important to think of yourself as a sexual creature, says Dr. Bartlik. You are capable of giving—and more important, receiving—pleasure.

Meander in the A.M. Before you get ready for work, throw on some sweats and take a quick walk outside, suggests Dr. Block. Nothing long or taxing; just one lap around the block. Five minutes. No huffing, no puffing.

While you're walking, you can increase your sensual awareness by taking notice of the things around you. Inhale deeply. Taste the air. Listen for birdsong. Smell the lilac—or pollution—in the air. Appreciate that bank of azaleas in your neighbor's yard. Feel like a creature absorbing the sights and sounds of your patch of jungle.

Get sensuous in the shower. Use a scented soap or body wash. "Spend a little extra time washing your erogenous zones and savoring how it feels," Dr. Bartlik suggests. Some research suggests that particular scents can modify our moods. In men, for example, the scent of vanilla has been shown to increase the circulation of blood to the penis.

For purposes of feeling more sexual, it doesn't matter which soap scent you choose, as long as you deliberately enjoy it. As you savor the water sluicing down your body, concentrate on the soap's perfume and your own sensuality. Appreciate your ability to smell the raspberry, strawberry, or lavender.

Be a licentious listener—or reader. Do you listen to audiobooks on the drive to work? Well, deep-six that tome on tax-free investing and try an erotic novel. You needn't stoop to trashy reading—in fact, you can take the literary high road. Try *Lady Chatterley's Lover* by D. H. Lawrence or the novels of Henry Miller. They're full of passionate prose and may get your engine running, says Dr. Block.

Even if sultry books aren't available on tape, try perusing the printed page for pep. Some of the books on Dr. Block's recommended reading list include *Little Birds* by Anais Nin and *Tart Tales: Elegant Erotic Stories* by Carolyn Banks.

Walk to work. If possible, park your car a half-mile from your office and walk the rest of the way. You don't have to set a fast pace. You're just using the machine here, keeping the instrument in good repair. Take advantage of chances to move, advises Dr. Bartlik.

Dare to stair. When you get to work, if you work on an upper floor, take the stairs, says Dr. Bartlik. Burn a few calories. You can bet Lady Chatterley didn't take the elevator. Neither did her lover.

Get nostalgic. Every now and then, take a five-minute meditation break from filling out those F-23 forms. Sit quietly at your desk and just invite your memory. Think back to the most sexual moment of your life—the patriotic postpicnic assignation that Fourth of July or that encounter in the coatroom at your cousin's wedding. To restore your vim, remember some of your greatest hits, suggests Dr. Bartlik.

Run and be randy. To boost your need for badda-bing-badda-boom, you'll need to grab some exercise whenever you can, says Dr. Bartlik. Research has clearly shown that exercise raises desire. So if you want to be a sex machine, it may help if you swim, walk, jog, ride a bike, or dance several times a week.

Put movies in your mind. There's no need to pace on that treadmill without any entertainment. As you're raising your heart rate a bit, add a little spice to the sweat. Pop a hot movie into the VCR, suggests Dr. Bartlik.

If you like romance to be soft-focus, go with Bergman and Bogey in *Casablanca*. If you like a little more visual stimulation, try *The Postman Always Rings Twice* with Jack Nicholson and Jessica Lange or *Body Heat* with William Hurt and Kathleen Turner. Both have high heat quotients. If you're more straightforward still, go ahead, work out while you watch an adult video. You'll be working your body and rousing your mind.

If you're a woman and have found that most erotic videos don't appeal to you, try one produced by Femme Productions. Their free catalog is available from Femme Productions, 302 Meadowland Drive, Hillsborough, NC 27278. "They're tailored to a woman's sensitivity and needs," Dr. Bartlik says. "They're respectful of women and their sexuality."

To take advantage of their spark, slip one of these videos into the VCR before your partner comes into the room and enjoy it by yourself for a few minutes. "It's a way of priming the pump," says Dr. Bartlik.

Bathe yourself bolder. Any sensual experience may help put you in the mood. Draw a hot bath and add some sensuous, slickening bath oils. Luxuriate in the feel of the water caressing your body, says Dr. Bartlik. As the warmth of the water spreads over you, picture the way your body looks when it is aroused.

Try a bon-bon and a beaujolais. When you get out of the tub, eat some chocolate. Research suggests that it may stimulate other appetites. Chocolate contains a substance that stimulates the production of a neuropeptide called phenylethylamine, which has been shown to boost desire, according to Dr. Bartlik.

Then take a sip or two of wine. Swoosh the vino across your tastebuds. Enjoy it. Leave a drop on your lip. Share it with a kiss and hope that your partner's in the mood for a workout.

INNER EAR PROBLEMS

Bring the World Back into Balance

Ask for electric shocks behind your ear. Drink whiskey before going to work. Attach a gyroscope to your hip. People sent these far-fetched folk remedies (and others) to astronaut John Glenn in 1964 after he withdrew from the race for a U.S. Senate seat because of chronic dizziness. In his case, the dizziness resulted from a fall in the bathtub that led to damage to his inner ear, the part that sends us signals about balance.

Inner ear damage is only one condition that can make you feel dizzy or faint. There are many others, including low blood pressure, migraine headaches, aftereffects of flu, or diseases such as multiple sclerosis. So if you've been feeling dizzy, the first thing to do is check with your doctor. If he tells you it's an inner ear problem, here are some motions that will help you get your balance back.

"The idea with these exercises is to give the brain balance information that turns on the compensatory mechanisms," says Brian Blakley, M.D., Ph.D., professor and chairman of the department of otolaryngology at the University of Manitoba in Canada. "These exercises work especially well for people who have traumatic or degenerative damage to the ear. Aging can involve the nerves of the ear, but commonly, the brain is the main culprit in causing dizziness due to aging."

Give Your Balance System a Workout

The exercises recommended by Dr. Blakley should challenge the balance system and make you a little dizzy, so the rule is to start slowly but build up to rapid movements. Repeat each of the exercises two or three times per session, Dr. Blakley suggests. If possible, repeat the sessions three times every day. Some people with neck or back pain may be unable to do these exercises. If you're unsure if you can do them safely, ask your doctor.

Bend like a bow. This exercise is for dizziness that is positional, which means that it is aggravated by changing the position of your head with respect to gravity. Bending over, looking up on a shelf, or rolling over in bed are examples of positional changes.

Sit on the side of a bed, look straight ahead, and keep your shoulders square. Then quickly bend from the waist as far as you can to the right and hold for 20 seconds. Next, sit up and bend to the left. Alternate until you've done each side two or three times.

Look both ways. This exercise is for more general forms of dizziness other than the positional type described above.

Sit on the edge of a bed and look straight

ahead. Slowly turn your head to the right, then return it to the starting position. Repeat slowly, then gradually increase the speed—without straining—for about 20 seconds. Repeat, turning to the left, for another 20 seconds.

Listen to your knees. While sitting on the side of a bed, turn your head 45 degrees to the right. This will position your head halfway between the forward-facing position and a position where your chin would be over your right shoulder. Then lower your head so that your left ear moves toward your left knee, bending your neck as far as is comfortable to bring your head close to horizontal. Alternate between the upright and bent positions as rapidly as is comfortable for about 20 seconds. Repeat the exercise for 20 seconds on the other side, bending your right ear toward your right knee this time. This exercise is also for general, not positional, dizziness.

Press Away Dizziness and Fainting

When you do acupressure, experts believe that you can press certain points on the body to relieve problems in other areas, and they have identified some acupressure points that help relieve the spinning sensation. Similarly, reflexologists work on specific points on the feet, ears, or hands that correspond to parts of the body.

Do the following exercises as often as you feel dizzy or faint, recommends Mary Muryn, a certified teacher of polarity (energy healing) and reflexology in Westport, Connecticut.

Press your palm. Find the exact center of your left palm and press it with the thumb of your right hand for about 30 to 60 seconds, then release. When you've "worked" the palm of your left hand, you can switch positions and apply pressure to your right palm, says Muryn. "You should press hard enough for it to hurt a little," she says. Alternate between pressing and releasing until the dizziness passes, but no longer than three minutes. This technique will get energy circulating throughout your body and help dispel feelings of faintness or dizziness, says Muryn.

This acupressure technique is also effective in an "emergency" situation. If you suddenly feel as if you're going to faint, apply heavy pressure to this point for a few seconds and it will bring you out of your dizzy spell, Muryn says. You may also want to sit or lie down if you feel faint.

Try a foot rub. "Any kind of strong foot massage will jolt you back to reality if you feel like you're about to faint," Muryn says. While seated, take off your shoes and use the thumbs of both hands to knead the bottom of one foot for a few minutes, she says. Or roll one foot firmly back and forth on a glass soft-drink bottle to apply pressure to the critical area of the sole.

If you find that this technique is especially effective in relieving dizziness, you might want to purchase a specially designed wooden foot roller that you can carry with you and use any time. You'll find them in natural products catalogs, at beauty supply stores that carry therapeutic products, or at foot-care stores.

INSOMNIA

Some Well-Timed Motion Can Help You Sleep

There is one upside to insomnia—the comedy of Groucho Marx. The cranky comic was one of the legions of folks who struggle night after night to fall asleep and stay asleep. It's easy to imagine him strutting about at 3:00 A.M., cigar dangling, coattails flapping over a pair of flannel pajamas, whipping up withering witticisms. Who knows? If he'd been a good sleeper, Groucho might have been much less sarcastic and far less amusing.

But Groucho is the only funny thing about insomnia. According to a national Gallup survey, insomnia affects 87 million Americans. And it can lead to serious health problems, according to Gary Zammit, Ph.D., director of the Sleep Disorders Institute at St. Luke's–Roosevelt Hospital Center in New York City and author of *Good Nights*.

The results of insomnia are often a catalog of bad news, since sleep deprivation can affect everything from your mood to your motor skills.

If you're always tired, you can't be a good worker, spouse, parent, or friend. It can lead to bad job performance and discord at home. If you're tired, you're not too efficient when operating machines at work or driving your car on the way home. People who struggle with on-going insomnia are at higher risk for car collisions and other serious accidents, says Dr. Zammit.

Insomnia has many different causes. Certain medical conditions, such as chronic pain or thyroid problems, can cause it. Some antidepressants, such as fluoxetine hydrochloride (Prozac) and related drugs; some cardiac drugs, such as propranolol (Inderal); and other medications may also interfere with sleep. If you think that one of these factors could be affecting you, talk with your doctor.

An uncomfortable sleeping environment can also be to blame. Noise, heat, cold, or even a lumpy bed can all keep you tossing and turning. Psychological factors, including stress and anxiety, are also sleep thieves, says Dr. Zammit. When the mind reels, sometimes sleep won't come.

But whatever the cause of your insomnia, it's important that you do something about it. Long-term sleep deprivation is bad for your health and your relationships.

Take two steps. First, call your doctor and tell him about your problem. Second, start exercising. An energetic workout—at just the right time of the day—is a safe and effective sleeping pill, says Dr. Zammit.

Warming Up to Cool Down

It's not entirely clear whether our bodies have to cool off before we can sleep or if they cool off right after we fall asleep. But in a delightful inversion of logic, the best way to drop our temperatures a bit at night is to raise them well before bedtime by exercising. Experts believe that this helps beat insomnia. Here's why.

Once our bodies get warm from exercise, they have a natural inclination to cool down. And as body temperature falls, momentum appears to make it more likely that we'll conk out once we hit the sack. When poor sleepers get a head start on chilling out by working out, they seem to have an easier time nodding off. "It's like a big push down the slippery slope to sleep," says Dr. Zammit.

Aerobic exercise—walking, running, biking, swimming, or dancing—seems to be the best workout prescription for insomnia. Whatever activity you choose, it should be done vigorously enough and long enough to make you feel warm. Dr. Zammit recommends at least 20 minutes a day.

But here's the important anti-insomnia idea. You've got to time your workout just right. Research suggests that if you exercise too early in the day, you don't get the exercise-induced sleeping edge. Why? Because your body will warm up, cool down, and reach equilibrium long before bedtime. You won't have the cooling momentum going for you when you hit the sheets. Conversely, if you work up a sweat too late in the day, your body will be warmest just when it's supposed to be cooling down.

To fight insomnia, the experts recommend exercising in the late afternoon or early evening. If you can squeeze in a visit to the local gym after work, that's great. But even if your day is just too busy to include another stop, there's hope. Here's an hour-by-hour anti-insomnia evening plan designed by experts to help you cool down, relax, and get on the glide path to dreamland.

The Sleep-Easy Evening

Take these steps to sweet, sweet sleep.

5:30 P.M.: Do the after-work workout. When you arrive home from work, stick dinner in the oven and start up the stereo. Pick out a half-hour's worth of your favorite dance music and get moving. While the roast is roasting, you should be rocking. Do the mashed potato, the pony, or even the twist.

Dance is a fun form of aerobic exercise. Moving around your living room for 30 minutes will vaporize stress and, more important, warm your body up. The cool-off effect will benefit you later when you hit the hay, Dr. Zammit says.

8:00 P.M.: Take an after-dinner dip. The dishes are done, the garbage is out. Now's the time to extend a more formal invitation to the sleep fairies.

A warm bath is a time-honored way to summon slumber. It works on the same principle as working out, says Dr. Zammit. Raising your body temperature encourages it to drop later on, setting the stage for sleep.

Try scenting your soak with a few drops of an essential oil known for its relaxing properties. "You might want to try lavender," says Dr. Zammit. "It's been shown to promote sleep in older folks." No matter what your age, the soft floral fragrance can let you down easily.

9:00 P.M.: Do a gentle yoga stretch to relax. Judith Lasater, P.T., Ph.D., a physical therapist and yoga instructor in San Francisco and author of *Relax and Renew*, recommends the legs-up-the-wall pose for heading off insomnia.

Sit on the floor facing a wall, about 8 to 10 inches away. Slowly roll back and put your legs straight up against the wall, keeping your knees relaxed. You can put a small pillow or folded blanket under your head to support your neck. Hold this pose for 5 to 10 minutes. Then bring your knees toward your chest, roll to one side, and sit up slowly.

Beyond Those Sheep: Mental Moves for Insomnia

Falling asleep is not something that you actively *do*, says Judith Lasater, P.T., Ph.D., a physical therapist and yoga instructor in San Francisco and author of *Relax and Renew*. "You simply have to let go and allow sleep to happen."

But letting go isn't always easy. When you find yourself too tied up to drift off, try the three expert-suggested methods below. They're designed to help sleep slip in when you're not even looking.

Become Brando. If you've ever considered a career on the stage, this tip is for you. As you're lying in bed, find the most comfortable position possible. Then "pretend you are in a play," says Dr. Lasater, "and you're acting the part of someone who's supposed to be asleep." Let yourself really relax into the role. In the same way that you'd fool an audience into thinking you're actually out cold, you'll pull the wool over your own eyes. Don't be surprised if you fall asleep for real.

Make mental room. Still awake? Maybe it's that list of worries being recited ad infinitum in your mind. When the wheels won't stop turning, turn a page—a notebook page, that is. Keep a journal on your nightstand to quickly jot down the stuff that's cluttering up your brain, says Gary Zammit, Ph.D., director of the Sleep Disorders Institute at St. Luke's–Roosevelt Hospital Center in New York City. "Tell yourself that you can worry about it tomorrow," he says. Since you'll know that everything's written down, there's no extra worry about forgetting anything that you may need to think about. Shut the notebook and see if that doesn't give you "permission" to sleep.

Escape to Eden. Guided imagery, or visualization, is the escapist's way to relax into sleep. In its simplest form, you mentally pick out a place (real or made-up) that you find calming, then imagine yourself there, says Dr. Zammit. Imagine everything, down to the tiniest detail.

Try to make the experience as vivid as possible. Focusing on your senses will bring the scene to life. For example, if you picture yourself walking in a forest, smell the pine needles and hear the crunch of dry leaves under your feet. Feel the gentle breeze on your face. Take your time and work on making your imaginary place as real as you can.

Losing yourself in a fantasy like this allows your mind to be free of worries. And your body will follow suit, becoming free of tension and allowing sleep to sneak up on you.

10:00 P.M.: Get close to the one you love. Sex is another traditional—and natural—sleep aid. And according to Dr. Zammit, preliminary animal research indicates that there may be some medical merit to the myth.

If you're the type who finds physical intimacy to be the ticket to complete relaxation, this time of day is the right time to do some loving. "If it works for you, you'll know it," says Dr. Zammit. "And if that's the case, I'd say do it as often as you need to."

Of course, if you find that sex delivers an energy boost, giving you the urge to balance your checkbook or fold a load of laundry. you'd be better off saving the seduction till morning.

11:00 P.M.: Let a breathing pose help you doze. Deep-breathing exercises allow your blood pressure to drop and your pulse rate to slow, says Dr. Lasater. When the body's inner processes are working in slow motion, sleep is only a breath away. Here's the way to float into dreamland on a cloud of fresh air, she says.

▶Lie on your back with a pillow under your head, and allow any tension in your body to release. Then take a long, slow inhalation followed by a long, slow exhalation.

▶Breathe normally for one breath.

▶Switch back to long and slow: Inhale completely, then exhale completely.

▶Take another normal, easy breath.

▶Keep breathing, gently and quietly, alternating deep breaths and normal breaths, for 5 to 10 minutes.

Little by little, allow your breathing to even out again and become light. If you're not already asleep, focus on the peace-inducing effects that your breathing can deliver. Sweet dreams.

INTERMITTENT CLAUDICATION

Turning Off the Pain

Experts have described intermittent claudication as a heart attack of the lower leg. It's a painful condition characterized by a cramp in a calf muscle. The cramp, caused by poor circulation, appears during some kinds of exertion when the calf muscles are crying out for more blood and the oxygen that it comes with it. If the cramp is accompanied by chest tightness, it may be an indication of coronary artery disease as well.

"With intermittent claudication, a person will get a deep, painful cramp that feels like it is at the bottom of the calf muscle right next to the bone," says Dan Hamner, M.D., a physiatrist and sports medicine specialist in New York City.

How to stop the cramp? Stop exerting yourself. The cramp will stop pretty quickly once you do, says Dr. Hamner. The only problem is that it will return when you resume exercising. And that poses a painful dilemma, because sometimes the best means of relieving intermittent claudication involves the same activity that triggered the cramp.

"Although intermittent claudication is aggravated by exercise," says Dr. Hamner, "exercise is also the best way to cure it. "

To deal with partially blocked blood vessels in your legs, you may be able to create blood-flow bypasses with aerobic activities like walking. When you go for a stroll, sometimes you can stimulate the production of new circulation routes in your legs, according to Dr. Hamner.

Before you add more blocks to your walking regimen, though, Dr. Hamner recommends that you get a thorough checkup and a stress test from your physician. Sometimes, circulation problems in your legs can be omens of more serious problems elsewhere. About half of the people with intermittent claudication also have some plaque that's begun to clog up their coronary arteries. "Generally, if you have plaque in your leg arteries—what we call peripheral vascular disease—you need to get a stress test, because your heart may be just as bad," he says.

Walking the Walk

Once your doctor has given you the green light to exercise, here are some helpful exercises that may help alleviate or prevent the leg pain. Don't forget to do a warmup routine before exercising and wear clothing that will keep your legs warm.

Chip off more blocks. The next time intermittent claudication strikes, slow down until the pain diminishes and then continue your walk at your original rate, says Dr. Hamner. Think of the new blood vessels that you are creating that will deliver the blood to your legs and feet.

"When you start feeling the muscle cramps, try to walk a short distance with the pain before you stop to rest," says Dr. Hamner. "The objective is to extend the distance that you walk. If you always try to increase the distance, you may be able to walk a few hundred yards farther each week."

With four weeks of steady walking, Dr. Hamner says that new capillaries will sprout in your legs to help ease the pain caused by the blocked blood vessels.

Head for the hills. Besides gradually building up the distance, Dr. Hamner suggests that you vary your walking routes to include hills as well as flat terrain.

"By varying walking surfaces, you stimulate the formation of new vessels that help blood bypass the blockages," he says. If the pain worsens when you walk uphill, check with your doctor about other possible conditions, such as arthritis of the spine.

Trek on the treadmill. Walking on a shock-absorbing treadmill and climbing on a stair machine can add variety to your workouts, especially during inclement weather, says Dr. Hamner.

Head for the dance floor. You can two-step to keep the pain away by doing the polka, tango, waltz, or even country line dancing, says Dr. Hamner.

"Dancing is a great exercise for this because it requires you to use your calf muscles more," he says. "If you can, try to go up on your toes every now and then while you dance. It will give you a little extra calf work."

Walk in the water. If your calves are really screaming in pain, a few laps of water walking may be in order, says Dr. Hamner.

"I recommend that some people try walking waist-deep in a pool with a water vest on for as long as they can tolerate," he says. "Water aerobics have both cardiovascular and psychological benefits."

JET LAG

Of Body Clocks and Hard Knocks

The crew of the *Enterprise* could beam themselves across galaxies in a blink—and without missing a beat. But for us less highly evolved *Homo sapiens*, even a flight from New York to Paris throws us way out of whack. It can take two to three days to recover from the muscle aches, fatigue, and general discombobulation known as jet lag.

Why? It's because our bodies operate on 24-hour internal clocks that count on regulated exposure to darkness and light. "This biological clock orchestrates the efficient functions of the body's heart, blood pressure, eyesight, mental ability, alertness, sleep-wake cycle, and even digestion," says Albert Forgione, Ph.D., a psychologist and director of the Institute for the Psychology of Air Travel in Boston. When we travel across several time zones, all of our systems get thrown off-kilter. Add the stress of air travel to the assault on our body clocks, and the result is the funk of jet lag.

There have been countless crackpot suggestions for tricking your body's timing mechanism, many of which included blindfolds worn at high noon. But the body is pretty tough to dupe. Here, however, are a few techniques that may limit the shock to your corporal clock and minimize the complicating aches and pains that often come with being on a plane.

You've Taken Wing: Try Some Jet-ercises

You're crammed into seat 16D. No aisle. No window. No room. You're fighting two feelings: one, your sister doesn't appreciate that you're traveling 2,500 miles for your nephew's birthday, and two, your right leg is about to explode into a charley horse. Here are some no-room tension-busters, recommended by Adriane Fugh-Berman, M.D., a Washington, D.C.–based medical researcher and a frequent flyer.

Try the red-eye shoulder shrug. With both feet flat on the floor, lift your shoulders up and down in a relaxed, I-don't-care shrug. Next, roll your shoulders forward and then back. Do this exercise 15 times and repeat these steps every 30 to 60 minutes.

Rise to the mile-high thigh lift. Sit with both feet flat on the floor and place your left hand firmly down on your left thigh. Lift your thigh as you press down with your hand. Do this 20 times, then do the same with your right thigh and your right hand.

Launch the DC-10 tilt. Tilt your head to

the left, moving your left ear toward your left shoulder. Hold that position for a few seconds, then raise your head back to the center. Then tilt your head to the right, forward, and back. Repeat 10 to 15 times in each direction.

Tips for after Touchdown

Once you arrive, you need some exercise tactics to help your body clock tick along. Here are some tips for short-term layovers and long-term visits.

Achieve terminal velocity. If you have landed at a sprawling airline terminal and have some time between flights, skip the barstool and head for the open aisles. You've just been shoehorned into a seat for four hours. If you walk a bit, you'll give your muscles a workout and help your circulation. Dan Hamner, M.D., a physiatrist and sports medicine specialist in New York City, suggests a cruise through the concourse.

Soak up some sun. Light helps your body's internal clock adjust more quickly to the time at the new location, says Dr. Forgione. If it's rainy, Dr. Forgione suggests spending some time indoors under bright lights.

Fight the urge to nap. If you leave New York's Kennedy International airport at 7:10 P.M. on a nonstop flight and arrive in Paris at 8:35 A.M. the next day, use your mental powers to avoid daytime napping. By staying awake, you will sleep better that night and reset your body clock faster, says Dr. Forgione.

Work out late. *Philadelphia Inquirer* sports columnist Bill Lyon offers this next-day jet-lag advice: Do some exercise the afternoon after you arrive. "A workout is the best sleep inducer," says Lyon, who has logged more than two million miles in the air.

Preflight Instructions

To jaunt around jet lag, take precautions before the flight. Try these easy "exercises."

Exercise wisdom. Be reasonable when making your travel schedule. Give yourself plenty of time to check in and relax before you board. Stress doesn't cause jet lag, but it does make it worse. "A schedule with tight timing between flights will only create irritation and anger," says Albert Forgione, Ph.D., a psychologist and director of the Institute for the Psychology of Air Travel in Boston.

Exercise the juice option. Flyers should load up on water, fruit juice, fruit, cheese, and carbohydrate-rich pasta and avoid alcohol, caffeinated beverages, and salty foods. "The airplane is a very dry place, so it is important to keep from being dehydrated," says Dr. Forgione. "Salt makes you thirsty; alcohol, especially beer, can make you lose water from your body by making you urinate often, and carbonated drinks send gassy fluids into your stomach, causing bloating and discomfort."

Exercise self-deception. As soon as you board the plane, set your watch to the local time of your destination. Dr. Forgione says that this will help you psychologically adjust to your new location.

Job Burnout

Re-Igniting Enthusiasm

Chicago Bulls guard Michael Jordan sidelined himself for a couple of seasons. "Far Side" creator Gary Larson voluntarily stopped his weekly comic strip at its peak.

You don't have to be an NBA superstar or a nationally syndicated cartoonist to be among the burnout bunch. Homemakers and CEOs also experience the singe of burnout when they feel bored, powerless, overlooked, or driven by a perfection-at-all-costs attitude.

"Burnout is a sign that you're not having fun with what you're doing," says Peter Wylie, Ph.D., an organizational psychologist and management consultant in private practice in Washington, D.C. "Most people ignore the subtle signs that it is time to make a change and put up with incredible discomfort until someone makes the decision for them—like firing them." It is important to actively address the sense of burnout. Trying to ignore those feelings can lead to more serious consequences, such as clinical depression or ulcers, according to Dr. Wylie.

You may not have a lot of control over things such as your workload and the kinds of tasks you're assigned to do. But you still can rekindle that burnt-out flame in a variety of ways. According to Dr. Wylie, sharing your frustrations with someone you trust and performing satisfying exercises aimed at developing self-confidence and compassion for others top the list.

Tilling the Soul

Planted in another boring company meeting? Feeling deep-rooted frustration that your boss never notices your accomplishments?

Gardening is a great anti-burnout activity, says Diane Relf, Ph.D., professor of horticulture and extension specialist at Virginia Polytechnic Institute and State University in Blacksburg. Whether it's a one-acre vegetable plot or a window box full of petunias, gardening relaxes you, stimulates your mind, and gives you a healthy workout, she says. Digging, pushing a mower, raking leaves, or spreading mulch can often burn as much energy as a brisk walk, she adds.

"Gardening is an ideal way to get back in touch with the rhythms of life and get away from the extreme stress and pressure and lack of satisfaction that many jobs hold," says Dr. Relf. She presents these earthy tips to rejuvenate you the next time you feel plowed under.

Weed out worries. One way to visually and physically feel a sense of accomplishment is to free your garden of unwanted weeds. The yanking and pulling works the fine motor muscles in your fingers plus the larger muscles in your arms, shoulders, and back. At the same time, your mind focuses on the task in front of you and frees you from thinking about your job. "When you see the soil getting cleaned up, you start feeling positive," says Dr. Relf. "In most cases, you become so totally engrossed in gardening that you stop thinking about the worries of your job and the worries of the world."

Spread yourself around. Dr. Relf recommends that you select three garden tasks that you want to accomplish in a given day and get them done in alternating chunks rather than spending hours on one chore. Maybe you want to weed, plant a few rows of green peppers, and water the rosebushes. To get a healthy workout without straining your knees, shoulders, or back muscles, Dr. Relf advises that you set a timer and take it along to the garden. After every 15 to 20 minutes, rotate to the next task until you've completed a few rotations and all the tasks are done.

"This way, you are using different tools and different body parts and you avoid becoming obsessed in your garden," she says.

Exercise a little charity. Typically, gardeners plant more fruits, vegetables, and herbs than they can consume. So why not offer your surplus to co-workers, neighbors, and friends? You'll feel a sense of pride that can overshadow any burnout blues, says Dr. Relf.

"One woman I know has a lot of older neighbors, and she worries that they are not getting the right foods," says Dr. Relf. "So she grows plenty of tomatoes, collard greens, and other vegetables and gives them to her neighbors. Being charitable to others brings out a feeling of personal satisfaction."

Stepping into Something New

Maybe each day on the job is the same-old, same-old. And your home life is equally predictable. Every Monday, you eat meat loaf at Aunt Mary's. Wednesdays, you shop for groceries. Fridays, you head for the fish fry at the Kiwanis Club. Weekends, you scour garage sales and pay bills.

"I encourage people to add variety to their lives, especially when they feel burned out," says Dr. Wylie. "By varying your stimuli, you break up the routine, the monotony of life. You perk up."

Be sure to cover all your senses when you venture into something new, he adds. Consider these habit-breaking activities.

Switch hands. Right-handers, try writing with your left hand; southpaws, do the opposite. The simple act of pen-switching can stimulate creativity and douse burnout, says Cam Vuksinich, a certified personal trainer and certified clinical hypnotherapist specializing in stress/pain management in Denver.

"With hand movements, the motor skills are driven by opposite hemispheres of the brain," she says. "So if you feel burned out, try nondominant writing. Even the sensory stimulation of having a pen in the opposite hand helps spur creativity."

Other hand-switching opportunities include brushing your teeth, stirring a pot of chili, and wearing your watch on the opposite wrist.

Change pursuits. If you're a swimming enthusiast who has been doing freestyle strokes at your local YMCA for the past five years, try breaking up the monotony by taking snorkeling lessons. You'll be invigorated by the challenges of this new activity, says Dr. Wylie. "All of a sudden, whole new worlds open up for you," he says. "You get to meet new people and exercise in a different way."

Other novel activities to consider are fencing, juggling, kayaking, and pitching horseshoes.

Doing for Others

It's easy to have a pity party when you feel consumed by work demands and ignored by your boss. Experts say that's a sign that you need to shift your thoughts toward others in need.

"When we get burned out, we become self-absorbed," says Dr. Wylie. "If you aren't getting any positive feedback from your boss, gain it by doing something good for someone else."

Head a hike. Consider becoming a volunteer at your local park and leading a pack of curious kids on a nature walk on a sunny Saturday morning, says Elaine Wethington, Ph.D., associate professor of sociology at Cornell University in Ithaca, New York. You're sharing your knowledge with others—an empowering feeling, she adds.

Lead the pack. Offer to help an ailing neighbor by walking his beloved beagle before you head to work and when you return. You'll benefit by being in the fresh air and enjoying the companionship of a tail-wagging new buddy, says Vuksinich.

Step in style. If you love country line dancing on Saturday nights, consider sharing your savvy steppin' with residents at your local nursing home once a month, says Vuksinich. You'll get your exercise in a fun, energizing way, and get a kick out of helping others.

Job Loss

Bouncing Back Physically and Emotionally

The company may call it downsizing, layoffs, or "we simply need to part ways." But the person who loses a job, no matter his income or stature, is more likely to call it unsettling, unnerving, unnecessary, and unfair.

Fortunately, few lose the same job twice. A notable exception was Billy Martin: The fiery Hall of Famer lost his job as manager of the New York Yankees a total of five times. Although he'll never make it into a hall of fame for that, it just might be a record.

Whether or not you're likely to get your old job back, there's a good chance that you'll be employed again before long. Not only that, the chances are very good that you'll find a better, higher-paying job, according to a survey by the Employee Benefit Research Institute. Half of workers who lost their jobs between 1993 and 1995 reported earning about 25 percent more in new jobs that they landed by 1996.

In fact, you can almost expect to lose a job sometime in your career. "Anyone starting a career ought to plan on eight different jobs in their lifetime—that's the norm these days," says Michael McKee, Ph.D., a psychologist and vice chairman of psychiatry and psychology at the Cleveland Clinic Foundation in Ohio. "Realize that the notion of job security is a thing of the past and always have a Plan B ready in case the company goes out of business tomorrow or your job is abolished."

Whether or not you have a Plan B in the wings, job loss is still likely to come as a shock. And recovering from that shock requires a two-step strategy. First, you must deal with the immediate avalanche of emotions—the anger, the hurt, and the fear. "When you lose something as important as a job, don't try to ignore how you feel or stifle the need to grieve, or it will come back and bite you in the butt," says Peter Wylie, Ph.D., an organizational psychologist and management consultant in private practice in Washington, D.C., who specializes in job issues.

Second, you must strive to get in top shape—physically and mentally. Whether you drop a dozen pounds or firm up your tummy, you're likely to increase your confidence level because you look and feel better about yourself. And self-confidence can usually be detected by prospective employers, experts say.

Applying Immediate First-Aid

Ever try to suppress a sneeze? Magnify that many times and you'll realize the emotional and physical toll that your body and mind suffer if you try to dismiss being laid off with a simple shrug. You'll heal faster if you flush it out of your system, say job experts.

"The difficult thing is not to take it personally even though it has a personal impact on you," says Dr. McKee. "It's bad enough that it happened. Don't complicate it by blaming yourself." Most often, people are laid off because the corporation wants to save money by cutting jobs, not because of some job-perfor-

mance deficiency, Dr. McKee reminds us.

Even so, Dr. McKee and Dr. Wylie advise that you spend at least a few days releasing your emotions in a productive, healthy manner. They offer these outlets.

Scream like a fiend. Go ahead—holler, rage, and rant. But only in the company of a caring, compassionate friend, Dr. Wylie says.

"Take the time to scream, yell, or cry in the presence of someone who cares about you and who will not judge you," he says. "Go ahead

---

Sweating the Details

After a 12-year tenure as a public relations specialist for a publishing company outside Philadelphia, Sarah, a woman in her midforties, was told that her job was being eliminated.

Stunned, she worried about what would happen to her and what she should do next. She eventually regrouped and discovered new ways to be self-employed. She shares these tips with others who have been downsized.

Keep sweating. A regular at morning workouts, Sarah stuck to this regimen after being laid off and during her job hunt. She pedaled for an hour in a cycle Spinning class and then lifted weights for another hour. By 8:00 A.M., she felt energized and confident.

"Daily workouts give me energy and keep me in shape," says Sarah. "They also keep me from sounding groggy, like I just woke up, when I make morning calls to prospective employers. I sound energetic. They can hear my confidence. That's

important to the impression I make."

Stick to a schedule. "I make a schedule of tasks to do each day, much as I did when I was working," she says. "I feel a sense of accomplishment when I get things done."

Speak the downsizing language. Sarah recommends not letting personal feelings of hurt or anger creep into the conversation during a job interview. Take out the emotion and speak the corporate language.

"I tell them that my company underwent a reorganization, and that resulted in it downsizing. And, of course, in these situations, public relations is often the first department cut," she says. "They can relate to this and understand."

Be in command of your day. For Sarah, the job loss has opened up new opportunities to do freelance work for her former employer and former clients, giving her more control over her time. She is now able to visit friends whom she once neglected because of job demands.

and scream, 'I feel like telling my boss off' or 'I can't believe they would do this do me.' It is important to vent your feelings, to talk them out with a close person for as long as those feelings persist."

If your anger doesn't seem to tone down after a few days or a week, Dr. Wylie recommends seeking a professional therapist or counselor to work out a resolution.

Hurl your anger. Get yourself a pair of darts. Draw a caricature of your ex-boss or former company on a piece of paper and mount it on a cork board. Take a few steps back and let the darts fly. Celebrate each time your dart hits the target. After a few games, you should find that you feel better, says Dr. Wylie.

"This exercise won't make the underlying feelings go away," he says, "but it will take away the uncomfortable, pent-up energy." Remember, it's the symbolism here that's doing you some inner good. Don't let any violent impulses spill over into reality.

Punch a bunch. Go a few rounds with a punching bag. You can also use big fluffy pillows. Vary your punches: a right jab, a left uppercut, a right hook, and a few rabbit punches. Work your hands and feet for 10 to 15 minutes until you're glowing with perspiration and less frustration, says Dr. McKee.

The Makings of the Comeback Job Candidate

You've released the final scream and let the last tear fall. Now you're ready to regroup and begin the job search. At this stage, experts stress the need to exercise daily, stick to a schedule, and tackle tasks that deliver a sense of accomplishment.

"Realize that you have been given the gift of time—the rarest commodity in the world today," says Dr. McKee. "Use it to make yourself physically fit and strong. That can translate into making you feel stronger emotionally and intellectually."

If you're between jobs, experts recommend these exercises to keep you in shape and help you prepare for the next job interview.

Head for the pantry. What if you can't afford to join a gym and the notion of jogging bores you? You can get a free, thorough workout by converting your kitchen pantry into a home workout center, says Dr. McKee.

"You can use cans of soup, sacks of potatoes, whatever, to do pushups, situps, and arm and leg lifts," he says. "After a couple of weeks, you should start to see muscle definition. You will find you can lift more as you get stronger. Do aerobics, too—walk, jog, bike, or jump rope. Keep a record of your progress on your refrigerator door." That record provides a great statement of self-esteem, he points out.

Start out at your own comfort level and, if it's okay with your doctor, work up to about an hour a day, suggests Dr. McKee.

Take the stairway to health. Dr. Wylie encourages his laid-off clients to "view the world as a gym." And the nearest part of that world, in his area, includes the steps and escalators inside the Woodley Park metro in Washington, D.C.

"The nonmoving stairway has 153 steps, and the moving one has 90 steps," he says. "You can treat the escalator like a treadmill, except that you don't have to pay for it and you get a great opportunity to do some people-watching while you exercise."

Karate your body. Take up karate, or any form of martial arts, at the local YMCA or community center. Karate helps you build flexibility and self-confidence and master job interview anxiety, says David Lee, a mind-body expert from Saco, Maine, who helped develop an accredited program in guided imagery for health professionals.

"Sparring can be a great confidence booster, especially if you're uncomfortable with confrontation," he says. The more you spar, the more you begin to lose your fear of confrontation. "The directness of sparring makes other, nonphysical forms of confrontation easier. It builds your 'warrior nature,' which can be used in all kinds of situations requiring assertiveness," explains Lee. When you step away from the mat with confidence, that feeling should carry over for the next job interview.

Tame the trim. Armed with a gallon of white paint, a brush, and some rags, make an appointment to finally finish the baseboard in your basement. By the end of the job, you will see a notable improvement in your self-confidence as well as in your basement, says Dr. Wylie.

Unclutter the garage. Suppose broken rakes, old, half-empty paint cans, and boxes of out-of-style clothes are filling up your garage. Take time to toss the debris and donate items that you think others might use.

"Use this time to get all those home projects done that you kept putting off because you were too tired when you came home from work," says Dr. Wylie. "You will work up a sweat and see a real sense of progress and accomplishment. That will make you more confident during the job hunt."

KNEE PAIN

When Your Hinges Hurt, Take a Stand

Women get weak in the knees more than men. That's not sexism but physiological fact.

Why? Because a woman's pelvis is broader than a man's. This places greater pressure on the insides of the knees, says Michael Ciccotti, M.D., orthopedic surgeon and director of sports medicine at the Rothman Institute at Thomas Jefferson University in Philadelphia and a team physician for the U.S. Women's National Soccer Team. Women are twice as likely to tear the all-important anterior cruciate ligament, according to Dr. Ciccotti.

The human knee is a marvelous machine. It consists of two sturdy bones, the thigh bone (femur) and the shin bone (tibia), joined together by thick, stretchy ligaments. The kneecap, officially called the patella, is the protective bony shield up front. The knee has only one design flaw. It's not great at moving from side to side. When it is strained laterally, sometimes the ligaments are damaged.

Knee pain can attack from several angles. Arthritis is one common cause. If you're significantly overweight, you can put additional wear and tear on your knees. If you're sedentary and don't exercise, the muscles and ligaments in your knees can atrophy and weaken before their time.

"As we get older, we often run into problems with cartilage tears in the knees. After we've been on our feet for so many years, the cartilage may get brittle and tear even without a specific injury," says Terry Nelson, M.D., an orthopedist at K Valley Orthopedics in Kalamazoo, Michigan, and a former marathon runner.

But none of these difficulties need bring us to our knees. Exercise can often alleviate the pain and thwart its return.

If you have knee pain, resist your natural inclination to stop, sit down, or get off your knee and do nothing. "The body's normal reaction to pain is to shut down and hold things still. You really have to work hard to overcome that tendency," says Marian Minor, P.T., Ph.D., a physical therapist and associate professor of physical therapy at the University of Missouri School of Health in Columbia.

Exercise can often assuage pain. Researchers at Wake Forest University and the University of Tennessee conducted a random study of 439 people over age 60, starting them on a regular aerobic and resistance exercise program. They found that the exercisers were able to reduce the disability and pain in their arthritic knees.

After you get an okay from your doctor, Dr.

Minor recommends gradually starting a stretching and strengthening program that keeps the muscles, tendons, and ligaments in your knees both strong and flexible. If exercising causes increased pain or swelling, check with your doctor or therapist to modify your program.

Stretching: Please Those Knees

For limber knees, you need to expand your focus to include workouts for the whole leg, say doctors. Here are some simple stretches to keep everything loose and lubricated.

Make waves. Head for the pool to stretch achy knees, suggests Jane Katz, Ed.D., professor of health and physical education at John Jay College of Criminal Justice at the City University of New York, world Masters champion swimmer, member of the 1964 U.S. Olympic performance synchronized swimming team, and author of *The New W.E.T. Workout*. She recommends the following exercise, especially for people with arthritic knees to improve their range of motion.

1

- ►Stand chest-deep in a warm pool (about 85°F) with your back against the wall.
- ►Grasp your right knee with both hands and bring it toward your chest (*1*).
- ►Release and straighten your right leg in front of you, with your toes pointing up (*2*).
- ►Return the leg to the starting position. Try five repetitions with your right leg and then five more with your left.

2

Hold up the wall. There's a long section of connective tissue that runs all the way from your hip down to your calf. It frequently gets tight, causing pain in the knee. This stretch will help ease the pain and keep that tissue toned, says Dr. Ciccotti.

- ►Stand at arm's length from a wall with your left side toward it.
- ►Place your left palm against the wall at shoulder level.
- ►Keeping both feet flat on the floor, cross your right leg over your left leg and lean your hip gently toward the wall.
- ►Hold for about 30 seconds and then uncross your legs.
- ►Switch sides and repeat. Try this stretch at least twice a day.

Grab an ankle. Strong, flexible quadriceps (the muscles at the fronts of your thighs) can help stabilize the kneecap, says Thomas Meade, M.D., orthopedic surgeon and medical director of the Allentown Sports Medicine and Human Performance Center in Pennsylvania. He offers this easy quad stretch.

Cycling Power

The stationary bike parked in your bedroom or living room can be a real knee pleaser, says Kim Fagan, M.D., a sports medicine physician at the Alabama Sports Medicine and Orthopedic Center in Birmingham. Working out on the bike can soothe an arthritic knee or most other types of chronic knee pain. To ensure a successful ride, Dr. Fagan recommends raising the seat as high as you can while still being able to reach the pedals and keeping the resistance level at a minimum. You should flex your knees as little as possible while riding.

Start with a 5-minute ride and gradually build up to 20 minutes, Dr. Fagan suggests. "Initially, you are using the cycle to increase range of motion and strengthen the knee. Once you build the time up, this also becomes a cardiovascular exercise," she says.

Stand with your back straight and your feet shoulder-width apart. Place your left hand on a tabletop or a chair for support. Pull your right foot straight up behind you, grabbing your ankle in your right hand. Keep your knee pointed toward the floor. As you pull your foot back and up, you should feel a stretch in your thigh muscles. Hold the stretch for at least 20 seconds and then relax. Try five, then switch legs and repeat. (See page 336 for an illustration of this exercise.)

Ham it up. Give equal time to stretching the hamstring muscles located on the backs of your thighs, adds Dr. Meade, which also help stabilize your knees.

▶Stand and prop your right foot on the seat of a chair. Keep your leg straight and lock your knee. Bend your left leg slightly (*3a*).

▶Place both hands on your right thigh just above the knee.

▶With your back straight, lean forward from the hip until you feel a stretch in the back of your thigh (*3b*).

▶Hold for 20 seconds and then relax. Try five of

3

these stretches and then switch to the other leg and repeat.

Strengthening: Steps toward Nobler Knees

Healthy knees depend on strong quadriceps and hamstring muscles, say doctors. They offer these muscle-toning exercises.

Take a 10-second pose. This isometric exercise can reduce pain and improve the ability of the muscles to absorb shock and cushion the knee joints, says Kim Fagan, M.D., a sports medicine physician at the Alabama Sports Medicine and Orthopedic Center in Birmingham.

While sitting or standing, straighten your leg completely, flex your foot back, and tense the quadriceps muscle around your knee joint. Hold for 10 seconds before letting it go. Try five squeezes.

"Ten seconds of this exercise five times a day will significantly strengthen the knee muscles and reduce pain and fatigue," says Dr. Fagan.

Sit and lift. This strengthening exercise doesn't look dramatic, but it will help tone your knee muscles, says Dr. Meade.

▶Sit on the floor with both legs straight out in front of you.

▶Put a pillow under your right knee, but keep it straight.

▶Bend your left knee and place your left foot flat on the floor.

▶Place both hands, palms down, behind you.

▶With the toes of your right foot pointed up, slowly and steadily try to push your right thigh down toward the floor. Hold for five seconds.

▶Raise your right heel a few inches and bend your toes toward your body. Hold for five seconds.

▶Try 10 raises with your right leg, then move the pillow under your left knee and do 10 more with the left leg.

Take one step at a time. Turn your stairs into a mini-gym for your knees, suggests Dr. Fagan.

With your left foot on the floor, place your right foot on the first step. Shift your weight onto your right leg and straighten your right

---

When to See a Doctor

Certain kinds of knee pain require medical attention. You should see your doctor if you have any of the following symptoms, says Dan Hamner, M.D., a physiatrist and sports medicine specialist in New York City.

▶Pain around the kneecap that occurs for the first time when you are climbing or coming down stairs and persists for two or more days

▶Kneecap pain that occurs for the first time after sitting for a long time and persists for two or more days

▶Pain that does not respond to the RICE (rest, ice, compression, and elevation) technique or a painkiller

---

knee. Your left foot will naturally rise off the floor. Counting to 10, slowly lower your left foot back to the floor, heel first. You'll feel a pull in the back of your lower right leg. Try to work your way up from 5 to 15 step-exercises for each leg twice a day.

Play knee-ball. The following exercise strengthens the muscles of the inner thigh and stabilizes the knee, according to Jeffrey Willson, certified athletic trainer at K Valley Orthopedics.

Lie on your back with your knees bent and your feet flat on the floor. Place a volleyball or soccer ball between your legs at midthigh. Squeeze your legs together and hold for 20 seconds, then relax without letting the ball drop. Repeat at least 10 times twice a day.

LARYNGITIS

Pamper Your Voice

If the world's divas would get out there and exercise, fewer of them would have voice problems in the middle of a matinee. So says James A. Koufman, M.D., director of the Center for Voice Disorders at Wake Forest University in Winston-Salem, North Carolina. Dr. Koufman advises the singers he treats at the center to run, walk, or paddle their way to aerobic fitness. It's good for their health in general and for their talented vocal cords specifically.

"Aerobic exercise builds lung volume," he explains. "And lung volume is what provides the horsepower for the voice."

Build up your lung volume and you'll get more vocal volume, without straining your voice. And that's the key to laryngitis prevention and cure: avoiding vocal cord strain and irritation.

Laryngitis is medical lingo for inflammation of the larynx, or voice box, the place in your throat where your vocal cords are located. When all's well with your throat, you produce sound by exhaling air from your lungs and up through your vocal cords. The passing gust makes the cords vibrate, and those vibrations turn into your voice—whether you're yelling at the children, cajoling the plumber, or singing the National Anthem. If you strain your voice by shouting at a louder volume than you can comfortably achieve, you inflame your vocal cords. Since swollen vocal cords don't vibrate as well, they may produce sounds that are different from those desired, says Dr. Koufman.

Laryngitis comes in two types. Acute, or short-term, laryngitis is the result of either a specific vocal strain—an afternoon spent screaming for the Sox at Fenway—or of an infection. It may be accompanied by a cough and a scratchy feeling or pain in your throat and will most likely disappear within days if you give your voice a rest. Chronic laryngitis causes the same symptoms, but, as the name suggests, it's longer-lasting. Chronic laryngitis has a few leading causes: reflux (commonly associated with heartburn); smoking, allergies, or chronic infections like sinusitis; and habitual vocal straining. This strain isn't due to the specific trauma of those extra innings against the Yanks. Most often, it's the result of limited lung volume and/or poor breath control that put a strain on your larynx every time you speak.

If your hoarseness lingers longer than a few weeks, says Dr. Koufman, you may have chronic laryngitis and should see a doctor. If it turns out that you've been abusing your voice, the following exercises—aerobic and vocal—can help you be gentle with your larynx.

Develop Breath Support

For anyone who has to speak or sing loudly, the trick is not to strain your vocal cords but to pump more air past them. Good air-pumping technique is what singers call breath support. For good breath support, you need good lung volume, good posture, and good breath con-trol. Here's how to get all three.

Get moving. Remember, aerobic exercise is what builds lung volume. "I know a number of well-known performers who have, over the years, become dedicated joggers," says Dr. Koufman. You don't like to jog? It's no tragedy. Any kind of aerobic exercise will do.

Stand tall with the Alexander Technique. Poor posture can undermine the best efforts of the most voluminous set of lungs. If you're hunched over, air can't flow freely from your lungs to your vocal cords. To perfect your pos-

---

Reflux, Laryngitis, and Exercise

A good number of the chronically hoarse and voiceless folks who turn up at the Center for Voice Disorders at Wake Forest University in Winston-Salem, North Carolina, have none of the classic symptoms of reflux, the disorder that most of us call heartburn. They don't feel a burning sensation in their chests. But they have reflux nonetheless, and it's what's causing their laryngitis, says James A. Koufman, M.D., who heads the center. "Two-thirds of my patients with laryngeal and voice problems have reflux as the primary cause," he says.

Stomach acid and enzymes can cause laryngitis. People who still experience reflux despite good dietary habits may need med-ical treatment.

Reflux is the backflow of the stomach's digestive fluids into the throat. It may be silent, occurring without any heartburn or other digestive problems, explains Dr. Koufman. In addition to hoarseness,

common symptoms of silent reflux are too much mucus in the throat, chronic throat clearing, coughing, difficulty swallowing, or the sensation of having a lump in the throat.

You can help prevent reflux by avoiding certain foods—particularly fried foods and other high-fat foods, chocolate, mints, and al-cohol—and by reducing your weight if you are overweight, says Dr. Koufman. Reflux can be aggravated by eating late at night, he adds.

The best way to maintain a healthy weight is through exercise. Shoot for 30 minutes of aerobic exercise, like walking, swimming, tennis, or jogging—or even 30 minutes of a moderate activity like golf or raking leaves—six or seven days a week, suggests Wayne Miller, Ph.D., professor of exercise science and nutrition at the George Washington Uni-versity Medical Center in Washington, D.C. Keep it up, along with a healthy diet, and you can expect to lose a pound or so a week.

Laryngitis Limiters

If a cold or the flu or a day of howling and otherwise abusing your vocal cords has left you hoarse, Peak Woo, M.D., associate professor of otolaryngology at Mount Sinai School of Medicine in New York City and clinical director of Mount Sinai's Grabscheid Voice Center, suggests the following.

Rest your voice. Give a yell only if your life depends on it.

Drink deep. Swallow at least eight glasses of water a day to quench the fire in your inflamed vocal cords. If you have a cold or the flu, you often have excessive mucus in your throat that can irritate your cords as well. Drink plenty of water, and you'll thin the mucus and wash some away. Avoid drinking alcohol, though, since booze can dilate the blood vessels in your throat and make your vocal cords even more swollen.

Puff not. Cigarettes just irritate the vocal cords, making a bad throat situation even worse.

Jump in the shower. And stay there for half an hour or so. The humidity will soothe your cords.

Suck candies. Sucking on hard candy stimulates your salivary glands, and saliva can help keep your vocal cords moist and more comfortable. But don't use mints, as they can be irritating.

---

ture, sign up for a course in the Alexander Technique, which will teach you good posture. "Some singers absolutely swear by Alexander Technique," says Dr. Koufman. (For more details, see "Alexander Technique" on page 11.)

If taking classes is not an option for you, the basic rule of good posture, according to Dr. Koufman, is to stand up straight with a relaxed neck and shoulders. It requires good balance with no slumping or leaning to one side.

Run for president. Effective speakers understand the value of pacing and pausing. Remember JFK's stump speeches? He never rushed. And he used lots of nice, short clauses—"Let the word go forth" and "The time has come"—and paused between them. Exercise proper breath control, and you won't run out of air in midsentence and be tempted to strain your vocal cords to be heard.

Skip the sigh. "Some people breathe in, then breathe out in a sort of sigh, and only then begin to speak," says Dr. Koufman. This bad habit robs you of breath and sound. If you've exhaled most of the contents of your lungs, you've got only a dribble of air left with which to make noise. If you want to be heard, you'll have to strain your voice. So breathe in before you start to speak, then skip the sigh and start spouting as soon as you start to exhale.

Slow down your speech. Ever-notice-how-some-people-talk-as-if-their-words-were-inextricably-linked-and-speeding-along-like-commuter-cars-on-a-Japanese-bullet-train? These fast talkers go barreling along until they're virtually out of air, then strain their voices to squeak out a few more words before gasping for breath. If you're a fast talker, slow down just a bit, suggests Dr. Koufman. If you pause now and then to take a breath, and refill your O$_2$ supply, you won't run out of air—or sound.

Reclaim your clarinet. If you played a wind instrument in the high school marching band, pull it out and give it a toot, suggests Dr. Koufman. (You don't have to wear the hideous uniform.) To play a wind, you have to use proper breath control, meaning that you have to pause regularly to inhale and stock up on air.

Work Out—The Kinks in Your Neck

Excessive tension in your neck can cause excessive tension in the muscles controlling your vocal cords. And that can strain the old vibrators, leaving you vulnerable to laryngitis. Here's how to relax things.

Massage your neck. If your neck muscles tense when you talk, gently rub and knead your neck with your fingers, suggests Dr. Koufman. Better yet, ask someone else to do it. Say it's a medical necessity.

Remember relaxation exercises. Some people tense their throat muscles when they're

---

Who Put the "-itis" in Your Larynx?

The suffix "-itis" comes from the Latin for "inflammation." Whenever you see it at the end of a medical word, you can bet that something is red or swollen. Appendicitis? Inflammation of the appendix. Bronchitis? Bad news for the bronchi, the tiny airways leading to your lungs.

---

under stress. To relax in trying times, try some deep breathing, suggests Dr. Koufman. Inhale slowly and deeply so that you completely fill your lungs and your lower abdomen pouches out. Exhale just as slowly, letting your abdomen sink as the air escapes. Meditation, massage, and yoga will also help you relax, Dr. Koufman says.

The Center for Voice Disorders at Wake Forest University offers extensive information on its Web site. To get there, use your Web browser to search for the organization's name.

LAUGHTER

The Hee-Haws of Good Health

WHAT IT IS

Seriously folks, laughter isn't necessarily a joking matter. Centuries before audiences bent over in hearty cackles at the physical slapstick of the Three Stooges or the everyday bemusement of Jerry Seinfeld, laughter was known to have its healing powers.

Henri de Mondeville, a fourteenth-century professor of surgery, declared, "Let the surgeon take care to regulate the whole regimen of the patient's life for joy and happiness, allowing his relatives and special friends to cheer him, and by having someone tell him jokes."

Laughter is the music of mirth. Do it more often, and you'll get a head start to good health.

WHAT IT DOES

Laughter is a low-sweat way to get lots of physical benefits. Here are a few examples of how guffaws do your body good.

Circulation: The more we laugh, the more efficiently the blood flows in our bodies. As we work the muscles in our faces and stomachs, the all-important oxygen supply to our body's organs increases, say experts.

Immunity: Laughter enhances our immune systems, say researchers. Chuckles increase the production of the body's natural disease-fighting cells.

Mood: Laughter also stimulates the release of endorphins, the mood-elevating brain chemicals that are responsible for making us feel good.

"Short bursts of adrenaline kick in when you laugh," says Steve Allen Jr., M.D., assistant professor of family medicine at the State University of New York Health Sciences Center in Syracuse, who, like his famous comedian father, sees laughter as tonic.

Stress control: Ever try to feel angry and happy at the same time? It's impossible for those two emotional states to exist simultaneously, say medical experts and humorists.

"When we laugh, our muscles contract and we decrease our levels of stress hormones," says Steven Sultanoff, Ph.D., a clinical psychologist in Irvine, California, and president of the American Association for Therapeutic Humor. "Laughter works by reducing the production of cortisol, the stress hormone."

Problem-solving: "Laughter energizes us and gives us more energy to do better work," says Dr. Sultanoff. "By diverting our attention from a problem with laughter, we allow our unconscious minds to work on the problem."

REAPING THE BENEFITS

Amid all the obligations in life, it's easy to get cheated on laughs. Experts offer these tips for maximizing your amusement.

Rise and smile. Start your day with a smile and thank yourself for waking up, suggests Loretta LaRoche, a national stress expert, president of The Humor Potential in Plymouth, Massachusetts, and author of *Relax—You May Only Have a Few Minutes Left*. If you can face the day with a sense of humor and gratitude, it will help you physically, emotionally, and spiritually to view your life in a more optimistic fashion, says LaRoche.

Be a humor hunter. Make a point of seeking out life's amusing little treasures every day. Read cartoon books like "The Far Side" by Gary Larson and "Dilbert" by Scott Adams. Dip into a funny book. Get a funny audiotape to listen to in the car. It doesn't matter which writer or comedian you choose. The point is to make humor a small daily goal. Put yourself in the path of funny people, says Dr. Sultanoff.

Magnify mayhem. The next time minor calamities strike, try to find the laughter in the folly, suggests LaRoche. The key is to exaggerate minor nuisances. This is what is called catastrophizing. If you are stuck in traffic, for example, you might first think, "Great! My boss will be mad." Then, "She'll fire me for missing her meeting." Then, "Won't she be sorry when she finds me still sitting here next week, still clutching the steering wheel, with cobwebs in my hair!"

"This game of exaggeration gives you distance from your little stresses and keeps them from becoming major stresses," explains LaRoche.

Create a joy journal. To convert doldrums into delights, consider scribbling down phrases, quotes, or observations that make you laugh. Stockpile them, tape them on your computer or refrigerator door, and read them whenever you feel down, suggests Dr. Sultanoff. Journaling helps people find simple pleasures in stress-filled times, he says.

Stock up on silliness. Keep a clown's red rubber nose or a pair of Groucho Marx glasses in your office drawer. The next time deadlines, telephone calls, or your boss's demands send your stress level soaring, open the drawer and slip on the fake nose or glasses. Stop and take a gander at yourself in the mirror, suggests Dr. Sultanoff. Watch how quickly your scowl turns into a grin.

TRIAL RUN

Need some tonic? Here are some one-liners from the master of the form, the late Henny Youngman.

On airline cuisine: "The food on the plane was fit for a king. 'Here, King!'"

On golf: "The other day I broke 70. That's a lot of clubs."

On horse racing: "The horse I bet on was so slow, the jockey kept a diary of the trip."

On hotel accommodations: "The room is so small, when I put the key in, I broke the window!"

LAZY EYE

One Side Sees All

Back in the days of ancient Greece, the mythical one-eyed cyclops terrorized and tormented villagers. Or so the legend goes. But from his point of view, he made his all-seeing eye more efficient because it did the work of two eyes.

That's what happens today to people who have lazy eye (medically called amblyopia): They also see the world from a one-eyed perspective. They have two eyes, but one dominates, forcing the other to retreat from active duty.

Doctors estimate that 2 or 3 out of every 100 people develop lazy eye during infancy, and it is usually hereditary.

Both eyes see an object, but they each send different images to the brain. The strong eye sends a crisp, clear image; the weak eye delivers a blurrier image. For someone with amblyopia, the brain can't blend two images into one picture. Forced to choose, it picks the clear image, and after a while, the weaker eye becomes lazy from disuse. If this problem isn't corrected, you could lose the ability to see out of that eye. The weak eye may also pull inward, outward, upward, or downward.

If you have amblyopia, you'll be more likely to help the condition if you avoid labels like "lazy eye," some doctors say.

"If you say you're in the process of visual change, not that you're nearsighted or have lazy eye, I believe something would shift in a positive way," says Jacob Liberman, O.D., Ph.D., a Colorado optometrist and president of Universal Light Technology, a company that researches and manufactures light and color technology for healing, in Carbondale, Colorado. "In many ways, the diagnosis or the label of lazy eye can do more to retard the healing process than the disease itself."

"Lazy eye is due not to disease but to disuse," adds Paul Planer, O.D., an optometrist in private practice in Atlanta and president of the International Academy of Sports Vision in Harrisburg, Pennsylvania. "If your two eyes have trouble working together, you will subconsciously close off one eye from seeing."

Fostering Teamwork

The most common treatment that eye specialists prescribe for lazy eye is wearing a light-blocking patch for periods of time over the strong eye. This forces the unused eye to focus and see. The eye patch is removed when a specialist determines that vision in the lazy eye matches the sight strength in the good eye. In other cases, ophthalmologists may recommend prescription glasses or surgery to correct the problem.

But behavioral optometrists (optometrists who have conventional training but also see the value of the mind and body working together in a healthy way) prefer to combat lazy eye with a variety of eye exercises. These exercises (which may be done with an eye patch, but check with your eye specialist first) are de-signed to strengthen the weak eye and get both eyes working in tandem. Here are some of their suggestions.

Capture your wandering eye. You can get your brain to read with the weaker eye by using two distractions on your stronger eye, says Meir Schneider, Ph.D., a licensed massage therapist, founder of the Center and School for Self-Healing in San Francisco, and creator of the

---

Energize Your Troubled Eye

Sondra Harris hesitates to show her baby pictures. Each captures the image of an infant with a lazy eye. Her right eye looks obligingly into the camera, but her recalcitrant left eye turns toward her nose.

She isn't much happier with pictures from her teen years. During adolescence, this Utica, New York, native wore thick prescription glasses that helped her vision but made her brown eyes appear mammoth. Before age 15, she had eye surgery twice, but the procedures didn't completely correct her condition.

"It was always embarrassing for me in school with my half-crossed eye and corrective glasses," says Harris, now in her forties and a corporate communications manager for a technology firm in San Rafael, California. But an unusual series of exercises that she devised for herself contributed significantly to getting the upper hand over lazy eye.

During what she calls her spiritual phase in her early twenties, Harris explored various options to correct her lazy eye. She was greatly helped by yoga. For one hour each day, she sat cross-legged with her arms relaxed. While doing whole-body yoga movements, she devoted at least five minutes daily to "working" her eyes.

She moved her open eyes clockwise and then counterclockwise slowly. Next, she looked up and held the stare for a few seconds. Then she looked down for a few seconds. She finished the routine by looking left and holding the stare for a few seconds in hopes of strengthening the weak muscles of her left eye.

"I worked out of a yoga book with no instructor and did this for a couple of years," says Harris. "My doctor pooh-poohed it, but it's a pretty permanent fix. My lazy eye finally corrected to near-perfect."

She relies on her yoga eye exercises whenever her eyes tire. She finds that she also needs to do them after a sauna. "For some reason, the heat from the sauna tends to relax the muscles, and I find my left eye moving toward my nose again," says Harris. "When I do the yoga exercises, I'm fine again."

Meir Schneider Self-Healing Method. Here's what he recommends.

Attach a piece of paper, one inch wide by two inches high, to the bridge of your nose. Hold a book off to the side of your weaker eye, far enough away so that your stronger eye is partially blocked from seeing it. Begin reading with the weaker eye while waving your hand in front of the stronger eye to distract it. Limit the exercise to five minutes; you need to avoid tiring the weak eye. Repeat it as often as you like.

Begin and end with this "palming" technique: Rub your hands together for about 30 seconds. Close your eyes and cover them with the cupped palms of your hands. Breathe slowly and deeply and try to feel the warmth of your palms penetrating your eyes. Take about 20 breaths. With each breath, feel yourself growing more relaxed as you let strain and tension flow out of your eyes. Slowly remove your hands, open your eyes, blink, and let in the light.

March your eyes in all directions. With this exercise, you can build teamwork with your eyes, according to Dr. Schneider.

▶Look with both eyes as far as you can upward, downward, left, and right. That's the easy part.
▶Next, try to combine movements. First look upward, then (while still looking skyward) move both eyes slightly to the left so that they peer upward diagonally. Hold that position for a count of three.
▶Blink several times.
▶Move your eyes—while still looking up—diagonally to the right and hold that gaze for three breaths.
▶Blink several times and then look down.
▶Repeat the previous steps while looking down. Stop if you feel any strain or discomfort.

Believe what you will see. This exercise

Play Beam Tag

You are never too old to enjoy the benefits of flashlight tag, says Don Teig, O.D., an optometrist and co-director of the Institute for Sports Vision in Ridgefield, Connecticut.

In a dark room, have a friend wave a flashlight so that you can see the beam on a wall. Your mission is to match your flashlight beam with your friend's beam pattern as it goes up, down, sideways, and in circles. To keep yourself from using your head to follow the light, try balancing a book on top of your head. This forces your eyes to work together to follow the beam.

---

taps your mental abilities, says Robert-Michael Kaplan, O.D., a behavioral optometrist, consultant in vision care in Gibsons, British Columbia, and author of *The Power behind Your Eyes.*

▶Find a comfortable, quiet place to sit or lie down.
▶Close your eyes, take a few deep, relaxing breaths, and release.
▶With your eyes shut, imagine looking through them and seeing the world two-eyed, looking straight ahead and with both eyes working together.
▶Blink and open your eyes wide. Let the light of the room or the sun refresh them.

Ignite your peripheral vision. This exercise, recommended by Dr. Schneider, teaches the brain to accept the input of both eyes by blocking the dominant eye from "stealing" the other eye's turf—each eye sees only its own separate periphery. The exercise also creates more relaxed, balanced vision for both eyes. You will need two dripless candles and a small, oblong piece of black construction paper about two inches wide by four inches long.

►Tape the paper across the top of the bridge of your nose so that it is in front of your eyes.

►Sit in a dark room holding a lighted candle in each hand.

►Start with the candles where you can easily see them, one on either side of your face.

►Looking straight ahead at all times and wiggling the candles continually, slowly move them outward until they are barely visible.

►Stop, close your eyes, and visualize that you can see the candles clearly and effortlessly while looking straight ahead.

►Open your eyes. Is it easier to see the candles? If so, move them a little farther outward and repeat. If one candle is clearer than the other even though it's an equal distance outward, close your eyes and visualize that you see both equally well.

►When you meet this goal, move both candles a little farther out and repeat.

LOVE HANDLES

Come On, Baby, Let's Do the Twist

Whoever coined the term *love handles* to describe the loose flesh hanging over the sides of our belts was one of the great spinmeisters of all time. It takes some style to claim that fat deposits, made worse by mushy muscles, are actually a handy convenience to help your sweetheart get a grip.

The best way to shrink love handles and keep your silhouette svelte is to work your oblique abdominal muscles, the muscles on the sides of your torso, says Michael Yessis, Ph.D., president of Sports Training in Escondido, California, and author of *Body Shaping*.

For Peak Obliques: Twisting the Day Away

The obliques are our Chubby Checker muscles—they get a workout when we do the twist. Here are two easy calisthenic exercises that will help you work them into shape, says Dr. Yessis.

Do the reverse trunk twist. Lie on your back with your arms out to the sides and your palms down. Bend your knees, keeping your feet together, and raise your legs until they form a 90-degree angle with your body (*1a*).

Lower both knees to one side and try to touch the floor. Try to keep your shoulders and

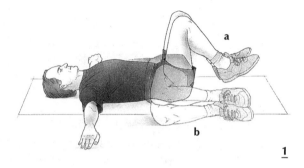

arms on the floor, and maintain the bend in your knees (*1b*). Then raise your legs back to center. The goal is to feel the twist in your torso.

Repeat the twist on the other side. Aim for 3 repetitions per side to start. Over the course of eight weeks, you should gradually build up to 20 repetitions per side, says Dr. Yessis.

Make like a wicked witch. A broomstick or a long pole can help you cast out your love handles. Here's how.

Sit on a bench or a stool with your feet resting on the floor about shoulder-width apart. Put the broomstick (or any other light bar) behind your head and rest it on your shoulders. Hold the broomstick as close to the ends as you can while keeping your arms slightly bent (2).

While keeping your hips stationary, twist your body as far to the right as you can. Hold

<u>2</u>

<u>3</u>

the position for a second (3). Then turn slowly back to the left as far as you can.

Once again, take care to hold the extreme position for a beat. Don't turn so fast that you bounce back and forth. Go slowly and steadily, without jerky movements.

Start with 10 turns in each direction and slowly build up over two months to 30 twists in each direction. Try to do two sessions a day.

Lay Into Your Lats

The second set of muscles that can help make your love handles less "handleable" are the latissimus dorsi. These large, fan-shaped muscles wrap around the back of your body, connecting your shoulder with your pelvis. Their relationship to love handles is sort of like that of a troublemaking kid sister: They're responsible for a lot, but they tend to be overlooked when justice is being meted out. In order to make sure they pay their dues, use this unique move recommended by Dr. Yessis.

Pull up your pelvis. Stand with your feet

hip-width apart, placing your left hand on a countertop or chair back for support, if desired (4).

Shift your weight to your left foot as you tilt the right side of your pelvis up toward your shoulders. If you're doing this subtle move correctly, your right heel should lift up from the floor (5).

Lower your right hip and foot. Repeat 15 to 20 times, then switch sides and repeat.

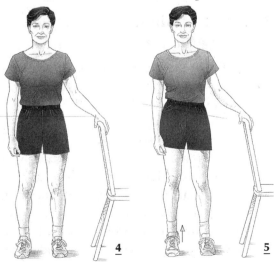

<u>4</u>

<u>5</u>

Spin Some Platters and Tune Up Your Fine Self

Try this trip down memory lane to help fight love handles, recommends Dan Hamner, M.D., a physiatrist and sports medicine specialist in New York City. Go into the attic and dig out that thing called a record player and one of those funny-looking discs called 45s. Look for a platter with Chubby Checker's name on it. Turn on the music and act like a kid again. Since twisting works the obliques, Chubby is a worthy workout coach.

For maximum muscle benefit, make sure you stretch the twist as far as you can to the left and back again to the right. You'll also want to modify Chubby's twist a bit. He had lots of hip movement, but you should keep your hips in place and twist your torso around them. Remember to alternate which foot you put in the forward position, so you work both sides of your obliques. This will not only help tone up your muscles but also keep you in touch with the youthful you.

Sports with a Twist

Some leisure-time activities are especially beneficial in attacking your love handles. Try twisty sports—they're oblique optimizers, says Dr. Yessis.

Paddle on. If you're lucky enough to have access to Lake Glimmerglass and a canoe, hop in and start stroking. The twisting motion of paddling works the obliques well. Just be sure that you rotate your shoulders outward and back as you pull the paddle toward the stern of the canoe.

Serve it up. Tennis hates love handles. Why? Because when played properly, the game requires that you twist your torso on many shots. Give tennis a try if you've got lateral bulges where you don't want them.

Be a swinger. Golf is yet another twisting sport. To get the anti-love-handle benefit, you don't even have to take up the game—just pick up a club and work out with it.

To make the proper motion on the backswing, turn your entire upper body so that your front shoulder turns back under your chin. It's important to keep your hips steady and your eyes on the ball. As you shift your weight onto your back foot, your club should move from straight in front of you to the top of its stroke.

On the downswing, make sure that you make a full swing, turning your hips and then your shoulders as you follow through so that eventually, you're turned with your chest facing the target. This twist will work your obliques.

LOW BLOOD PRESSURE

Help with a Dizzying Deficiency

In *Gone with the Wind*, Scarlett O'Hara's perpetually fainting Aunt Pittypat comes across as a fairly hysterical character, swooning as often as she does. But who knows? Maybe she simply had low blood pressure.

One of the most common symptoms of low blood pressure is feeling lightheaded, dizzy, or even faint when standing. While public health messages tend to focus on the hazards of high blood pressure, low blood pressure can also be troublesome. Not only can it leave you feeling woozy when you get to your feet, it can also cause serious complications if you have other health problems, such as artery disease.

"In general, the lower blood pressure is, the better," says Gerald Fletcher, M.D., professor of medicine at the Mayo Clinic in Jacksonville, Florida. "But truly low blood pressure can also cause problems."

If you have low blood pressure, standing up or standing for a long time can make you dizzy because it temporarily diverts blood from your brain. Normally, gravity pulls the blood in your midsection down toward your feet, so the blood pressure in your upper body drops momentarily. Usually, this prompts your heart to beat faster. Simultaneously, the blood vessels in your legs contract, boosting your blood pressure back to normal and pumping blood to your cranium.

If you have low blood pressure to begin with, though, the changes in your heart rate and blood vessel diameter may not raise your blood pressure quite high enough to get sufficient blood to your head. The same thing can happen if you have an illness or nervous system problem that prevents your heart from beating faster or prevents your blood vessels from contracting properly. Whatever the cause, if your brain doesn't get enough blood, you feel faint.

Unfortunately, the consequences can be far more serious if blood flow to your brain is already limited by atherosclerosis, the presence of fatty plaque deposits in the blood vessels leading to your head. If the drop in blood flow is drastic, it could lead to a stroke, warns John Hall, Ph.D., professor and chair of the physiology department and director of the Center for Excellence in Cardiology Research at the University of Mississippi Medical Center in Jackson.

The Causes—And an Effective Remedy

A variety of things can cause dizzying low blood pressure, a condition that doctors call orthostatic hypotension. Some of us inherit it. Health problems like diabetes and heart failure can cause it, and so can certain medications, including cardiac drugs known as beta blockers. Dehydration can cause it as well. So if you feel dizzy or faint when you stand, you should see your doctor. She should check to see whether you have an underlying medical problem that needs attention and whether any medication that you're taking might be the culprit.

If everything else is ruled out, there are things you can do on your own that will help—and exercising is chief among them. When you exercise, your muscles contract, putting the squeeze on nearby blood vessels. "The squeezing action helps return blood back to your heart," explains Dr. Hall. And once the blood is back in your heart, it's just a short sprint up to your head.

In a study at the Mayo Clinic, men and women with low blood pressure reported far fewer bouts of dizziness after they started practicing muscle contraction exercises. Tests showed that the exercises boosted their blood pressure. So the next time you feel woozy when you try to stand, sit or lie down and wait until the feeling passes, says Dr. Hall. Then try the following exercises.

March to nowhere. While standing, simply lift your knees up and down and march in place. In the process, you'll contract and relax your thigh muscles, pumping blood up to your noggin. Repeat 50 times, suggests Dan Hamner, M.D., a physiatrist and sports medicine specialist in New York City.

Tighten your legs. While standing, tighten the muscles in your legs. Hold for a few seconds, then relax for a second. Repeat several times, suggests Dr. Fletcher.

How Low Is "Low"?

When you have your blood pressure checked, you get a two-number readout. The first number, your systolic pressure, is the pressure inside your blood vessels when your heart beats. The second, your diastolic pressure, is the pressure when your heart rests between beats. Your blood pressure is low if your systolic pressure is under 90 and your diastolic is less than 70, explains Gerald Fletcher, M.D., professor of medicine at the Mayo Clinic in Jacksonville, Florida.

A reading of 86/68, for instance, is low. Normal blood pressure is anywhere between 90/70 and 140/90. Anything higher than 140/90? You guessed it—high blood pressure.

Get a Boost with Regular Workouts

Regular exercise will also help you keep your blood pressure up in the safe range. Among other things, it will get your muscles in shape so that they can do a better job of contracting and squeezing blood back up to your brain. Here are some great ways to give your blood pressure a boost.

Walk in the water. Pace up and down a shallow lane in a pool. Water exercise keeps your leg muscles and other muscles in good shape so they're fit to contract and return blood to your brain. When you're submerged up to your neck, the water exerts pressure on your legs and stomach and consequently on the blood vessels inside them, say experts at the University Hospitals of Cleveland in Ohio.

Cover some ground. If it's tough getting to a pool, do some walking, cross-country skiing, cycling, fencing, or inline skating. Shoot for 30 to 40 minutes of this kind of exercise at least six days a week, says Dr. Fletcher.

Carry your weights. Resistance training strengthens your muscles so they're able to contract more forcefully. Exercises that build your calf and thigh muscles are particularly helpful, says Dr. Hall. Try the following calf-building exercise, as well as exercises that build your other muscles, twice a week.

Stand facing a wall or a chair with your feet together and your knees slightly bent. Holding onto the wall or chair, rise up on your toes, then slowly lower your heels back to the floor. Repeat 10 times.

Drink water heavily. Exercising without pausing to drink can leave you dehydrated, and that can make you dizzy, since dehydration is itself a cause of orthostatic hypotension, says Dr. Hall. You lose a quart of water after just 45 minutes of moderate exercise, so replenish it. Always have a quart bottle of water with you when you exercise, and stop for a slug every 10 minutes or so.

LOW SELF-ESTEEM

Nourishing Your Athletic Ego

It's a sunny, humid Sunday afternoon in Chino, California, and 49-year-old Jenny Hanley is sweating it out on a heat-scorched soccer field. She's not cheering on a Midget League son or daughter. She's playing—rough, if necessary.

"I play hard," Hanley says. "I love the competition." Hanley, the director of special education in nearby Ontario, California, didn't start playing soccer until she was 37. Although she was a little nervous about joining the local league, she hoped that soccer would build her athletic skill. It did much more than that.

"This was something I'd never done before, and I was a little lacking in confidence," she explains. "So I practiced and practiced. The more I practiced and the better I got, the more confident I was. I tell other women that this is a great way to build your self-esteem."

Achievement on the soccer field—or the tennis court or the walking trail—*can* give your self-worth a boost, studies suggest.

Even if you're no paragon of speed or agility, athletic accomplishment and the self-esteem boost that goes with it are still within your grasp, says Nico Pronk, Ph.D., director of the Center for Health Promotion at HealthPartners in Minneapolis. If you struggle with feelings of inferiority or incompetence, try exercising for an ego boost. The trick is choosing a form of exercise that you enjoy and setting exercise goals that are a bit of a stretch but are still within reach.

Picking the Right Workout

Exercise can and should be fun. It's not just huffing and puffing on a treadmill. Dancing is exercise. So is paddling across a blue-gold lake under the harvest moon. So step one is to choose a sport or exercise that you really enjoy and one that's suited to your style and personality. Here's how to find your perfect exercise match.

Accept who you are in terms of exercise. If you're just starting to exercise, it's best to choose something simple like walking, says Edward McAuley, Ph.D., professor of kinesiology at the University of Illinois at Urbana-Champaign. "Pick something that's appropriate for your skill level and that's going to give you some immediate success," he says. Early successes mean immediate deposits in your self-esteem account. They also increase the odds that you'll keep on exercising and continue to make deposits.

Decide whether you're a lone wolf or a social butterfly. If you feel self-conscious exercising in front of other people, or if, like Garbo, you simply want to be alone, try a solitary pursuit, says Robert Motta, Ph.D., a psychologist and director of the doctoral program in psychology at Hofstra University in Hempstead, New York. Go walking in the morning, before the chickens get up, and enjoy the quiet. Grab your bike and head out for a solo spin around the neighborhood in the twilight after work.

You say that you'd rather have company? Sign yourself and a partner up for tango classes. Or join a hiking group and wander the woods. A group adds peer pressure, which can help you get out of a warm bed on frosty mornings. "If other people expect you to be there, you're more likely to show up," says Dr. McAuley.

Figure out your learning style. If you're trying a new athletic pursuit, a teacher can show you how to do it right and help you avoid frustration and injury, which can undermine the best-laid exercise plans, says Dr. Motta. Sign up for a half-dozen lessons. That should cover the basics of forehand, backhand, and serve or the fundamentals of the golf swing.

If, on the other hand, you know from experi-ence that you don't take instruction well, choose a sport that's relatively easy to learn without a teacher, like mountain biking, says Dr. Motta.

Go for a Goal

Setting and reaching a goal—athletic or otherwise—can give your self-esteem a big lift. The trick is to set goals that challenge you but don't require winning the Boston Marathon. "A lot of people get turned off by exercise because they set their goals too high," says Dr. Motta. Remember these rules for inventing your ambitions.

Forget your neighbor. Don't compare yourself to others. If you're working your way up to jogging a mile, don't get discouraged because your neighbor Wilson does 10 miles every Saturday. Just slowly but surely increase your exercise level, says Judith Beck, Ph.D., director of the Beck Institute for Cognitive Therapy and Research in Bala Cynwyd, Pennsylvania, and clinical assistant professor of psychology in psychiatry at the University of Pennsylvania School of Medicine in Philadelphia.

Pumping Pride: Why Exercise Boosts Self-Esteem

Exercise gives self-esteem a lift for a number of reasons, says Robert Motta, Ph.D., a psychologist and director of the doctoral program in psychology at Hofstra University in Hempstead, New York.

For starters, exercise can improve your appearance, your health, and your sense of self-discipline and control, says Dr. Motta. Research suggests that exercise also triggers key biochemical changes—increasing the quantity of feel-good neurochemicals bathing your brain cells—that play a role in self-esteem improvements.

But overarching all of these individual potentially helpful esteem-building changes is a more global one, says Dr. Motta. "The mind and body are one, so if you make one healthier, you make the other healthier," he says. "When you exercise, your body feels better, so you feel better."

Wrestling Self-Defeating Thoughts to the Ground

What's that you say? You can't exercise because you're *not athletic*? Unfortunately, many of us may accept that label just because the playground gang always picked us last for touch football. Be alert for self-sabotaging thoughts. The next time you hear these self-defeating one-liners in your head, challenge them, suggests Judith Beck, Ph.D., director of the Beck Institute for Cognitive Therapy and Research in Bala Cynwyd, Pennsylvania, and clinical assistant professor of psychology in psychiatry at the University of Pennsylvania School of Medicine in Philadelphia.

▶When you tell yourself: "I'm not athletic," try this encouraging self-talk: "I don't have to be a world-class athlete to go for a walk. I think I'll head over to the park right now and walk once around the track. I'm going to try to walk every other day—a little farther each time." You're not trying to win at Wimbledon, you're just trying to stay in shape.

▶When you tell yourself, "I've tried to exercise before, but I can't stick with it," counter with a realistic response such as, "Maybe I was too much of a perfectionist. I gave up because I promised myself that I'd exercise every day and I felt like a failure because I couldn't manage it. This time, I'll shoot for a more manageable goal. I'll try to exercise three or four times a week, and I'll block out time for exercise in my appointment book."

Get stronger slowly. Last time out, you ran for 5 minutes? Good going. This time, try for 6 minutes, says Dr. Motta. Don't shoot for 10 minutes right away—that's too much. You don't have to raise the bar much each time in order to make considerable progress over a short time. If you start by walking or running a mile and add a mere 20 yards to your distance each time you go out, you'll be covering two miles at a shot within three months.

Don't overextend. A word of caution: If you're too sore to exercise the day after you raised your sights, you've probably raised them too high, too quickly. Throttle back a bit, says Dr. Motta. Remember: Exercise should be enjoyable. If it's not enjoyable, you're more likely to quit and do your self-esteem some injury.

Revel in writing. Keep an exercise diary, suggests Dr. Motta. Every time you exercise, jot down what you've done. Writing down your accomplishments makes them tangible. You can flip through your diary and see all the progress you've made, right there in black and white. You'll feel good.

Run across New England. Find a map of the United States. Using a highlighter, circle your hometown. Then pick a destination—some faraway place that you love, like Cambridge, Vermont—and circle it. At the end of each week, total the number of miles you've run, walked, or cycled, suggests Dr. Motta. Then use the highlighter to trace the path from your place to Cambridge, coloring in as many miles as you've logged. See how long it takes you to "get there." When you do, celebrate your accomplishment. Buy yourself something that makes you feel good, like a pair of shoes or a new outfit, or have a massage.

MACULAR DEGENERATION

Aging Eyes See Hope

It's like there are little holes in my vision. I can't see straight on very well. But around the edges are little holes where I can see quite clearly.
—**American artist Georgia O'Keeffe (1887–1986)**

As it was for Georgia O'Keeffe, so it is for others with macular degeneration: The world of sight is best viewed peripherally, where all images are clearest. Macular degeneration is a disease that destroys the light-sensitive cells in the macula, the central part of the retina, or innermost layer of the eyeball. This area houses image-capturing receptor cells in the central vision that are needed for reading, close-up tasks, and some color and distance vision.

What triggers this disease is unknown, but the main risk factor is simply too many birthdays. Macular degeneration strikes about 14 percent of Americans between the ages of 55 and 64. The number escalates to 37 percent among those over age 75.

A fatty diet, overexposure to sunlight, and having a family history of the disease also raise the risk stakes, and your chances increase if you smoke. A 12-year study of nurses ages 50 to 59 who smoked showed that they were at greater

---

Rename the Disease

"People say, 'Doctor, I have macular degeneration.' I call it macular *regeneration*. We must look at the possibility of being regenerated. You can increase that healing ability by concrete suggestions and by creating an image of wellness rather than one of destruction and disease," says Robert-Michael Kaplan, O.D., a behavioral optometrist, consultant in vision care in Gibsons, British Columbia, and author of *The Power behind Your Eyes*.

---

risk for macular degeneration than their non-smoking peers.

Avoiding as many risk factors as possible will help reduce your chances of getting macular degeneration. In addition, eye exercises may help. In fact, some behavioral optometrists (optometrists who have conventional training but also see the value of the mind and body working together in a healthy way) recommend that people with macular degeneration perform exercises that relax their eyes and concentrate on maintaining their peripheral vision.

"The bottom line with macular degeneration is relaxation and more relaxation," agrees Meir Schneider, Ph.D., a licensed massage therapist, founder of the Center and School for Self-Healing in San Francisco, and creator of the Meir Schneider Self-Healing Method. To get improve-

The Eccentric View

Earthquake survivor Bethea Wilson of San Francisco wasn't prepared for what she saw while driving her car in the Bay Area one day in 1982.

Buildings suddenly appeared to jog, break, and arch skyward. She shook her head and blinked several times, but the haunting images remained.

"It was like the buildings just slipped," recalls Wilson, now in her mid-eighties. "I said, 'Good night, what's going on here?'"

It wasn't an earthquake; it was her eyes that were shaking things up.

The next day, an ophthalmologist told her that she had macular degeneration. Five more ophthalmologists echoed the following comment, she says: "No known cause, and no known cure."

The news jolted Wilson, a skilled embroidery artist of large-scale commissioned hangings. Her work is draped on hospital meditation room walls as well as on the shoulders of members of the clergy. For six years, big, black blots haunted her central vision. She could no longer see to stitch. She misjudged curbs and tripped. She had trouble inserting her house key into the front door lock and handling coins and currency.

A friend urged Wilson to see an optometrist who uses the holistic approach (an optometrist who has conventional training but also sees the value of the mind and body working together in a healthy way). He determined that she had good vision cells left in her macula and introduced a technique called eccentric viewing to help Wilson see more clearly. Instead of looking at objects straight on, Wilson moves her eyes upward and over to the right to use the good cells in her macula. She uses the same technique with magnification in order to read.

Twice a day, she gives her body a complete workout to improve her circulation and get the blood flowing to her eyes. She begins her exercise by tensing and relaxing her feet for a count of eight. Then she moves her way up her legs, thighs, abdomen, chest, shoulders, arms, and hands to her jaw and then her eyes. At each body location, she tenses and then relaxes her muscles. Early on, Wilson made a commitment to stay active, and she uses walking as her mode of transportation.

Wilson believes that the optometrist, daily eye exercises, and her feisty spirit kept her from surrendering to a no-hope prognosis. "I'm not cured, but I have harnessed my disease and learned to live with it," she says. "My eyes have not degenerated more since I began my exercises. In fact, they got a lot better. I'm functioning in life again, and functioning very well."

ment, he says, you need to bring serenity into the eye exercise session. Keep the eye exercises short, with lots of rest periods. Make it a point to look at the things you enjoy seeing and just let them appear to you, however fuzzy they may be.

Let the sun shine in. This exercise is designed to ease your eyes. Before you begin, protect your face with sunscreen, says Dr. Schneider. Unless it's a problem for you, do this exercise while standing.

►Find a sunny place outside or in front of an open window sometime before or after the peak hours of the sun.

►Close your eyes. Turn your face slightly up so that the sun's rays pour directly onto your closed eyelids. (On cloudy days, aim your face where the sun would be if it were out and do the exercise with your eyes open.)

►Start turning your head slowly from side to side. Do this for two to three minutes. To keep yourself from furrowing your brow and squinting, lightly massage around the eye on the side that's turning away from the sun.

►Stop and use this "palming" technique: Rub your hands together for 30 seconds or so. Close your eyes and cover them with the cupped palms of your hands. Breathe slowly and deeply and try to feel the warmth of your palms penetrating your eyes. Take about 20 breaths. With each breath, feel yourself growing more relaxed as you let strain and tension flow out of your eyes. Slowly remove your hands, open your eyes, blink, and let in the light for one or two minutes.

►Repeat the steps. Slowly work your way up to 15 minutes of "sunning" each day.

Make waves. Since macular degeneration strikes central vision, we must rely more on peripheral vision. This exercise is designed to give

---

Try Eye Sweeps

Here is an exercise that concentrates on what we see out of the sides of our eyes. Although this exercise is great for people with macular degeneration, it can be used by anyone, says Meir Schneider, Ph.D., a licensed massage therapist, founder of the Center and School for Self-Healing in San Francisco, and creator of the Meir Schneider Self-Healing Method.

While walking down a quiet sidewalk, pick a point just in front of your feet. Sweep your eyes from one edge of the sidewalk to your chosen point and then to the opposite edge of the sidewalk. Sweep your eyes back again and blink a few times. Do this exercise every few blocks to help stimulate and strengthen your peripheral vision.

---

your peripheral vision a healthy workout, says Dr. Schneider.

►Find a comfortable, quiet place to sit or stand.

►Extend both arms straight out to the sides and parallel to the ground. Keep your arms just within your range of vision.

►Moving from your wrists, wave your hands up and down.

►Keep your head straight and unmoving, facing forward.

The goal is to see your hands waving without moving your head and without using your central vision. Move your arms back slightly as your peripheral vision improves.

Focus on colors. Some behavioral optometrists recommend color therapy for some

Avoid Gridlock

Sometimes in the early stages, people don't know that they are developing macular degeneration, says Jacob Liberman, O.D., Ph.D., a Colorado optometrist and president of Universal Light Technology, a company that researches and manufactures light and color technology for healing, in Carbondale, Colorado.

The following exercise is a good way to keep regular tabs on your vision and detect any possible vision loss. It uses the Amsler grid, which contains equally spaced intersecting horizontal and vertical lines and a solid dot in the midpoint.

►If you wear glasses or contact lenses, keep them on. Hold the Amsler grid about 12 to 15 inches from your eyes in good light.

►Cover one eye.

►Look directly at the center dot on the grid with your uncovered eye.

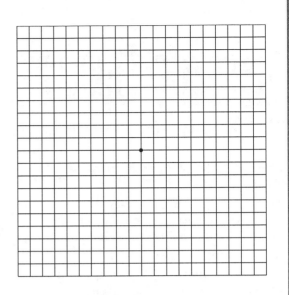

►While staring at the dot, note if any areas on the grid are distorted, blurry, or dark.

►Repeat these steps with your other eye.

Consult your eye doctor if you detect any signs of wavy, blurry, or dark spots on the grid.

people with macular degeneration to help them make the best use of their remaining sight. This exercise requires looking at certain colors of light, moving the eyes away from the light, and then moving them back to the light again. The key is to keep the eyes moving.

"They are empowering themselves to take charge," says Robert-Michael Kaplan, O.D., a behavioral optometrist, consultant in vision care in Gibsons, British Columbia, and author of *The Power behind Your Eyes*. "I help them choose colors. Cooler ones slow down inflammation and warmer colors activate blood flow."

This exercise should be performed in an optometrist's office the first time, so you will need to find an optometrist who also works with color therapy. Once your doctor has chosen the colors that will benefit your eyes, you can attach various colored light filters over a light source and continue the exercise at home.

MASSAGE THERAPY

Rubbing for Royals and Regular Folks

WHAT IT IS

What's the first thing you do when you bump your elbow or stub a toe? If you're like most people, you instinctively reach down and rub the little hurts of life.

Massage therapy—rubbing on a grand scale—has been touted as a healing tool since antiquity. In fact, Julius Caesar got daily back rubs in an attempt to control his epilepsy. While we don't know if massage helped Caesar stop seizing, there's no doubt that therapeutic rubbing, kneading, and touching can offer relief from a long list of ailments.

WHAT IT DOES

"Massage is clearly a good adjunct therapy that has the healing potential to help many conditions," says Shawne Bryant, M.D., a gynecologist and massage therapist in private practice in Virginia Beach, Virginia. She cautions that people with medical conditions need to consult with a health professional about their treatment and learn all the available options.

Although it's most strongly associated with the relief of stress, muscle tension, and backache, studies show that beyond its mere manipulation pluses, massage can help with a range of ailments far more complex than a stretched hamstring, from asthma and depression to chronic fatigue syndrome and even serious immune deficiencies like HIV. Preliminary research has also shown that massage can temporarily relieve tremors caused by Parkinson's disease and may work against high blood pressure and reduce agitation in Alzheimer's patients.

Exactly how therapeutic rubbing works isn't entirely clear, but this much is known: The stimulation of the nerve endings in the skin sends messages zipping up through the spinal cord to the brain and sets our biochemical pot to bubbling. Specifically, massage increases the levels of healthful hormones like serotonin, which helps the body cope with pain and boosts mood, says Maria Hernandez-Reif, Ph.D., director of the massage therapy research program at the Touch Research Institute at the University of Miami School of Medicine. It's even possible that regular sessions of therapeutic massage could help reduce the need for prescription pain drugs, she says.

Massage also has a positive effect on the release of other beneficial substances in the body. A study involving HIV-positive men showed that massage can increase the number of natural killer cells—immunity soldiers—patrolling the blood. And other research suggests that massage may boost insulin levels, says Dr. Hernandez-Reif, resulting in better digestion, better food absorption, and thus, better nutrition.

Massage not only accentuates the positive, it also cuts down on the negative. It has been shown to shrink the body's balance of norepinephrine and cortisol, stress hormones that are found in above-average amounts in people with chronic illnesses like depression, Dr. Hernandez-Reif says. Within just 20 minutes, the soothing effects of massage can bring stress hormones back to normal levels.

REAPING THE BENEFITS

If you've never had a massage, you don't know what you're missing. The following expert tips can help you get rubbed the right way.

Check those references. To be sure you have a good first experience, it's important to choose your therapist carefully. Referrals from friends who have had massages or doctors who believe in its effects are the best places to start, says Marianne Bergmann, a certified massage therapist in Bethlehem, Pennsylvania, and a member of the American Massage Therapy Association.

Keep it personal. Consider your personal preferences. Bergmann mentions that some women, for example, may be more comfortable getting a massage from a female massage therapist. And if you have allergies to certain scents or oils, be sure to speak up.

Try the M&M technique. Massage and meditation may be the sweetest M&M combination since the candy. At the end of a massage session, give yourself a few minutes to soak up the feeling of total relaxation, says Bergmann. A moment or two of meditation may make your transition back into the world a little easier. And what's more, you'll form a memory of serenity that you can summon up the next time you feel tense.

TRIAL RUN

The Asian technique known as shiatsu—a form of acupressure that may use massage—is based on the theory that health relies on the flow of life, or chi, through the body. When the easy movement of chi gets blocked, health problems can result, according to Bergmann.

Shiatsu uses firm pressure on various points of the body to restore the flow of chi. This gentle point-and-press method can be performed on a friend or on yourself. And it works through clothing. "It may be an easier introduction to massage for modest people," says Bergmann.

Try any or all of these simple anti-headache shiatsu routines to push your pain away.

▶Resting your fingers on your head, use your thumbs to press into the indentations just at the base of your skull on either side of your spine. Hold for 10 to 15 seconds (*1*).

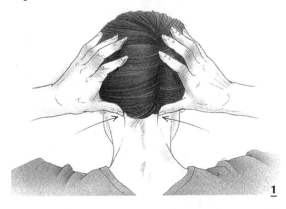

1

2

▶Use one thumb to press into the indentation formed by the two tendons at the base of the other thumb. Hold for 5 to 10 seconds, then repeat on your other hand (2).

▶Rest your fingers on your head and gently

3

press into your temples with your thumbs. Take a deep breath while rotating your thumbs for 5 to 10 seconds (3).

MEDITATION

Building Mental Muscles

WHAT IT IS

Meditation has a misleading reputation. Lots of people think of it as "zoning out," going into a trance, or becoming zombielike. In fact, it's exactly the opposite of getting spacey.

Meditation is the art of learning to pay attention and the craft of controlling your thoughts. Too often, our thoughts drift in an endless stream of random thinking, moving here and there like a bottle in the surf. Meditation stops that bobbing about and lets us focus our awareness.

Although formal meditation is usually done while seated, either on a cushion on the floor or in a chair, informal meditation can happen anywhere and anytime—even when you're in motion. There's nothing mysterious or magical about meditation. It is a skill that can be learned, like driving a car with a stick shift or using a sewing machine.

WHAT IT DOES

Gaining control over our thoughts through meditation generates many health blessings. Here are just a few of the good things meditation can do.

Reverse the effects of stress. Meditation re-duces levels of harmful stress hormones. In a study at the Center for Health and Aging Studies at the Maharishi University of Management in Fairfield, Iowa, meditators showed an impressive, healthful drop in one type of stress hormone after only four months of practice. Reducing stress hormones is a good bet for health.

Lower blood pressure. Going the peaceful route is great for people with high blood pressure. In one study, meditation was shown to bring down blood pressure significantly after just three months of daily practice.

Control nervous system disorders. Regular meditation seems to relieve the symptoms of some kinds of tic disorders, according to Janet Messer, Ph.D., a psychologist in private practice in Eugene, Oregon. These disorders result in the characteristic muscle contractions that cause facial twitches or other symptoms such as compulsive explosions of inappropriate words (Tourette's syndrome). Although it is not a solution to these disorders, meditation seems to alleviate the symptoms for several hours.

Help break bad habits. Meditation bestows its boons on our minds as well as our bodies. In a case study in the department of psychology at Pepperdine University in Culver City, California, researchers found that meditation could help end chronic nervous problems such as nail biting.

Develop healthy self-awareness. Over time, meditation improves the ability to watch your thoughts without reacting to them in an emotional way, even when you're not meditating. You may find yourself feeling calmer overall and less likely to react to life's minor annoyances with angry responses.

Meditation can work its magic on any ailment for which stress is a cause or a contributing factor. Some of the health problems that may improve when people practice meditation include premenstrual tension, irritable bowel syndrome, insomnia, high blood pressure, allergies, angina, erectile dysfunction, indigestion, and many aches and pains.

REAPING THE BENEFITS

Fitting focus time into an already busy schedule is perhaps the biggest obstacle for would-be meditators. These tips will help you find time you didn't know you had. Anything done with awareness can become a meditation in itself.

Find delight in dishwashing. Opportunities for a moment of meditation are everywhere. For centuries, monks have found serenity in the quiet obligations of labor. A mountain of dirty plates and cups is actually a chance to tune out worry and pay attention to what you're doing. Don't think back on the day or plan for tomorrow. As you're soaping and rinsing, try to keep your mind entirely in the moment. This tedious work can actually be relaxing, says Mark Epstein, M.D., clinical assistant professor of psychology at New York University and author of *Thoughts without a Thinker* and *Going to Pieces without Falling Apart*.

Meditate in motion. Instead of reading a dog-eared, sweated-over issue of the same magazine every time you work out, sharpen your exercise focus with meditation. Tune in to the soles of your feet while you're running or walking, for example, or simply concentrate on your breathing while you exercise. "You'll be putting your mental energy back into your physical activity," says Dr. Epstein. This gathering of energy may be what's behind the boost that meditation gives to athletic performance.

Make your menu mindful. Three times a day, every day, you have another opportunity for meditation in action. When you sit down to a meal, take time to savor the first few bites. It will bring peace to your table.

The experience of eating starts well before the first taste. Start by bringing your attention to your breath. When you feel quiet and calm, begin eating slowly. Let your senses come into play. If you're eating finger foods, feel the texture. Inhale the aromas. Appreciate the colors on your plate. While chewing and swallowing, stay focused on what is happening. Concentrate on each bite just as you would concentrate on your breath.

TRIAL RUN

All of the following meditation variations begin in the same position. Either sit in a chair with your feet on the floor or on a cushion with your legs crossed. Place your hands in your lap. Close your eyes or lower your eyelids just enough to limit visual distractions. Your head, neck, and spine should be straight. Relax your shoulders and let your body become still as you begin to pay close attention to your breath, says Dr. Epstein.

When disruptive thoughts arise—and they will—don't get discouraged. Meditation reveals how busy our minds can be. With time and practice, you'll notice more and more internal silence.

Here are three basic methods of keeping your concentration. Although you may want to try them all, you'll probably find one that works best for you, says Dr. Epstein.

Be aware of your breath. The Buddha—probably the most famous meditator of all time—was said to have used this technique to reach the ultimate state of awareness known as enlightenment. For the rest of us, it's one of the simplest methods for attaining deep relaxation and concentration.

To start, focus on your breath. Notice each inhalation and exhalation. It helps to place your mental attention either on your belly as it rises and falls or on your nostrils, where the breath is gently flowing in and out.

Don't change or force your breath—just pay attention to what it does. If you notice your mind wandering, return your attention to your breath. Just for now, let your breath be the most important thing in the world.

Count for clarity. If you're a beginning meditator, counting can make watching your breath easier. Begin counting your breaths, either on the inhalation or on the exhalation. Focus your attention on the numbers and your breath. When you reach 10, start over.

Meld with a mantra. When many people think of meditation, they immediately picture a grizzled old swami sitting cross-legged and chanting the same word over and over again. Mantras—words or sounds used as tools for focusing the attention—are used by many religions, including Christianity and Hinduism. The words used in the mantra may be different, but their effect is similar in all religions. Mantras help keep your mind from wandering back to your everyday worries—mortgage payments and wayward children wandering the mall.

It's easy to come up with your own mantra. Just choose a word, phrase, prayer, or sound that has a special, positive meaning for you. *Peace*, *amen*, *om*, and *love* are just a few possibilities.

Once you've decided on a mantra, focus on the words or sounds just as you focus on your breath. Repeat the mantra silently to yourself as you meditate. When thoughts intrude, try to have your mantra fill your mind instead.

Meditating between 20 minutes and an hour is ideal, says Dr. Epstein, but even 10 minutes is beneficial. The important thing is to get into the habit of doing it daily, he says.

MENOPAUSE

Exercising through Changes

A century ago, proper women never even whispered the word *menopause*. Thirty years ago, menopause was treated as a dark secret, not as a natural stage of womanhood. Today, a generation of Baby Boomers is entering menopause filled with new-found empowerment and confidence.

"Women are no longer ashamed of menopause, of speaking the word, going through the phase, or having hot flashes," says Sadja Greenwood, M.D., assistant clinical professor of obstetrics, gynecology, and reproductive sciences at the University of California, San Francisco. "I now see women in their late thirties seeking information to prepare for it."

In fact, 52 percent of American women ages 45 to 60 view menopause as the beginning of a new and fulfilling phase of womanhood, according to a national Gallup survey conducted for the North American Menopause Society.

"These findings reveal a real change in attitudes about menopause," says Wulf Utian, M.D., Ph.D., executive director of the North American Menopause Society in Cleveland. "American women are becoming more proactive in educating themselves about it."

Strictly speaking, menopause begins after your final menstrual period, the moment your ovaries stop releasing eggs. Although the average age for menopause is 51, some women go through it in their thirties, and others don't stop menstruating until their midfifties.

But it's helpful to think of menopause in a broader sense, as the changes that a woman goes through in the years just before her final menstrual flow, a time known as perimenopause. Some women breeze through these years virtually symptom-free, while others have many physical and emotional problems that can last for up to 12 years.

Exactly why the passage to midlife is easier for some women than for others is under debate. It probably depends on how much hormonal change you go through and how suddenly you go through it. The issue of hormone replacement therapy (HRT) is controversial. It may make sense for some women but pose unacceptable risks for others. Fewer than half of American women take hormone replacement therapy to help them navigate menopause.

But although HRT may not be suitable for all women, being fit has menopause-managing advantages. According to research at the University of Newcastle in Australia, there is evidence that women who are physically active have fewer problems during menopause than women who are sedentary and out of shape.

"Exercise is a key component of optimal health at all ages, but particularly through midlife and beyond," says Dr. Utian. Here are some exercises to counter common menopausal maladies.

Fending Off Blue Moods

With hormone levels shifting like a seesaw, some women experience bouts of depression and anxiety during menopause. One remedy is as simple as breathing. Here's an easy anti-stress breathing exercise to try every day, suggests Dr. Miller.

▶Lie flat on your back with your knees bent and both feet flat on the floor.

▶Close your eyes.

▶Inhale deeply from the diaphragm, breathing in through your nose for five seconds.

▶Exhale completely, breathing out through your nose for 10 seconds.

▶For complete relaxation, continue this easy in-out breathing for three to five minutes.

To make sure that you are breathing from your diaphragm, place your hands on your stomach. Your hands should rise and fall each time you inhale and exhale, says Dr. Greenwood.

Maintaining Bladder Control

Some women entering the menopausal years find that they must get up in the middle of the night to relieve an insistent bladder. Others find that they urinate tiny amounts whenever they sneeze and laugh. What's happening?

"When estrogen levels go down during menopause, the bladder tends to contract, and some women find that they must urinate more frequently," says Dr. Greenwood.

You can bolster your bladder by doing daily Kegel exercises, which strengthen the pelvic-floor muscles that control your urine flow. Hold for three seconds, then release. Do 10 repetitions three to five times per day, advises Dr. Greenwood. (For a more detailed explanation of these exercises, see "Incontinence" on page 196.)

Weight Control and More

Menopause frequently leads to weight gain or to changes in their figures that many women lament. Exercise is the best defense. Steady, aerobic exercise for 30 minutes a day three or four times a week can keep your muscles toned and your bones strong. And it's easier to stay on a healthful, low-fat diet that keeps your weight down if you're exercising regularly.

Doctors urge women in the menopausal years to select an activity that they like so they'll stick with it. "Energetic activities like brisk walking, swimming, and tennis get your heart rate up and allow your lungs to take in more oxygen," says Dr. Greenwood. "Aerobic exercise both elevates people's moods and helps them prevent weight gain."

Cut the rug. Hit the dance floor with your partner, a neighbor, or a friend. Nudge him into joining a dance club where you can both enjoy an evening of country line dancing, polkas, or whatever is your pleasure. "Dancing is a wonderful exercise and a wonderful way to bring you and your husband together," says Dr. Greenwood.

Sidestep the night sweats. To avoid insomnia, night sweats, and morning-after irritability, Dr. Greenwood recommends brisk walking or some steady pedaling on a stationary bike for at least 30 minutes. Just avoid heavy exercise two hours before you go to sleep. That can throw off your sleep-wake cycle.

Cooling Hot Flashes

Hot flashes are probably the most common complaint during menopause. You're walking along, minding your own business, when suddenly you feel an out-of-nowhere warmth coming over you. Often it starts in your face and your head and spreads quickly to your chest and then all over your body.

What causes hot flashes? The precise mechanism isn't clear. Experts believe that the drop in estrogen levels that comes with menopause compromises the ability of the hypothalamus, the brain's thermostat, to regulate your body temperature. Once your body gets hot, it tries to cool itself off by sweating, and you end up drenched. To top it off, sometimes hot flashes are followed by chills. Try these exercises to handle hot flashes.

Blow out the fire. Deep breathing can help stop a hot flash, says Dr. Greenwood. "When you feel a hot flash coming on—usually there is a little aura feeling, like heat is rising in your body—you can go into deep, slow breathing, and often that can abort the flash," she says. "The key is to take six to eight breaths per minute, five seconds in, five seconds out."

Take a hike. For some women, a vigorous 20-minute stride through the neighborhood park three times a week may help decrease the frequency of hot flashes, says Mona Shangold, M.D., director of the Center for Women's Health and Sports Gynecology in Philadelphia. An Australian study of 220 women over age 40 showed that those who exercised reported having fewer hot flashes and night sweats than those who did not.

Turn off the hot button. When you feel a hot flash coming on, you may be able to halt it in its tracks by using the Chinese therapy called acupressure, says Susan Lark, M.D., director of the Menopause and PMS Self Help Center in Los Altos, California, in her book *Menopause: Self Help Book.*

▶Sit straight in a chair.

▶Wrap both hands around your neck, with your fingers resting on the backs of your upper shoulders and your thumbs on both sides of your neck on top of your collarbone.

▶Using your thumbs, press gently into both sides of your neck for one to three minutes (1).

▶Then press your upper shoulders and shoulder blades gently with your fingers for one to three minutes (2).

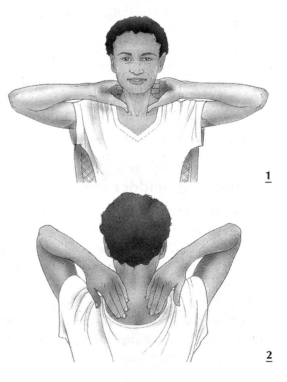

1

2

To bear with menopause, become a yogi. The art of yoga can be a woman's best friend during menopause, says Richard Miller, Ph.D., a clinical psychologist and co-founder of the International Association of Yoga Therapists in Sebastopol, California. Studies have shown that the stretching and relaxation that are central to yoga help with many of the annoyances of menopause, including hot flashes. If you're in—or getting near—the menopause years, consider taking a beginner's class.

Keeping Romance Red-Hot

If you enjoy sex, menopause does not have to curb your desire. Although blood flow to the vagina may decrease during menopause, some women are able to maintain the same orgasm level, says Dr. Greenwood. But the sexual response to menopause is varied, and other women may completely lose interest in sex and channel their energies into other passions like volunteer work. That's okay, too, says Dr. Greenwood.

"It is important to note that some women are glad that they are not thinking about sex all the time and want to do other things with their lives. That is normal, too," she says. "But if you are interested in sex, there is no reason to think that you can't continue once you reach menopause." Her best bedroom tip? "Make sure your partner gives you plenty of time in lovemaking to become aroused and lubricated."

MENSTRUAL DISCOMFORT

Getting Free from "Female Complaints"

Way back in 1876, an entrepreneur named Lydia Pinkham pitched her Vegetable Compound to fight monthly blues. Women came from far and wide to order the ornate glass bottles of dark liquid that promised a "sure cure to female weakness."

Now, more than a hundred years later, we've got shelves full of monthly medicines, from Extra-Strength Midol to Premsyn. But you don't have to resort to pills to relieve period woes. You can put the kibosh on monthly maladies with exercise, stress-reduction methods, and a little feminine ingenuity.

Cures for Killer Cramps

It's no wonder that menstrual cramps hurt so much. These internal spasms occur when hormones called prostaglandins overload your system. They reduce the blood supply to the uterus, and the muscle responds by clenching tightly. In short, cramps are actually similar to a mini-heart attack.

Sound scary? Well, while cramps are definitely no picnic, most of the time they're not much more than a common nuisance. Half of all menstruating women will be bothered by cramps at least once in while, but an unlucky 5 to 10 percent will be incapacitated by their pain.

While you may feel just plain "cursed" at that time of the month, cramps actually happen for some interesting reasons. Since blood and tissue are leaving the uterus through such a tiny opening—the cervix—it's inevitable that there will be some odd sensations going on, says Jill Maura Rabin, M.D., associate professor of obstetrics, gynecology, and women's health at Albert Einstein College of Medicine in the Bronx. "The cervix itself is being stretched a bit," she says, "and the nerves there are being stimulated."

While cranky cramps aren't usually anything to worry about, we'd all be better off without them. Check out these four suggestions.

Walk it off. Thirty minutes of physical activity three to five days a week, all month long, is a good bet for avoiding period pain and boosting health overall. But exercise is also good for on-the-spot treatment, says Dr. Rabin.

When cramps have you immobilized, resist the urge to curl up into a ball. Get up, get out, and get moving for 15 minutes or more. Walking is gentle enough to do right through the pain. And the easy leg- and arm-swinging motion of strolling helps in more ways than one. "Exercise will boost your circulation, get rid of the lactic acid buildup that cramps can leave behind, and relax tense pelvic-floor muscles, " says Dr. Rabin.

Pad your pain. Tampon use may crank up cramp intensity for some women, says Dr. Rabin, possibly because tampons affect the free flow of blood. Try using a pad, at least at the very start of your period.

Go for the O. Orgasm—either alone or with a partner—can bring quick cramp relief. The rhythmic muscle squeezing that happens during climax increases blood flow to the uterus and pelvic area. This counters circulatory congestion and relieves feelings of heaviness and pressure, says Barbara Bartlik, M.D., a researcher in the human sexuality program at the New York Hospital–Cornell Medical Center in New York City.

Snuff the smokes. Tossing the tobacco offers a list of rewards that just keeps getting longer. Two studies suggest that quitting smoking could help put you on the road to becoming cramp-free. In both cases, researchers found that smoker's periods were more troubling in general. According to one of the studies, smokers' cramps lasted longer than those of nonsmokers. While it isn't clear just why this is so, going smoke-free is a no-risk method that might just ease your monthly pain.

Fending Off a Back Attack

As if cramps weren't bad enough, there's another thug to worry about—and this one attacks you from behind. Back pain that accompanies menstruation is due to blood congestion and muscle tension radiating outward from the pelvic region. Relieving the tightness that tends to build up there can work wonders in erasing the pain. These techniques will bring relief to your aching bustle.

Turn up the heat. Moist warmth, like what you'll get from a hot-water bottle, is especially good for relieving lower-back muscle tension, says Dr. Rabin. A warm shower can do the trick, too, especially if you can aim the stream right where it hurts. Or try babying your lower back with a compress. Soak a thick towel in moderately hot (not boiling) water. Wring it out, lie on your tummy, and drape the warm wetness over your lower back and hips. Relax until the towel cools off.

Fix it with flexibility. When pain hits, the natural response from your muscles is to tense up, says Diane Pennock, a certified aerobics instructor at the Marsh Center for Balance and Fitness in Minnetonka, Minnesota. Her three-step stretch routine counteracts a tight lower back and works for everyone (except those with herniated disk problems). Since the ending point of each step is the starting place for the next one, you can move through these stretches in a flowing sequence

The single-knee squeeze. Lie on your back on the floor with a mat or blanket underneath you. To get a little more comfortable, bend one or both knees and keep your feet flat on the floor. Take three slow, deep breaths to relax. On the next inhalation, draw your right knee to your chest. As you exhale, draw your knee closer to your body. Concentrate on your low back area expanding and dropping toward the floor. Relax your knee and repeat two more times—inhaling, exhaling, and drawing your knee to your chest. Lower your right leg and repeat with the left. Move on to the next step.

The double-knee hug 'n' roll. Bend both knees and gently bring them toward your chest. Inhale, then exhale and use your hands to draw your knees a little closer. Picture your spine lengthening and loosening. Then wrap your arms around your knees or thighs and gently rock forward and back along the length

of your spine. Rock back and forth three times. Come to a stop and then lower your legs, placing your feet flat on the floor and keeping your knees bent. Continue to the final step.

The hammock lift and lower. Inhale. As you

Peeling Away the Layers of Pelvic Pain

While discomfort and even some pain aren't uncommon just before and during menstruation, chronic pain throughout the month is another story. Pelvic pain that doesn't disappear with menstruation should be checked out by your doctor.

Chronic pelvic pain is tough to diagnose because there are so many possible causes for it. Endometriosis, urinary tract infections, and even gastrointestinal problems are just a few. Pelvic pain that doesn't quit has even been associated with a past history of sexual abuse.

Even more troubling is that sometimes, despite examinations and tests, doctors can't identify what's causing pelvic pain. And that doesn't make it any easier to bear—in fact, the mystery of your condition may even seem to make the situation worse.

To help you manage pelvic pain for which your doctor has found no medical cause, guided imagery is well worth a try, says Dennis Gersten, M.D., a psychiatrist in private practice in Solana Beach, California, and author of *Are You Getting Enlightened or Losing Your Mind?*

Imagery will help you feel more in control when you're dealing with the pain. And while imagery seems like nothing more than a simple relaxation technique, it has been

shown to have complex effects on the body. Imagery lowers blood pressure, releases muscle tension, and slows heart rate, thus counteracting some of the most troubling effects of chronic pain, explains Dr. Gersten.

To try the imagery exercise that follows, find a quiet spot where you won't be disturbed, get into a comfortable position, and close your eyes before beginning.

Focus first. Focus on your solar plexus, the spot just a few inches below the center of your chest, right where your ribs and stomach meet. Imagine that this place is the center of all of your energy.

Move it down. Let your gathered energy move slowly down until it rests low in your abdomen, in your reproductive organs. As the energy descends, pay attention to any places where the energy feels stuck or blocked.

Make your way in. Imagine a door leading into the blocked area. Picture yourself opening the door and discovering the problem. See yourself walking into the room and repairing what needs to be fixed.

Wash it clean. Next, allow a cloud of bright, white light to wash out your entire pelvic area. Imagine the light cleansing and clearing out any pain or darkness.

exhale, push into the floor with your feet and use your buttocks to lift your hips and pelvis up. Don't rise too high; your waist should stay slightly lower than your hips. Inhale deeply. As you exhale, lower your body, leading with your upper and middle back and finishing with your tailbone. Move slowly, lowering as you exhale. Relax. Then roll onto one side to protect your back as you stand up.

Ease it with imagery. The mind is a powerful tool for defusing pain. Try this three-step imagery exercise and watch your discomfort drip away like water, says Dennis Gersten, M.D., a psychiatrist in private practice in Solana Beach, California, and author of *Are You Getting Enlightened or Losing Your Mind?*

Start by sitting or lying down comfortably in a quiet place where you won't be disturbed for 10 minutes or so.

▶First, focus. Turn your attention to your pain. Picture it as having a certain size, shape, and color. Imagine the texture of your pain. Is it smooth or rough? Is it moving around or staying still? Try to get a clear image of just what your pain looks like.

▶Next, liquefy. Allow your pain to become liquid. It's still the same size, shape, and color as in the first step, but now imagine that it's made of liquid.

▶Finally, roll. Watch your liquid pain roll down the arm or leg closest to it. Let it pick up speed, flowing effortlessly out of your fingers or toes. Then watch as your pain keeps rolling, leaving the room. Maybe you can imagine it rolling into the sea or down a drainpipe.

Now check in with your body. Assuming that your pain was at 100 percent before doing the imagery exercise, see what level it's at now.

Mood Swings: Getting Off the Crabby-Go-Round

Pain isn't the only cause of menstrual discomfort. Irritability, anxiety, and mood swings are also par for the course for some women. When weepiness or fits of anger start filling your days, it's important to counter the trend because mental stress can compound menstrual pain, and vice versa.

Moods and emotions—bad or good—are direct products of our thoughts, says Dr. Gersten. So to rein in negative moods, get to the root of the problem by taking control of your thoughts.

Meditation and visualization are the two best steps to generating peaceful, healthy thoughts. Start with meditation to clear some mental space, then use positive imagery to encourage healthy thoughts to begin to grow and flourish.

You can practice this dual-level approach for 10 minutes two or even three times a day to see improvements in your monthly experience over the long term. But you can also call up a mantra any time you need it for some instant soul-soothing, says Dr. Gersten. "These techniques are perfect to get you through crisis situations."

Start with mind-weeding meditation. Imagery is akin to planting seeds in a garden, says Dr. Gersten. To ensure that the "flowers" you plant will take root and blossom, first you need to clear some mental space with meditation.

Dr. Gersten recommends using a meditation tool called a mantra to help you quiet your mind. Personalize your path to peace by choosing a word or phrase that you find uplifting. This will be your mantra. It's best to select a mantra that is two to nine syllables long, says Dr. Gersten. Some possibilities are *great spirit*, *higher power*, *peace and love*, or your chosen name for God, but it's really up to you.

Close your eyes and repeat the mantra of your choice silently to yourself for five minutes at a time. Let the sound and the sentiment of your word or phrase fill your mind, drowning out any stressful thoughts.

Sow the soothing seeds of imagery. After meditating on your mantra, take another five minutes to establish a healing image in your mind through imagery, says Dr. Gersten. An exercise called the wave can teach you to relax in just one minute.

Close your eyes. Imagine yourself lying on the seashore, cradled in warm, soft sand. Listen to the waves as they roll in, moving farther and farther up the beach, closer and closer to your body. Feel the waves start to gently wash over you. As the water touches you and then rolls back to the sea, imagine that all of your tension and anxiety are being pulled into the waves and washed away. As each wave recedes, taking your stress with it, you feel a little more relaxed.

You can practice this exercise for as much or as little time as you have. And you don't have to wait for the monthly blues before letting the waves tame your tension.

MIDLIFE CRISIS

Searching for Satisfaction

Spanish explorer Ponce de Leon searched the New World for a rumored Fountain of Youth. He never found it.

Centuries later, Baby Boomers are discovering a new Fountain of Youth: the middle years of their lives. It's a time to treat the forties and fifties more like a midlife opportunity than a midlife crisis.

We are well-aware of the danger signs of midlife crisis: By age 50, some of us lament never achieving our goals to be a CEO or climb Mount Everest. We fret that our new boss is 15 years our junior. We realize that the daughter whose first baby steps we remember so vividly is suddenly graduating from college. We spend a pleasant Sunday afternoon weeding and hoeing in the garden and face the possibility of waking up to the unwelcome surprise of aching muscles the next day.

And then it hits us—"Hey, I'm about halfway through this thing called life."

"Recognizing that half of your life is over can be a jolting, frightening thing," says Peter Halperin, M.D., clinical instructor of psychiatry at Harvard Medical School and staff psychiatrist at Massachusetts General Hospital in Boston. Or it can be a calling card for you to sort through the real choices and options in your life and make this a midlife opportunity time.

Reconnect with Your Past

Discard the message from Barbra Streisand's "The Way We Were" and view your forties and fifties more like Bruce Springsteen's "Better Days," in which he proclaims, "There's better days shining through."

The key to midlife is to acknowledge your youthful past but also to rejoice in what's ahead.

"Relax and be comfortable with who you are now and don't try to re-create who you were," says midlife expert Elaine Wethington, Ph.D., associate professor of sociology at Cornell University in Ithaca, New York. "You get a chance to have fun doing something good for your health." Here are some ways to make midlife memorable.

Slide back home. Maybe as a 21-year-old, you ruled the local softball diamond, and you have good memories of the cheers and pats on the back you received when you crossed home plate with a winning run. Even if 20 or so years have passed since a leather glove adorned your left hand, why not return to the field of your dreams?

This time, however, play on an over-40 league where playing—not winning—is the name of the game. Shift your focus from high batting averages and double plays to the simple pleasures of smelling the outfield grass, hearing the dugout chatter, and noticing a reddish orange sunset beyond centerfield.

If softball isn't your game, seek a senior league for volleyball, tennis, or golf and play for enjoyment, not for victory, suggests Dr. Wethington.

Flock back to nature. You fondly recall those early Sunday mornings spent with Uncle Joe, tiptoeing a wooded trail in hopes of spotting a shy blue jay perched on a high branch of an oak tree. Uncle Joe would patiently explain what markings to look for and offer delightful bird gossip.

Don't worry that decades have passed since your last bird walk. Just grab your binoculars, a brimmed hat, and a well-thumbed bird guide and resume this past passion on Sunday mornings. Take a hike in a local woods or state park, suggests Dr. Wethington. Recapture the joy you felt as a child as you spotted a busy robin feeding her brood.

Reconnect with Your Community

One great way to take the crisis out of midlife is to offer your time to others in your community. "The best cure for feeling sorry for yourself because you're getting older is to help someone less fortunate than you," says Dr. Wethington. "Helping others helps you put your own problems in perspective."

Striding through Midlife

Most days, Victoria Preuss of West Palm Beach tackles daily deadlines as an editorial department editor for a major South Florida newspaper. At night, she finds fine places to dine. On weekends, she cheers the Florida Marlins or renovates her 1920s home.

But at age 45, she opted to shake up her comfortable routine by accepting a challenge made by a co-worker: complete a 26.2-mile marathon. The last time she had run more than a block was 10 years earlier, but partnered with another running novice from work, Preuss trained for 11 months for a marathon in Kiawah Island, South Carolina.

Among a field of 900 runners, Preuss finished dead last. She was 30 minutes behind a 73-year-old grandmother and crossed the finish line only moments before race officials started rolling up the tape and taking down the banners.

A comedown? Not at all. For Preuss, the real prize came when she took that last step across the finish line, she says. She was cheered on by veteran runners who had showered, eaten, and then returned to the race site to give a supportive yell for final stragglers.

"When I crossed the finish line, I felt great," says Preuss, who did the marathon in six hours, 31 minutes. "I told myself that I entered to finish, not to compete like an athlete."

The marathon now serves as a reminder to Preuss that anything is possible in her middle years. "Whenever I happen to be at a low point, I just think of that marathon and know I can face other challenges," she says.

In fact, any way that you contribute to the public good has its own rewards. "You have more to bring to the community than you ever did because of your years of accumulated knowledge and wisdom gained from past struggles," says Dr. Halperin. Here are some specific ways that you can share and help others.

Reclaim the area. Organize your neighbors or co-workers and conduct a beach cleanup in your community. There's no water nearby? You can volunteer to help out with a park, trail, or wildlife area. A few hours of filling garbage bags with trash can deliver sweaty satisfaction when you create a litter-free park or shoreline, says Dr. Wethington.

Make habitat a habit. Devote one or two Saturdays a month as a volunteer member of a Habitat for Humanity building crew in your community. You'll learn new skills, such as hammering roof shingles and installing Sheetrock. You'll widen your circle of friends who share a passion for helping others. And you'll get the chance to work side by side with the family who will occupy the home, says Dr. Halperin.

Run for a cause. Certainly, your heart goes out to people with breast cancer, multiple sclerosis, or AIDS—and if you're a runner, your feet can get in the act as well. To rejuvenate your midyears, doctors recommend participating in charity runs. If that 5-K race for a good cause is more than you can run, walk it instead.

Become a soccer mom. Or soccer dad. Leave the comfort of the bleachers and coach a Bumble Bees team of eight-year-olds. Whether or not your own child is on a sports team, recreation leagues are always looking for coaches and helpers. This is a chance to pass on your skills to a young, eager generation and bring you a feeling of satisfaction in helping others, says Dr. Halperin.

Be a prof. If you don't know kicks about soccer, maybe you can tutor one night a week. Many communities have programs to help non-English-speaking residents learn the language, and there are literacy programs that constantly need volunteers. Check out the community needs at your local library—and as long as you're there, ask whether they need someone to read fairy tales to youngsters on Saturday mornings.

Reconnect with Yourself

If you find yourself sinking deep into a midlife rut, there are many ways to leap free of old habits. "Enriching and expanding yourself by trying new activities keeps your mind sharp and your attitude healthy," says Dr. Halperin. Here are some suggestions for trial runs that might turn into new pursuits.

Yield to yoga. If you've always liked physical exercise, maybe you're wedded to an activity of choice—say, weight lifting, jogging, or cycling. Could it be that a certain amount of boredom has crept in as you've pursued your favorite "bread-and-potatoes" exercise over the years? Consider celebrating your next birthday with some new pursuit that literally stretches your horizons. If you haven't done yoga before, for instance, the time may be nigh to give it a try.

"Yoga is great for the aging body. It preserves flexibility," says Dr. Wethington. You can check with your local YMCA for classes in your area.

Get teed off. As a 20-year-old, maybe the notion of lugging a bag of metal-handled clubs up and down a manicured landscape seemed b-o-r-i-n-g. A few decades later, golf may be just the game to cure midlife blues, says Dr. Wethington, who took up the game a few years ago.

But if you follow Dr. Wethington's cue, you might also find ways to customize the game to your liking. On a long par-5 course, her husband, David, starts each hole at the tee and booms a shot down the fairway. Dr. Wethington bypasses the tee, drops her ball about 100 yards from the hole, and plays a chip-and-putt-style game. She doesn't keep score, but she says she enjoys being outside and working on her short game.

"I don't have the upper-body strength to swing like Tiger Woods, but I do have enough athletic talent to play the short game," she says. "I've discovered golf to be a great game."

Try aquatic aerobics. Perhaps you're no longer motivated to sweat on rubber mats in gym aerobics. Dr. Wethington suggests that you climb out of your midlife rut by trying water aerobics. The movements are easier on your joints and help you keep in shape at any age. "I'm struck by the number of women in their sixties and seven-ties who do pool aerobics," says Dr. Wethington.

Leap into new arenas. With the kids away at college and your job skills honed, you may be ready to try other challenges offered at area adult education centers. This could be your golden opportunity to learn French, play an acoustic guitar, hang wallpaper, or sing beyond the confines of your shower walls.

"We're in a culture where education has become competitive and goal-oriented," says Dr. Halperin. "But pure learning for the sake of learning can be very rewarding physically and psychologically."

Never pictured yourself singing the lead in your town's community theater production of *Cats*? Well, just remind yourself that "Dirty Harry" himself —Clint Eastwood—delivered a jazzy, raspy rendition of "Accentuate the Positive" for the soundtrack recording from the movie *Midnight in the Garden of Good and Evil*. So go ahead—make your day.

MOTION SICKNESS

Soothing Moves

As captain of the *Cypre Prince*, a fishing boat that plies the choppy waters off British Columbia, Don Nohr knows a thing or two about motion sickness: First, you can prevent it, and second, if you get it, a few key moves will help you give it the slip.

"It's really not a big problem if you know what to do," Nohr says. Motion sickness—with its hallmark headache, sweating, and nausea—usually makes its unwanted appearances on boats and planes and in automobiles. It can show up whenever your senses bombard your brain with the message that you're moving somewhat erratically.

Motion sickness most often occurs whenever your brain gets different messages from different senses. Say that you're sitting in the backseat of a car with your gaze fixed just outside the window, watching the telephone poles flashing by. Your eyes tell your brain that you're moving fast. But since you're not being jostled around much inside the car, your inner ear, which keeps track of how your body moves in space, tells your brain that you're more or less immobile. The conflicting messages touch off motion sickness, explains Kenneth L. Koch, M.D., professor of medicine at the Pennsylvania State University Hershey Medical Center.

Your brain interprets the mixed messages as a danger signal. "The nausea, sweating, and headaches are your body's way of telling you to get out of there," says Dr. Koch.

Stop the Spinning

A few over-the-counter drugs, such as dimenhydrinate (Dramamine) and meclizine hydrochloride (Bonine), will help quell motion sickness when you can't escape, explains Dr. Koch. So will these key moves.

Press point P6. Pressing a particular point on your wrist—1½ inches below the base of your thumb and just slightly closer to your thumb than dead center—can prevent and ease motion sickness, says Dr. Koch, who has studied the technique.

According to the tenets of Traditional Chinese Medicine, needling and pressing acupoints (there are hundreds all over the body) helps relieve various health problems. Different points help with different health problems, and the point on your wrist, known as P6, is the one to press for nausea relief. In a study at Humboldt State University in Arcata, California, selfless volunteers who had their P6 points pressed while they were being spun around in a large revolving drum were less likely to suffer motion sickness than those who didn't.

Press the point firmly with your thumb or fingers and hold until symptoms lift, suggests Dr. Koch.

Wrap your wrists. A specially designed bracelet may help fight motion malaise. A product called Sea-Band, a bracelet equipped with small beads that press on just the right spot, has been shown to help alleviate seasickness, says Dr. Koch. It's also effective against carsickness, airsickness, and morning sickness.

Hit the deck. If you're feeling queasy aboard ship, it will help to get out of the cabin and up on deck, says Nohr.

The cabin is the wrong place to be because you can't see where you're going down there. In the cabin, your eyes tell your brain, "We're standing still," while your inner ear, registering your body's up-and-down, wave-induced movement, tells your brain, "We're moving." That kind of communication problem triggers motion sickness, explains Dr. Koch. So get back up on deck so that the messages your brain gets from your eyes jibe with what it's sensing from your inner ear.

By the same token, if you're traveling by car over somewhat bumpy terrain (the kind that signals your inner ear that your body's in motion), moving from the backseat to the front and looking out the window can help alleviate carsickness.

Once you're up on deck or in the front seat, you should also remember the following tips.

Look to the future. Looking at the horizon when you're aboard ship or focusing a considerable distance down the road when you're traveling in the car is another key anti-motion-sickness move, says Dr. Koch.

When you focus on things nearby, such as the waves dashing against the stern or the telephone poles whipping by, your eyes are inundated with evidence that you're moving fast, so they bombard your brain with "We're moving" messages. If you stare at the relatively unchanging horizon, your eyes will pick up fewer "We're moving" signs and transmit fewer signals to your brain.

Ease up. Tension makes motion sickness, and other ailments, worse, says Dr. Koch. To make everything easier, try these relaxation exercises before you board a plane or sloop or if you get motion sickness once you're aloft or at sea.

▶Use deep breathing to calm you and ease your symptoms. To do it, inhale slowly, drawing air deep into your lungs so that your stomach pouches out with each inhalation. Exhale slowly and completely, so your stomach flattens as you let the breath out. Repeat indefinitely.

▶Tense, then relax your muscles systematically, starting with your feet and moving up to your head. Tense your feet, hold momentarily, then relax. Then tense your calf muscles, hold for a couple of seconds, then relax. Work all the way up to your forehead muscles. By the time you get that far, you should be feeling far more at ease and noticeably less queasy.

MUSCLE CRAMPS

Relief for the Thighs (and Backs and Shoulders) That Bind

Who gets muscle cramps? Anyone who overworks a muscle. Too many hours raking leaves or weeding the flower bed—even marathon hours spent pressing the gas pedal on a long drive to the vacation cottage—can surface hours later as clenching muscle contractions. Although legs are the most common targets, cramps also stalk our feet, thighs, backs, shoulders, and necks.

"Any muscle that's used repeatedly is likely to get fatigued and more likely to cramp," explains John Cianca, M.D., assistant professor of physical medicine and rehabilitation and director of the sports and human performance medicine program at Baylor College of Medicine in Houston. "It just so happens that since we are on our feet all day long, the calf muscles are usually the first to go."

Any muscle that is overstretched or strained can react by cramping. Or you may get that reaction if you hit a muscle directly—by banging your thigh into the corner of the kitchen table, for instance. Sudden changes in temperature can also trigger cramps, says Dr. Cianca.

Although cramps can be plenty painful, they usually subside within a few minutes. To deal with them, you may want to begin with ice and massage, but the full complement of treatment and prevention strategies includes stretching and strengthening exercises, Dr. Cianca explains.

Cramps That Go Ouch in the Night

Almost as scary as nightmares, painful calf cramps jolt many of us innocent and unsuspecting souls from a deep sleep, say doctors.

These nocturnal nemeses can strike anyone at any age, but we're more likely to get nighttime cramps as we get older, according to experts.

Here are some quick remedies provided by Dr. Cianca to reduce the pain of a calf cramp and return you to a sound sleep. These techniques also work for daytime calf cramps.

Take a stand. What sounds simple is also effective. When a nighttime cramp sets your calf throbbing, get up and stand tall. Raise your hands straight over your head and hold this pose for at least 10 seconds. This mini-stretch reverses the direction of the muscle contraction.

Grab a towel or a T-shirt. If leg cramps have wakened you before, be as prepared as the Boy Scouts for the next assault. Keep a bath towel, a T-shirt, or a piece of rope within reach of your bedside.

1

When a muscle cramp hits at 3:00 A.M., loop the cloth or rope around the arch of your foot. You can lie on your back or your side, but keep the cramped leg straight and bend your other knee slightly. Slowly pull both ends of the towel or rope toward your chest, tugging the top of your foot toward your shinbone (1). Hold that stretch for 20 to 30 seconds and repeat a few times until the cramp vanishes into the night.

Massage the muscle. One effective cramp buster is a five-minute massage on the calf muscle. "Work the calf muscle up and down with the heel of your hand for about five minutes or until it feels good," says Dr. Cianca.

Take a slow stroll. You can also try to walk off the calf cramp. With each step forward, gently stretch your toes up toward your shin to lengthen the cramped muscle.

Some Anti-Cramp Exercises

Here are a few exercises to head off cramps before they have you moaning in pain.

Pedal off the pain. A good cramp fighter is that trusty stationary bicycle that you keep in your bedroom or recreation room, says Leonard A. Levy, D.P.M., professor of podiatric medicine and past president of the California College of Podiatric Medicine in San Francisco.

Cautions about Muscle Cramps

If you are starting to regularly experience muscle cramps when you're 5 to 20 minutes into a workout, you may want to check with your doctor.

This type of muscle cramp could be an early sign of an enzyme deficiency that can ultimately cause kidney failure, says Yadollah Harati, M.D., chief of neurology at the Houston Veterans Affairs Medical Center and professor of neurology at Baylor College of Medicine in Houston. The muscle may not be converting glycogen, the sugar that's stored in muscles, into lactate, the sugar that's needed for strenuous muscle activity. Ignoring the condition can lead to muscle breakdown and severe kidney problems, he adds. Another alarm signal is dark red urine, says Dr. Harati.

Although these enzyme deficiencies are rare, Dr. Harati says, they are serious. With prompt medical attention, however, you can prevent the condition from becoming life-threatening. "The sooner people realize the seriousness of muscle cramps during exercise, the better," he adds,

It is normal to have a few cramps here and there if you are new to exercise. The kind of cramps that signal an enzyme deficiency come frequently, beginning shortly after you start exercising. They can cause lingering muscle aches and pain after you stop exercising, Dr. Harati says.

Five to 10 minutes of steady pedaling at the lowest resistance setting may improve the circulation to the calf area and untie that muscle knot. Adding a regular cycling routine—even 15 minutes' worth—to your day may also help keep future cramps away, advises Dr. Levy.

Soak first. Treat yourself to a 10-minute relaxing soak in a warm tub before heading for bed. It will soothe your mind and calm your calves as well, says Dan Hamner, M.D., a physiatrist and sports medicine specialist in New York City. As added insurance, try doing foot circles in the tub. While you're still seated in a comfortable position, pivot your ankles. Do 10 circles clockwise and then 10 counterclockwise with each foot. The warm water will increase blood flow and help the muscles relax, he explains.

Relieving Tight Feet

Second only to the calves, your toes and feet are other hot spots for muscle cramps. Doctors offer these cramp calmers.

Roll in the relief. When a cramp puts your foot in a spasm, hobble over to the kitchen for relief. Kick off your shoe and grab a wooden rolling pin from the drawer. (An empty soda or wine bottle, tennis ball, or golf ball will also do the trick.)

Plant the arch of your cramping foot on top of the rolling pin and hold onto a chair or table for stability. Slowly move the bottom of your foot back and forth over the rolling pin for a few minutes. Most of your weight should be on your good foot, but try putting some weight on the painful foot. "You're helping to get the muscles more relaxed," says Dr. Levy.

Let your sheets breathe. Mom may have been a stickler about tucking your bedsheets tightly, but loose sheets are actually healthier for your toes and legs, say doctors. Dr. Hamner explains that sleeping on your back under tightly tucked sheets can leave your toes flexed toward the tops of your feet. You may get cramps in the arches of your feet while sleeping in this position, he says. So pull the top sheet out from under the mattress and give yourself some slack before you go to sleep.

Do some lifts. Try the following two exercises the next time your toes or feet get crabby, says Dr. Hamner.

First, stand so the weight of your body is on your heels and your toes are slightly off the floor, supporting yourself by holding onto a chair or table. Maintain that pose for at least five seconds, feeling the stretch in your ankles and calves. Lower your toes until they're flat on the floor, rest for a few seconds, then repeat the toe lift.

Next, lift your heels off the floor while your toes stay planted. Hold for five seconds, then relax for a few seconds.

Reining In Charley Horses

A charley horse is a painful contraction in the thigh muscle. It can last a few seconds or a few minutes and is usually associated with athletic activity, says Dr. Hamner. Here's some help for getting through one.

Try a thigh tamer. To treat a muscle cramp in the hamstring (back of the thigh), Dr. Hamner recommends that you try this stretch.
► Lie flat on your back on the floor.
► If the cramp is in your right leg, raise your leg by placing both hands just above or below the back of your knee and pulling it up, keeping it as straight as possible. Your left leg should be extended straight out on the floor.

2

►Pull your leg toward your chest as far as you comfortably can (2).

►Hold your leg firmly with your hands for at least 20 seconds until the cramp relaxes.

►If necessary, relax for a few seconds, then repeat to work out the cramp.

Harness the hams. For another hamstring stretch to fight a charley horse, Dr. Hamner suggests this exercise.

►Stand straight and clasp your hands behind your back.

►Keeping your head up and your knees and back as straight as possible, bend forward until you feel tightness in your hamstrings, but not pain.

►Hold this stretch for 20 seconds.

►Relax and try four more stretches.

Quell those quad cramps. Dr. Hamner recommends this stretch for cramps in the front of your thigh (the quadriceps muscle).

►Stand sideways next to or facing a wall and place your left hand against it for support.

►Bend your knee, grab your right foot with your right hand, and raise it behind your back, with your knee pointing down.

►Pull your leg back about six inches and tighten your buttocks, then pull back another two to three inches.

►Hold the stretch for 20 seconds.

MUSCLE WEAKNESS

Getting More Flux in Your Flex

Montana sheep and cattle rancher Elizabeth Lillegard put in a strenuous day's work almost every day of her life. Her ranch, in the foothills of the Rockies, had no running water, so Lillegard pumped water from a well and toted it in 10-gallon jugs. She hauled firewood and animal feed on her back, and she herded her livestock on foot. And when the time came to take her animals to market, Lillegard drove them, on foot, to the nearest town—23 miles away.

"My mother was not large but was always strong because of the work she did," says her son, Wade Lillegard, M.D., co-director of the sports medicine division of the department of orthopedics at St. Mary's Duluth Clinic in Minnesota.

While his mother didn't need to work out to keep her muscles strong, most of us do. Our plugged-in, gas-powered culture simply doesn't demand enough from our muscles, says Dr. Lillegard. Because they're underused, they become undersized and weak.

Losing muscle strength isn't just a cosmetic concern. Even if you have no intention of showing off your biceps at the shore, you need strong muscles. Without them, it's harder to heft groceries or to climb flights of stairs. Strong muscles also support your spine, pro-

tect your posture, and keep you looking youthful longer.

Exercise is a bona fide cure for muscle weakness—but not aerobic exercise. While running, walking, and other varieties of aerobic exercise will build endurance, they won't really build much strength, says Dr. Lillegard. To do that, you have to do resistance exercise—working out with free weights or weight machines or those stretchy elastic exercise bands.

Only resistance exercise builds the type of muscle fibers that make you stronger, says Dr. Lillegard. Known as type two or fast-twitch fibers, these are the ones that give you short-term, high-intensity power—the power you need to lift something heavy or give someone a powerful hug.

The Anatomy of Strength

A resistance training program needs to include a steady increase in the challenge level. If you want to build strong muscles, you must work against slightly more resistance than your muscles can easily handle, Dr. Lillegard explains.

"This kind of training actually strengthens muscles by causing microscopic damage to muscle fibers. The fibers respond by getting stronger to protect themselves from future damage," says Dr. Lillegard.

Before you start a resistance exercise program, check with your doctor. If you get the go-ahead, one of the easiest ways to get going is to start working out at home with elastic exercise bands, says Maryellen Bowman, R.N., a certified group exercise instructor and the group exercise director at the Multiplex Fitness Club in Deerfield, Illinois. You can buy packaged sets of bands at most sporting goods stores. Sold under a variety of names, such as Xertube and DynaBand, they cost less than $30 a set.

To get started with the bands, Bowman recommends the exercise routine that follows. Do the exercises in order, since each is followed by an appropriate stretch. Some of the exercises require you to anchor your band. You can loop an end around one of your feet, tie a knot in it and close it in a doorway, hook it to a sturdy ring, or loop it around a well-anchored post. Make sure that the band is securely anchored before you start exercising, says Bowman.

With the following routine, repeated at least three times a week, you'll be able to strengthen your muscles. If you do the entire routine in one session, allow 48 hours or more between workouts, since your muscles need that much time to rebuild after the microdamage of the previous session. If you break up the routine by muscle group, working your legs one day and your arms the next, for example, you need only 24 hours between sessions. Just remember to alternate muscle groups so that your muscles have time to rebuild between sessions.

You can expect to start seeing results—in the shape of more shapely, better-defined muscles—after about six weeks, says Dr. Lillegard. Long before then, though, your muscles will be gathering strength.

Before you pick up a single band, put on some dance music and spend five minutes or so

Daily Routine Resistance Training

You don't have to move to a ranch in the Rockies to get some extra resistance training into your daily routine. You just have to change the way you do a few things around your own ranch and grounds, says Wade Lillegard, M.D., co-director of the sports medicine division of the department of orthopedics at St. Mary's Duluth Clinic in Minnesota. For instance, try:

►Mopping your floor instead of using a buffer
►Carrying your laundry up and down the stairs rather than letting the laundry chute do the work
►Shoveling snow from your sidewalk yourself, by hand
►Using a manual lawn mower instead of one of those ride-on jobs

These amendments to your daily chores will help keep your muscles strong. But they probably won't do the job alone, unless you've got miles of sidewalk to shovel and dozens of acres to mow. You'll still need to do regular resistance training to keep your musculature tough.

cha-cha-ing at a moderate pace, says Bowman. A little dancing will warm up your muscles, which is important because warm muscles are more flexible and less prone to injury. If you're not in the mood for the Merengue, spend five minutes marching and swinging your arms or tread the stairs at a moderate pace, she suggests.

Pain? No Gain

When you do resistance exercises, you should feel a slight burning sensation in your muscles, especially toward the end of a set.

That feeling is the hallmark of what's called muscle fatigue. "If you feel the burn, it means that you're working your muscles sufficiently," says Bowman. "If you don't feel it, you don't have enough resistance."

That said, there's a difference between the mild burning sensation that you may feel after several repetitions of an exercise and sharp or stabbing pain. Typically, "good" pain will be bilateral, says Dr. Lillegard. If you feel pain only on one side, it may indicate an injury.

If you're in pain, you're doing something wrong, and you're bound to hurt yourself, warns Bowman. Since injuries can bench your training program for weeks, there's no long-term gain in outright pain.

If you feel pain while you're doing a strength-training exercise, stop. Check to make sure you're doing it properly, she says. If you're following directions and are still in pain, talk to a certified trainer at the local Y or gym or an adult education exercise class, Bowman suggests. A trainer should be able to recommend an alternative resistance exercise that doesn't hurt.

If you experience unilateral pain (pain on one side) or joint pain that does not improve when you avoid the activity that caused it, and you've tried using ice and anti-inflammatory medications for one to two weeks, see a doctor, advises Dr. Lillegard. You should also see your doctor if you have pain that becomes more severe or interferes with your daily activities.

The Repetition Plan

Begin with one or two sets of 10 to 15 repetitions of each exercise that follows and progress to two or three sets after a few weeks, says Dr. Lillegard. Each successive repetition should require just a bit more effort, and the 15th should be very hard. If you can zip through all 15 reps without the least struggle, you need more resistance. (To increase resistance, use a thicker band—most sets come with a range of sizes—or double up by using two bands at a time.) If you can't get through 10 repetitions, you need to lower the resistance, says Bowman.

Over time, as your muscles get stronger, you'll find the regimen easier and easier to do, and you'll need to progressively increase the resistance, says Dr. Lillegard. Each time you do, you'll build more strength. But you don't have to increase indefinitely.

"Once you've reached the point where you're satisfied with your strength, you can maintain it by sticking with the number of reps and the resistance you're used to," he says.

Here are the exercises, by muscle group.

Chest: Anchor one end of an exercise band in a doorway or around a sturdy pole at chest height and grasp the free end of the band. Then move sideways away from the door or pole until your arm is almost fully extended, with your elbow slightly bent (*1a*). Plant your

feet on the floor, shoulder-width apart.

With your arm still bent slightly at the elbow, move it to the front of your body at chest height, stretching the band as you go (*1b*). Slowly return to your original position, relaxing the tension in the band. Then repeat the process, holding the band in your other hand.

Stretch. Still standing, drop the band and hold your arms straight out in front of you with your hands just slightly lower than your shoulders. Moving your arms sideways as if you were swimming the breast stroke, bring your arms behind you as far as you can and push your chest forward. You should feel a stretch in your shoulders and chest. Hold for 10 to 30 seconds, then relax.

Back: Tie the center of one exercise band around a sturdy pole at shoulder level so that both ends are free. Holding one end of the band in each hand, back up until your hands are together in front of you. Continue backing away until the bands become taut (*2a*). Slowly open your arms as though opening a pair of

French doors, stretching the bands, then bring your hands back to a position just in front your shoulders (*2b*).

Stretch. Still standing, raise your right arm and move it across the front of your body. Cross your left forearm over your right forearm below the elbow and use it to gently pull your right arm farther across your chest. Hold for 10 to 30 seconds. Repeat, stretching your left arm with your right (for an illustration of this stretch, see page 382).

Shoulders: Hold one end of a band in each hand. Place the center of the band on the floor and step on it with one foot to anchor it. Place your other foot just behind the one anchoring the band.

Start with your arms straight down at your sides and your palms facing your thighs (*3a*). Slowly raise your arms in front of you so that they form a "V." Widen the distance between your hands as you raise your arms. Stop lifting when your hands are at shoulder level and

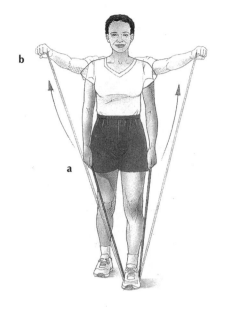

<u>3</u>

<u>4</u>

about three feet apart (*3b*). If you can't raise your arms all the way to shoulder height, lift them as high as you can and gradually work your way up to shoulder height. (If you don't feel resistance at this level, use a thicker band or stand with both feet on the band.) Then slowly lower your arms to the starting position.

Stretch. Standing with your arms at your sides and your palms facing backward, reach behind your back and grasp your left wrist with your right hand. Bend your left elbow slightly, then use your right hand to pull your left arm to the right until you feel a stretch in the front of your left shoulder. Hold for 10 to 30 seconds. Switch positions and stretch your right arm.

Arms: To develop your biceps, the important muscles located on the tops of your upper arms, tie the center of a band to a pole or sturdy ring at shoulder level. Hold one end of the band in each hand. Facing the pole or ring with your arms straight in front of you and your palms facing up, step back until the band is taut (*4a*).

Then slowly bring your arms up, bending your elbows to stretch the band. Pull your forearms as close to your chest as you can (*4b*), then slowly return to the starting position.

Stretch. Repeat the steps for the chest stretch, above. Then, with your arms straight behind your back, rotate your hands several times back and forth so your palms face up, then down, then up. Hold each new position for 10 to 30 seconds. You should feel a stretch in your biceps each time you rotate your hands.

To work the triceps, the muscles in the backs of your upper arms, fasten the center of a band to a pole or ring at chest level. Facing the pole or ring, take one end of the band in each hand. Keep your arms close to your sides and bend your elbows so that they form right angles. Back up until the band is taut, then move your elbows back so they're just beyond your back.

Position your feet so that your left foot is just behind your right. Keeping your back straight and your shoulders in their normal position

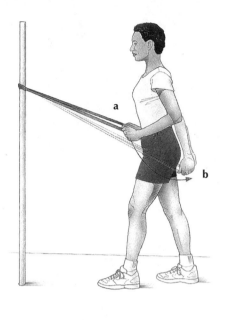

<u>5</u>

<u>6</u>

(not hunched up), bend forward slightly so that your torso tilts forward a bit at the hips (*5a*). Slowly straighten your arms at the elbows until your hands are behind your back (*5b*) You should be working against tension in the band.

Stretch. While standing, raise your right arm straight over your head as though you were reaching for something directly above you on the ceiling. Bend your right elbow so that your right hand touches the back of your neck. With your left hand, grasp your right elbow and gently pull it to the left. You should feel the stretch in the back of your upper arm. Hold for 10 to 30 seconds. Repeat with the other arm.

Legs: To tone the muscles in your outer thighs, make a loop with one end of a band and slip it around your left ankle. Position your feet about six inches apart and step on the band with your right foot, pulling out the extra band so it's held taut by the weight of your right foot (*6a*).

Slowly extend your left leg to the side, pulling against the resistance of the band (*6b*). Then slowly return to the starting position. Repeat with the other leg. (If you have a hard time keeping your balance, stand near a wall and press your hand against it for support.)

Stretch. Lying on your back, bend both legs so that you can comfortably rest your feet flat on the floor. Lift your right leg and rest your ankle on your left thigh, with your right knee bent to the side. Grasp your left thigh and pull your leg back toward your chest. You should feel a stretch in your outer right thigh. Hold for 10 to 30 seconds, then change positions and stretch the opposite leg.

To work your inner thighs, make a loop with each end of a short band. Slip one loop around your right ankle and the other around your left foot. There should be about six inches of band between your feet.

Lie on the floor on your right side with your left leg on top of your right, and prop your upper body on your right forearm. Bend

7

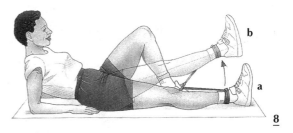

8

your left leg at the knee so that your leg makes a 90-degree angle, and put your left foot flat on the floor behind your right leg (7a). Slowly raise your right leg, working against the resistance of the band, until your leg is about a foot off the floor and the band has stretched to about 18 inches long (7b). Slowly lower your leg, then change positions and work the opposite side.

Stretch. Lying on your back, raise both legs so that they're perpendicular to the floor. Then move your feet apart slowly so your legs form the letter "V." Put your hands on the insides of your knees and gently push your legs farther apart until you feel a stretch in your inner thighs. Hold for 10 to 30 seconds.

To strengthen the muscles in the fronts of your thighs (quadriceps), start by sitting on the floor with your knees bent and your feet flat on the floor in front of you. Make a loop in each end of a band. Slip one loop under the arch of your left foot and put the other loop around your right ankle. Then bend your arms and lean back so that you're resting your weight on your forearms. Slowly straighten your right leg (8a) and then raise it toward the ceiling, pulling against the resistance of the band (8b). Switch the band and repeat with other leg.

Stretch. Lie flat on the floor, then roll onto your right side and rest your left leg on top of your right leg. Bend your left knee and bring

your ankle up toward your back. With your left hand, pull your left heel toward your buttocks while pushing your hip forward slightly. You should feel the stretch at the front of your thigh. Hold for 10 to 30 seconds, then change position and stretch the other side.

Buttocks: To firm up your gluteal muscles and the backs of your thighs (hamstrings), loop a band around your left ankle and stand on the other end with your right foot. Position your feet so that they're three or four inches apart (9). Keeping your left leg straight, slowly kick

9

it backward, feeling the resistance in the band (*10*). Slowly return to the starting position, then switch legs.

   *Stretch.*Lie flat on your back on the floor with your legs straight. Slowly raise your right leg as close to perpendicular as you can manage without straining. Then place your hands just above or below the knee and pull your leg toward your chest. You should feel the stretch in your buttocks and the back of your thigh. Repeat with both legs. (For an illustration of this stretch, see page 270.)

10

MUSIC

The Healer That You Hear

WHAT IT IS

What makes plants grow more vigorously, lets dentists drill without using anesthetic, and even boosts IQ scores? Music, a traditional healing tool that's never gone out of vogue, does all this and more.

Music fills our lives as a matter of course, starting with our mother's heartbeat and ending with the funeral dirge. Now, researchers are starting to document music's very real abilities to enhance relaxation and increase concentration.

WHAT IT DOES

People use music to alter their brain and body chemistry all the time, even if they don't realize it, says Paul Nolan, director of music therapy education at Allegheny University of the Health Sciences in Philadelphia. Whether you tap your foot in time to a musical rhythm or feel a rush of nostalgia at the soaring strains of a violin, music definitely gets under your skin.

Researchers don't yet understand how music raises spirits, soothes the soul, or makes you want to jump and shout, but they agree that the measurable effects are undeniable. Listening to your favorite music can reduce blood levels of the stress hormone cortisol while in-creasing amounts of painkilling hormones called endorphins. Music can normalize heart rate and blood pressure. And with the right music playing in the background, you might even do better on a quickie intelligence test.

One theory is that musical patterns help align brain waves in a beneficial way, says Don Campbell of Boulder, Colorado, author of *The Mozart Effect*. "Some rhythms of music can have a healthy effect on some rhythms of the body," he says.

The music of classical composer Wolfgang Amadeus Mozart is particularly potent. "Mozart is like the nutritional 'sports bar' of music," says Campbell. "It's very organized and uncomplicated, and it assists in organizing time, space, movement, and language." This simplicity seems to speak to multiple functions of the brain, including emotions, physical movement, and thinking skills.

REAPING THE BENEFITS

Music can be so much more than just a toe-tapping background. "A balanced diet of music can improve the body through exercise, concentration, and emotions," says Campbell. Make it a point to bring more music into your world with these tips.

Pick it up at any time. You're never too old to learn to play music, says Nolan. But you do need the time to practice. If you didn't get the chance to take music lessons when you were a kid, be assured that it's never too late. Right now could be the perfect opportunity to let your musician's soul soar. "Taking lessons will give you more than just musical abilities," says Nolan. "Music classes make great social clubs, too."

Belt it out. The easiest instrument to learn is one that you already know how to use—your own voice. Singing has the least mechanics to master and requires absolutely no special equipment. Join a beginner's choir or take voice lessons to increase your cantata confidence, Nolan suggests.

Singing increases blood flow throughout the body. You will notice that you are more alert and awake after vibrant singing, even while driving your car, adds Nolan. You'll be expressing yourself and filling your blood with health-giving oxygen at the same time.

Become an avid audience member. Scan the local newspaper listings for musical performances near you. An interview survey of 12,000 people at the University of Umea in Sweden suggests that regular concert-goers may live longer than people who are serenaded less often.

TRIAL RUN

Your very own CD, tape, or record collection is a first-aid kit of mood- and health-enhancing "medicines." Selecting the right music for you is an effective way to change attitude, calm down, or learn better, says Elizabeth Miles, an ethnomusicologist (someone who studies music in culture and human life) in Los Angeles and author of the book and CD series

Tune Your Brain: Using Music to Manage Your Mind, Body, and Mood.

Musicologists have identified certain styles of music that have an impact on your mood as well as your physical well-being. Here are Miles's musical recommendations for types of music and specific selections or pieces.

To energize: For popular music, some of the most upbeat categories are rock, hip-hop, samba, steel band, or even bagpipe. Many types of classical music can also be energizing. Miles recommends "Gaite Parisienne" (also known as the can-can) by Jacques Offenbach.

To relax: From Gregorian chants to lullabies, music of many kinds soothes souls the world over. In the category of jazz, Miles suggests *The Great Vocalists of Jazz: The Voice of the Soul*, a Blue Note recording featuring Billie Holiday, Nat King Cole, and others.

To focus: When you're trying to boost your intelligence, music can help you focus. The Mozart piece that has been scientifically shown to boost IQ scores is "Sonata for Two Pianos in D Major, K. 488." This makes great preparation for higher-level reasoning. But complex music can be distracting when you're concentrating on a task such as reading or writing. For this kind of focusing, try slower and simpler background music such as the Adagio from Mozart's "Divertimento" in B flat major.

To uplift: To get your spirits to soar, choose stimulating music with positive lyrics. If you're not familiar with "I Will Survive" by Gloria Gaynor, Miles suggests that you listen to it.

To heal: When you need to heal, it's important to choose music that's both familiar and relaxing. Many favorite vocal or classical selections are in the running, but Miles favors *The Four Seasons* by Antonio Vivaldi.

NEARSIGHTEDNESS

The Far-Reaching Effects of Eye Exercise

Life for the nearsighted among us is coziest up close. We clearly see everything within arm's reach.

Beyond that reach, the world becomes a blurry blob. With myopia (the medical term for nearsightedness), the eyeballs are longer than normal, and the corneas (the transparent coverings in front of the eyes) are too steep. As light rays pass through the cornea, they create an image in front of the retina (the innermost layer of the eyeball) that's fuzzy. As a result, close objects look clear, but faraway objects appear blurred.

Myopic people represent a powerful minority. Slightly more than one in every four Americans is nearsighted. Your family history can set you up for myopia. The chances increase if you're doing too much close-up work like reading or gazing at a computer every day. Stress can be a factor, too, says Meir Schneider, Ph.D., a licensed massage therapist, founder of the Center and School for Self-Healing in San Francisco, and creator of the Meir Schneider Self-Healing Method.

Doctors commonly prescribe eyeglasses or contact lenses to improve the vision of people who are nearsighted, and ophthalmologists can perform surgery to correct the cornea's shape.

In radial keratotomy (RK), the surgeon makes spokelike incisions on the cornea, and in photorefractive keratectomy (PRK), he relies on a laser to sculpt the cornea's surface.

But whatever corrective approach you choose, exercise can play a role, too. Behavioral optometrists (optometrists who have conventional training but also see the value of the mind and body working together in a healthy way) say that people can improve their nearsighted vision naturally through eye-strengthening and eye-relaxing exercises.

"The most important visual component is the brain," says Dr. Schneider. In his opinion, "clear vision happens when the brain allows the vision to be clear."

For more than two decades, Dr. Schneider has helped people shed their glasses through mental and physical exercises, starting with himself. "As a child, I had cataracts, glaucoma, and other problems," he says. "After five unsuccessful surgeries, I was declared legally blind." He could read only Braille until age 16, when he decided to do specific eye exercises to help his vision.

"An instructor taught me the importance of relaxing my eyes," says Dr. Schneider, who credits regular exercises for restoring some of his eyesight. "Today, although my eyesight isn't perfect, I read, write, and drive. I can see through tiny pinholes in the scar tissue of my eyes. I am proof that our eyes can improve if you give them the right conditions."

Dr. Schneider and some behavioral optometrists believe that perceptions and emo-tions play major roles in what we see. Robert-Michael Kaplan, O.D., a behavioral optometrist, consultant in vision care in Gibsons, British Co-lumbia, and author of *The Power behind Your Eyes*, says that he experienced double vision whenever he felt stressed, ate refined or fatty foods, failed to get enough sleep, or did not ex-

Goodbye, Fuzzy Mornings

The digital clock radio displayed the time in clear, red, glowing numerals, but when she awoke every morning, Barbara Anan, O.D., had to put her face within inches of the clock to see the time.

Not anymore. The clock is stationed six feet away on the far end of the dresser, and she can read the time from that distance without eyeglasses or contact lenses.

"I was nearsighted since age eight, and I could not see anything clearly more than five inches away," says Dr. Anan, an optometrist in her fifties who has a private practice in Washington, D.C.

Her vision worsened in optometry school from late nights of reading. "I started having trouble remembering what I read," she says. "That would slow me down on tests, and I would make mistakes."

She recalls that when her eyesight was at its worst, she had to climb off the chair in the eye doctor's office and get within 4 feet of the eye chart to read the big "E" on the top line. She can clearly see that "E" now while sitting in the chair, about 10 feet away.

Dr. Anan credits vision therapy activities for playing a major role in improving her eye-sight. She also wore special glasses and spe-cial rigid contact lenses to progressively reshape the curvature of her cornea over time.

Success did not come overnight, however.

Dr. Anan started doing vision therapy at her office when she was in her early forties. She would put chalk in both hands and try to draw circles simultaneously on a black-board—first clockwise and then counter-clockwise—while standing at arm's length from the blackboard.

During the exercise, you try to be aware of the movement of your hands and arms to sense whether they are moving at the same time and whether you are drawing the same size circles with each hand, she explains.

Dr. Anan urges nearsighted people to work with a behavioral optometrist to develop a customized vision-care program that includes office and home activities.

"The program must be doctor-directed to figure out exactly what you need—and everyone is different," she says. "In vision therapy, the cells in the brain get repro-grammed and learn a new, better way to function."

ercise. He says it is critical for people to adopt healthy eating habits, select a career that brings contentment, and regularly relax their eyes. He encourages people to treat themselves to massages, nature walks, or classical music.

With a Few Minutes' Effort, Stronger Eyes

But that's not all. Here are some strategic exercise tactics that could help you if you're nearsighted.

Give your eyes equal time. You wouldn't go a day without combing your hair or brushing your teeth. So why skip daily care of your eyes? asks Dr. Kaplan. He suggests that you close them and relax.

Here's how: Take deep, soothing breaths for at least five uninterrupted minutes. Visualize a gushing waterfall on a sunlit day. Try to see all the colors of that scene in your mind. Then blink and open your eyes slowly.

"Your eyes are not cameras; they are the doorway to your mind," says Dr. Kaplan. "I practice vision therapy every day. My vision and yours are well worth the few extra minutes each day."

Eat without lenses. At mealtime, remove your glasses or contact lenses. "When you remove your glasses, it puts you more into an intuitive, feeling state," says Jacob Liberman, O.D., Ph.D., a Colorado optometrist and president of Universal Light Therapy, a company that researches and manufactures light and color technology for healing, in Carbondale, Colorado. "You will eat more slowly and be more relaxed. What you see may be blurry at first, but after a while your eyes will feel wonderfully relaxed."

Play pencil tag. You can build up your eyes'

Try a New View

If you are nearsighted and want a good workout for your eyes, grasp this book and flip it upside down. Then try to read this sentence, says Meir Schneider, Ph.D., a licensed massage therapist, founder of the Center and School for Self-Healing in San Francisco, and creator of the Meir Schneider Self-Healing Method.

The upside-down text forces you to study the shape of each letter instead of reading the words for meaning. It should also make you more aware of what your eyes are doing when you read. Concentrate on the shapes of the letters—and don't forget to blink.

---

teamwork powers by using a pencil, says Don Teig, O.D., an optometrist and co-director of the Institute for Sports Vision in Ridgefield, Connecticut. Here's how.

▶Pick up a pencil in your right hand and extend that hand in front of you to full arm's length.
▶Slowly move the upright pencil toward the center of your nose.
▶Try to move the pencil as close as you can before you start seeing a double image.
▶Do this a few times each day, and don't forget to blink during the exercise.

The goal of this exercise is to get both eyes working together while keeping the moving pencil in focus at all times.

Follow the flying E. This exercise strengthens your side vision, says Dr. Teig. You will need a piece of cardboard about two inches wide and six inches long. Holding the cardboard sideways like a ruler, draw a capital letter "E" in the center, just below the top of the cardboard. The "E" should be the size of magazine print.

▶Hold the cardboard in your left hand about 12 to 18 inches directly in front of your eyes.

▶Cover your right eye with your right hand.

▶Keep your left hand at eye height and slowly move the cardboard in an arc around your left side.

▶Keep your head still and focus straight ahead as you try to see the "E" in your peripheral vision.

▶Repeat these steps with your left eye covered and the cardboard in your right hand.

Think blink. Blink when you're reading. Blink when you're tapping on a computer keyboard. Blink when you're driving. Blink when you're playing a video game. Blink when you're on a nature walk. The goal behind this simple exercise is to keep your eyes from staring too long at one spot, says Dr. Schneider. Blinking also prevents your eyes from getting too dry, which is also helpful if you want to improve vision. People with poor eyesight tend to blink less than people with normal vision.

Read in nature. This exercise gives you the chance to improve the focusing powers of your eyes and to enjoy the sunny outdoors at the same time, says Dr. Schneider.

First, photocopy the Snellen eye chart on the opposite page. Attach the chart to a tree, fence, or post. If you wear glasses or contact lenses, remove them. Now you're ready to begin.

▶Stand close enough to the eye chart so that you can easily read the top three or four large-print lines.

▶Read the smallest line that you can see clearly.

▶Interrupt yourself to relax, breathe deeply, and blink.

▶"Draw" the letters with your eyes by following the outline of each letter.

▶Occasionally raise both hands to your ears and wiggle your fingers to stimulate your side vision.

Meditation on the Color Black

Dark thoughts are not usually something you would aspire to. But meditating on blackness—the color, not the mood—can actually provide a soothing vacation for your eyes.

"You can rest your eyes by pausing for a visual relaxation exercise like this meditation using the color black," says Meir Schneider, Ph.D., a licensed massage therapist, founder of the Center and School for Self-Healing in San Francisco, and creator of the Meir Schneider Self-Healing Method.

Darken the room, play soft music, and find a position in which you can sit quietly and comfortably. Warm your hands by rubbing them together. Close your eyes and gently cover them with your cupped hands; rest the heels of your hands on your cheekbones, and don't press on your eyes. Breathe deeply and slowly. Imagine an ever-deepening blackness—a dark place or object. You will see lots of colors for a while, but stay focused on the color black. Don't try hard—just relax and enjoy the meditation. Dr. Schneider suggests doing regular 15- to 20-minute sessions.

- After reading each letter, close your eyes and picture it. Trace its outline in your mind's eye, then open your eyes and see if it's clearer.
- Look two lines below the previous line. Don't try to read the letters, but treat them like works of art, letting your eyes explore their shape and color. Again, stimulate your side vision with some finger wiggling.
- Interrupt your chart reading with this "sun-ning" or "skying" technique: Close your eyes and lift your head so that you face the sun for a few minutes. Keeping your eyes closed, turn your head from left to right. Prevent frowning or squinting by massaging around your eye as you move your head, without blocking the sunlight from your eyes.
- Blink a few times, then go back to the Snellen chart line that you started with and see if it's clearer. If it is, move to the line below it and again "draw" with your eyes, visualize, and check for increased clarity. If it isn't clear, repeat the previous line.
- Look off into the distance as far as you can. Pick out the smallest detail and move your attention from one detail to another. Stop and picture what you just saw, imagining it clearer. Now look a little nearer, lightly scan the details, and visualize again. Continue until you're looking at details nearby.
- Blink. Now look at the eye chart again and continue to work your way down the chart, line by line. Accept the blurs and enjoy the details. Continue to treat the letters like works of art, and remember to breathe, blink, and wiggle your fingers from time to time. With practice, you may be able to increase the clarity and read smaller lines of type.

NECK PAIN

Nix That Crick

Imagine trying to balance a 15-pound bowling bowl on your fingertips all day long. That's essentially what the seven bones and dozens of ligaments, muscles, and nerves in the neck have to do to properly perch and pivot your bowling—er, your head.

When our neck bones, shock-absorbing disks, and muscles are flexible and strong, the head generally moves painlessly in all directions. But necks are under a lot of pressure. They're often undermined by tension, muscle spasms, arthritis, loss of bone mass, and even poor posture. The phrase "a pain in the neck" became a cliché because neckache is both common and debilitating.

Fortunately, low-sweat exercises can help prevent and ease many types of chronic neckaches, say doctors.

Experts suggest two types of exercise that you might want to try: stretching exercises to improve the range of motion in your neck and strengthening exercises to help your muscles hold your head up proudly.

A word of caution: Your neck can be delicate. Doctors say that you should stop exercising immediately and see your physician if you experience numbness or sharp, intensifying pain during or after exercise.

Stretching for Full Mobility

A healthy neck is flexible enough to bend and twist easily in six directions: forward, backward, left, right, down toward the left shoulder, and down toward the right, says Kim Fagan, M.D., a sports medicine physician at the Alabama Sports Medicine and Orthopedic Center in Birmingham. Here are some easy stretching exercises that will give you good range of neck motion and relieve tension-induced pain. They can be done either sitting or standing. Dr. Fagan recommends moving in slow, controlled motions to get the most out of these stretches. She cautions not to stretch to the point of pain.

Head up and down. Start with your head erect, facing an imaginary spot on the wall at eye level. Slowly tilt your head forward until your chin touches your chest, as if you were about to nod off during a boring movie. Hold that stretch for 5 to 10 seconds. Then slowly tilt your head backward as far as you comfortably can without feeling pain. Hold that tilt for 5 to 10 seconds. Think of this action as a slow-motion nod of approval, says Dr. Fagan.

Do these head tilts twice a day, in the morning and evening. Start with 5 tilts and gradually build up to 15 per session.

Try the slow no-no. Start with your head

facing forward, then slowly rotate it to the left as far as you can. Hold that pose for 5 to 10 seconds. Relax, then slowly turn your head to the right as far as you can. Hold for 5 to 10 seconds, then relax. Try to start with 5 rotations and gradually build up to 15 twice a day. As your neck muscles warm up, you should be able to turn farther to the left and right, says Dr. Fagan. This stretch not only helps relieve neck pain but also improves flexibility, making it easier to do things such as looking over your shoulder when you're backing out of the driveway.

Be a head rocker. Facing front, slowly tilt your head to the left, bringing your left ear toward your left shoulder. You should feel a warm stretch in your right neck muscles. Hold that stretch for 5 to 10 seconds, then relax and slowly tilt your head back to the right, bringing your right ear toward your right shoulder. Hold for 5 to 10 seconds. Start with 5 of these complete head tilts and build up to 15 twice a day, recommends Dr. Fagan.

"You will actually feel a stretch, and as the muscles warm up, you should be able to get a better stretch with each one," says Dr. Fagan. "These stretches can help you deal with muscle pain and spasms." This stretch also helps relieve pain that comes from cradling a phone too long, she adds.

Face the clock. If you make time for neck stretches, activities such as working at a computer or reading a book should be less painful, says Thomas Meade, M.D., orthopedic surgeon and medical director of the Allentown Sports Medicine and Human Performance Center in Pennsylvania. Here's a stretch that should help, he says.

Lie on your back with your knees bent and your feet flat on the floor. Imagine that you're looking upward into the face of a large clock. Using your nose as a pointer, try to move your head as if you were following the hour hand, clockwise from 1 o'clock to 12 o'clock. Then move your head counterclockwise with the

Say No-No to Neck Pain: A Stance of Prevention

The first step in avoiding neck pain is proper posture. We have to avoid that head-forward hunch that we favor when facing our computers or scrunched-up behind the wheel.

Holding your head forward often strains muscles and triggers spasms in the neck, says Robert Markison, M.D., a hand surgeon and associate clinical professor of surgery at the University of California, San Francisco. "When we put ourselves in a head-forward, limb-forward posture, we transmit up to three times the usual force through the neck, the spine, and muscles around it," he says. To avoid neck pain, it's important to hold your head back, directly over your shoulders.

How's your posture? Check your alignment: Keep your spine as straight as you comfortably can while keeping your muscles relaxed. Your shoulders should be back, and your head and neck should be upright and directly over your shoulders, not tipped forward in front of your body. (See "Posture Problems" on page 313 for some posture-promoting exercises.)

same motion. Try five slow head circles in each direction. You can also try this standing up.

Neck-Strengthening Exercises

Once you have your neck muscles stretched, it's time to help make them stronger by building them up a bit. The stronger your neck muscles are, the less neck pain you'll feel. "Strengthening exercises can help prevent recurrence of muscle spasms and chronic pain in many cases," says Dr. Fagan. She recommends three "heads-up" exercises for strong, healthy neck muscles.

Rise up. To do this neck toner, lie on your back with your knees bent and your feet flat on the floor. Place a pillow under your head. While keeping your shoulders on the floor, tuck your chin and slowly lift your head toward your chest. Hold that position for 5 to 10 seconds and then lower your head. Try 5 of these at first and gradually increase to 15 twice a day.

Head sideways. Lie on your right side on the floor with your legs straight and your head on a pillow. Lift your head up slowly toward your left shoulder as if you were peacefully waking up from a sound sleep. Hold the lift for 5 to 10 seconds and then gently lower your head back to the pillow. Repeat four more times, then lie on your left side and do five more head lifts. Move slowly for optimum muscle-toning effect, says Dr. Fagan.

Look skyward. The next time you lie on your stomach to read a novel or watch television, take a break and try this one-minute neck muscle builder, says Dr. Fagan.

Lie face-down on a pillow with both hands on top of the pillow and your elbows pointed out. Lift your head straight up, tipping it backward and looking at the ceiling. Use your neck muscles, not your arms. Hold each lift for 5 to 10 seconds. With each successive lift, try to focus on a spot on the ceiling that requires greater range of motion backward. Do 5 lifts and work up to 15 twice a day.

NERVOUS HABITS

A Two-Step Exercise in Quitting

Your next-door neighbor may have trichotillomania. There's no need to lock the windows and doors, though. Trichotillomanics are no menace to society—only to their hairdos. *Trichotillomania* is the psychological term for the nervous habit of pulling out your hair. Like all other behavioral quirks—biting your nails, jiggling your foot, sucking your thumb, or fidgeting with your pen—trichotillomania can be a tough habit to break.

There's plenty of evidence that regular physical exercise lessens the stress that's at the root of any nervous habit. So start with a regular low-key workout of some kind. But beyond some exertion, the key to beating a nervous habit is a two-stage process, according to Ray Miltenberger, Ph.D., professor of psychology at North Dakota State University in Fargo and an expert on nervous habits.

Step one is to change your awareness. Most nervous habits are unconscious acts. Odds are that you're usually unaware that you're beginning to bite your nails, twist your hair, or chew your pencil, says Dr. Miltenberger. Before you can stop doing it, you have to bring the habit into your conscious mind and become acutely aware every time you do it.

Once you've become aware of the habit, you've got to move on to step two, which is to replace it with another, less annoying habit—what is known as a competing response.

Step One: Becoming More Aware

Here are four ways to raise your nervous-habit consciousness.

Hire spies. Ask family and friends to keep an eye on you and blow the whistle every time you twist your hair or bite your nails, suggests Judith Beck, Ph.D., director of the Beck Institute for Cognitive Therapy and Research in Bala Cynwyd, Pennsylvania, and clinical assistant professor of psychology in psychiatry at the University of Pennsylvania School of Medicine in Philadelphia. "Sometimes the act of monitoring alone will help you cut back," says Dr. Beck.

Your friends can simply say, "Nails," or, for secrecy, use an agreed-upon hand gesture when you begin to chew. That way, no one else need know. Of course, you should try to catch yourself in the act as well.

Keep count. Keep track of the number of times your spies bust you or you catch yourself, suggests Dr. Miltenberger. "Make a tally mark on your calendar each time," he suggests. Keeping tabs will not only help you become more aware of what you're doing, it will also give you a way to chart your progress at cutting down. Over time, you should find that you're biting or twisting less often.

Do some play-by-play. Twice a day, spend a few minutes alone during which you deliberately, consciously indulge in your habit, suggests Dr. Miltenberger. Really. Spend the entire time simulating the act of biting your nails and—this is the important part—*analyzing pre-*

Tension Tamers: Managing Your Mind and Body

An effective way to deal with an unruly nervous habit is to cut high anxiety down to size. You can do that with a few simple exercises—both mental and physical—that help you relax. Here are three techniques.

Grill yourself. If the prospect of leaving for work starts you gnawing on your nails, ask yourself, "What's going through my mind right now?" suggests Judith Beck, Ph.D., director of the Beck Institute for Cognitive Therapy and Research in Bala Cynwyd, Pennsylvania, and clinical assistant professor of psychology in psychiatry at the University of Pennsylvania School of Medicine in Philadelphia.

Perhaps you're afraid that the boss will chew you out because your department lost the Wigglesworth account. Once you've identified the thought that's giving you the heebie-jeebies, question it. Ask yourself, "What's the evidence that the boss will be critical of me?" suggests Dr. Beck. Were you responsible for that account, or did you just do the photocopying? If you were responsible, ask yourself, "What's the worst that could happen, and what will I do if it does?"

Now, the worst possible (although unlikely) outcome is that the boss will fire you. That's rough, no doubt about it. But even if the worst comes to pass, you'll figure out a way to handle it, says Dr. Beck. "Remember, you'll survive," she says. While being fired may seem like a catastrophe, it's actually calming to think that you can survive even if the worst happens.

Here are a couple of physical activities that might also help you relax.

Flex your muscles. Exercise can help calm you down when you break out in a nervous sweat, so take a short walk around the block or down the hall, says Dr. Beck.

Tighten, then loosen up. Stuck at your desk, dreading the boss's call? Try progressive relaxation, a calming exercise that involves alternately tensing and relaxing the major muscles in your nerve-wracked body. Start at the bottom and work up, Dr. Beck suggests. Tense the muscles in your feet and hold for a moment, then relax. Then move up to your calves, thighs, buttocks, stomach, hands, arms, shoulders, and neck, and end with the muscles in your furrowed brow. These exercises can be done in any private spot, usually sitting or lying down and with your eyes open or closed.

cisely what you're doing. Watch yourself in the mirror and describe aloud every step you take, Dr. Miltenberger says. You might sound like a sports announcer doing a running commentary like this: "I'm raising my right hand and bringing it up to my mouth. I'm opening my mouth slightly. I'm putting my finger in the corner of my mouth between my incisors. Now I'm biting down."

The idea is to become even more acutely aware of what you're doing, particularly when you're just beginning to do it.

Schedule regular rehearsals. Get rigid about practicing your "indulgence" sessions. Plan to practice at the same times every day—while you're waiting for your hair to dry after your morning shower, for example, and again while you're waiting for dinner to come out of the oven. Otherwise, you may forget, says Dr. Miltenberger. And that's no good, because breaking a habit requires consistency.

Once you become adept at catching yourself in your nervous exploits, you're ready to move on.

Step Two: Making the Switch

The habit-thwarting trick is to come up with another habit that's incompatible with your nervous one. You can't bite your nails and fold your hands in your lap at the same time, right? So folding your hands in your lap is incompatible with nail biting. In psychological lingo, it's a competing response. The best competing responses are unobtrusive, so they travel well. (Eating taco chips is also a competing response, but it's frowned upon at most business meetings.)

Try folding your arms across your chest, or carry one of those sand-filled "stress balls" in your purse or pocket and squeeze it whenever you feel like chewing. Develop any other habit that makes it impossible to indulge in the one you're trying to break.

Practice, practice, practice. Once you've decided on a competing response, add it to your twice-daily practice sessions. Start each session the way you always have. Begin to engage in your nervous habit and then (this is the new part) switch to your competing response. As soon as you start to bring your finger up to your mouth for a chew, for example, drop your hands into your lap and fold them there. Do play-by-play on what you're doing. Keep doing this over and over.

"Eventually, you should find yourself supplying the competing response at the very moment you feel your hand rising—before it even reaches your mouth," says Dr. Miltenberger. Don't limit your use of your competing response to your sessions, of course. Whenever you start engaging in your habit, substitute the competition. It's important to carry over what you've learned during your sessions. When you begin applying your competing habit to your everyday life, it becomes a natural response.

Keep your spotters. Once you've started practicing your competing response, alert your spies that you've moved on to phase two. Ask them to continue keeping an eye on you, but now ask them to tell you when you neglect to substitute the competition. "Ask them to say something like 'Do your exercise' or 'Remember your exercise' whenever you forget," says Dr. Miltenberger.

With patience and persistence, this repetition should help you modify your behavior. Instead of nibbling on your nails, you'll just fold your hands gracefully in your lap.

OSTEOPOROSIS

Bone Up on Weight-Bearing Exercise

James McNeill Whistler immortalized his mom's sullen face in his famous 1872 portrait. So why no smile, Mrs. Whistler? Perhaps it was the bowing of her upper back and the rounding of her shoulders—telltale signs of osteoporosis, the bone disease that can ruin your posture and diminish your life.

Osteoporosis—from the Greek meaning "porous bones"—slowly and silently steals calcium from our bones, making them more weak and brittle as we age. Osteoporosis picks up power as we age, especially after our midthirties, when bone mass development peaks. The worst cases leave some of us badly bent over and limited by frailty and fear of fractures. Brittle bones break easily. Fractures in the hips, spine, and wrists are quite common among people with advanced osteoporosis.

There is no known cure for osteoporosis yet. But weight-bearing exercise is a great way to slow down the bone fade, says Eric Orwoll, M.D., chief of the endocrinology and metabolism section at the Veterans Administration Medical Center and professor of medicine at Oregon Health Sciences University in Portland. And balance-building exercises are good protection from the falls that can have such debilitating consequences.

A Clean House and Healthy Bones

The next time you get ready to grab a vacuum, broom, or mop, heed these bone-safe tips from the National Osteoporosis Foundation.

► Imagine that your upper arms are strapped to your chest from the shoulders to the elbows and cannot move. This can help you keep your body in proper alignment.

► Face your work directly to keep your back from twisting.

► Keep your feet apart, with one foot ahead of the other.

► Shift your weight from one leg to the other to move the vacuum, broom, or mop.

► Always bend at the knees, not the waist.

Outflank Osteoporosis: Give Your Bones Some Burdens

Walking, running, aerobic dance, cross-country skiing, and other weight-bearing exercises are especially good ways to keep your skeleton spry, according to Dr. Orwoll. Why? It's the old use-it-or-lose-it wisdom: The more you demand of your bones, especially before age 35, the stronger you'll be, say doctors.

Move some metal. If you have access to a health club, try regular light workouts on the resistance-training machines, says Dr. Orwoll. Just easy sessions with light loads—three times a week for 30 minutes—will fortify you all over, from rib cage to wrist bones. The amount of weight you use really depends on the individual, explains Dr. Orwoll. Just start slowly and build up gradually. There's no need to overdo it.

Try dainty dumbbells. If you're far from fancy equipment, buy yourself some very light dumbbells and lift them—bicep curls are good—for 15 minutes a day, suggests Dr. Orwoll. (You can do this while watching television.)

Take a hike. You don't have to do anything too stressful, just walk, walk, walk. Get those legs moving, says Dr. Orwoll.

Have some letters to send? Walk to the mailbox. Meeting your grandson at school?

She Flexes with Ease

Libba Foshee tossed out her throw rugs and carpeted the floors in her Meritta, Ohio, home to reduce the risk of tripping and falling. She took those precautions after a physical exam a few years ago showed that she had lost 30 percent of her bone mass and 1½ inches in height. She first took estrogen supplements to help deter bone loss, but her reaction to the supplements was so negative that she had to stop. She then turned to exercise to stem the osteoporosis tide. She now walks at least 30 minutes a day, and here's her favorite rise-and-shine morning exercise.

She lies flat on her back in bed with her arms at her sides, a rolled towel under her lower back, and a thin pillow under her head. Then she bends her knees and lifts her buttocks off the bed with both hands. In that pose, she slides her left leg down so that it touches the bed and holds for a count of three. She returns her left leg to the bent-knee position and repeats the movement with her right leg. She does this 20 times before getting out of bed.

The important exercise principle is to avoid any sudden, jerky, or twisting movements. Here's how Foshee follows it. To get out of bed, she slowly swings her feet to the floor and then brings her body up. "If I drop something," she says, "I know I need to lower my body down so that one knee is on the carpet and my back stays straight. I don't twist.

"When I get into a car, I put my bottom in first with my feet still on the pavement," says Foshee. "Then I slowly move my feet inside the car, being careful not to twist my back."

Foshee's attention to exercise has paid off: Her last annual physical exam showed no further height loss. "I'm living proof that you can still exercise if you have osteoporosis," she says. "And, if you exercise on a regular basis when you're young, don't stop. It's important to keep your bones and muscles strong."

Hey, it's only a mile. Library-bound? Stroll to troll for the hot best-seller. Playing golf? Go cartless. Make sure you start at a comfortable level and increase your walking gradually. Shoot for 30 minutes of brisk walking three times a week.

Become a stair master. Avoid elevators. Make a daily habit of climbing stairs and your bones will thank you, says Dr. Orwoll. "Walking helps prevent bone loss," he says, "and going up stairs is a great way to increase your overall fitness."

Come and join the dance. Dancing is terrific weight-bearing exercise, explains Dr. Orwoll. Be the first one on your block to master the Merengue. If you start before your bones get porous, a folk-dancing habit will help give you especially strong shins and thigh bones. Why not try the hora or the highland fling?

Tap dance on your bones. Here is a creative exercise to increase blood circulation and help prevent bone loss. The following suggestions are from Meir Schneider, Ph.D., a licensed massage therapist, founder of the Center and School for Self-Healing in San Francisco, and creator of the Meir Schneider Self-Healing Method.

Bone-tapping massage works on any bone that's close enough to the surface to be easily felt; the spine, bony areas of the pelvis, ribs, fingers, knuckles, wrists, forearms, elbows, and shoulders.

Try this wrist-rap. Using the fingertips of all five fingers of your right hand, tap in a firm, steady, drumming cadence (about three taps per second) on your left wrist bones. Imagine that your fingertips are penetrating deeply and then feel them bouncing back up, as if your fingers were jumping on a trampoline. Keep your wrists

Osteoporosis and Amour

People with osteoporosis can enjoy sex, but they should consult their doctors or physical therapists for guidance, says Peggy Anglin, P.T., a physical therapist who practices at Duke University Medical Center in Durham, North Carolina. She advises keeping these loving thoughts in mind.

▶ Don't be shy about telling your partner what feels good and what is painful during lovemaking.

▶ Avoid any position that requires bending forward or twisting the spine.

▶ Avoid holding the full body weight of your partner.

▶ Try spooning, a position in which both partners lie on their sides with the man behind the woman.

loose and fluid. If you feel like your fingers are jabbing you, it's because your wrists have tightened up. Stop and shake them loose. To get the most effect, the tapping should be pleasant and relaxing, never painful, and your fingers should feel warm and alive. After about 30 minutes, tap your left fingertips on your right wrist.

You can use this technique anywhere, massaging each area for about 30 minutes. Because osteoporosis often affects the spine, and most of us can't reach our own backbones to drum on them, you may want to enlist a friend or family member or see a professional massage therapist. If you know someone who has the same problem, try swapping bone-tapping sessions.

The Art of Balance

If you have osteoporosis, falls are a big worry, says Dr. Orwoll. Consider these low-risk, easy-does-it exercises that will help you stay fit as they improve your balance.

Try tai chi. This ancient Chinese martial art is an excellent exercise for anyone with osteoporosis, says Lana Spraker, a Los Angeles master instructor who has taught tai chi for more than 25 years.

Tai chi involves slow, controlled movements that emphasize maintaining good balance. Several studies show that it helps reduce the risk of falling and maintains strength in the elderly.

"Tai chi is excellent for coordination, concentration, and peace of mind," says Spraker. "It strengthens muscle, brings joint flexibility, lowers blood pressure, and improves breathing and circulation." Aside from personal instruction, books and videos can be a good way to learn tai chi.

Become a water walker. Try walking in waist-deep water for 30 minutes three times a week, says Dr. Orwoll. The water supports your body, takes stress off your bones and joints, and helps improve your balance.

OVERWEIGHT

Keep Yourself Thin

Here's the bottom line: Dieting will *get* you thin, but medical studies show that you need to add exercise to *keep* you thin.

"We've found that exercise is actually the best predictor of weight maintenance," says Glen Blix, Dr.P.H., associate professor of health promotion and education at the School of Public Health at Loma Linda University in California. "People who are overweight should change their focus from weight loss to fitness."

No, you will not have to become an exercise fanatic. But if you're inclined to be sedentary, you'll want to consider the benefits of being moderately active. The National Weight Control Registry, an ongoing national study of people who have lost weight and kept it off for at least one year, found that 89 percent of the participants changed both their eating habits and their physical activity to lose or maintain weight. The types of exercise they selected included walking, bicycling, aerobics, and stair-climbing.

Exercise Will Feed You

"Exercise affects your weight in two ways," says Richard Cohen, M.D., a physician in private practice who specializes in nutritional medicine in Hanover, Massachusetts. "First, it will burn off the excess energy in your body, which is stored as fat. Second, exercise regulates the brain chemicals, or neurotransmitters, that often contribute to overeating."

The neurotransmitter dopamine is responsible for your alertness, memory skills, and feelings of hunger, explains Dr. Cohen. And serotonin affects your feelings of well-being, security, and craving. Numerous things can vary the levels of these brain chemicals in your body, including the foods you eat, the exercise you do, and the medications you take. Specifically, eating high-protein foods such as eggs, fish, and chicken will raise dopamine levels, while eating high-carbohydrate foods such as pasta and bread will raise serotonin levels.

"If you sit around doing nothing all day, your levels of dopamine will not be elevated and you'll feel sluggish," says Dr. Cohen. "So if you don't raise the neurotransmitter levels with exercise, you'll probably end up raising them with food." Too much food, to be exact. But if you choose to exercise, *voilá*! You may not have the urge to overeat.

Therefore, the key to weight control is to keep your neurotransmitters happy through healthy, active habits rather than food, says Dr. Cohen.

Victoria's Motivation Tips

You don't want any little frustrations to throw your exercise program off track. Victoria Johnson, a certified personal trainer and president of Victoria Johnson International, a fitness consulting group in Portland, Oregon, offers these tips for sticking to your weight-loss regimen.

Let go of exhaustion. Many people say that they don't exercise because they're too tired. If you feel too tired to exercise, consider changing your diet. "Food is fuel. Higher-quality fuel will give you greater performance," says Johnson. "Cut down on sugar and fat in your diet, and drink more water," she adds.

Get used to sweat. "A lot of people hate to sweat," says Johnson. Remember, though, that it's helping to cool your body down—it's not heating you up more. "Don't stop when you start sweating. That doesn't mean that you're finished exercising," says Johnson.

Be conscious of body alignment. Good posture can really make the difference between a worthwhile workout and a waste of time. "Your ears should be over your shoulders, and your shoulders should be over your hips," says Johnson. "The energy should emanate from the core of your body, which comprises the abdominal and back muscles. Proper posture increases the range of motion in all the joints of your body." Note that each exercise described in this chapter includes a "posture check."

Remind yourself to breathe. Yes, it's annoying when all those aerobics teachers remind you to breathe. But the truth is, a lot of exercisers forget to exhale, and they don't inhale deeply, either. Oxygen is an important part of the energy formula that powers your muscles. So when you're working out, "breathe two counts in and two counts out," says Johnson.

Choose the right weight. How heavy should your weights be? That depends on how strong you are now. Start out with light weights—three to five pounds. You know that you're using the right amount of weight if the muscle you're exercising feels tired, almost exhausted, during the last couple of repetitions.

Different muscles may require different weights. Large muscles, such as those in the back and legs, can take heavier weights. You may only need to use 3- to 8-pound dumbbells for your biceps, for instance, but need 5 to 10 pounds when working your back.

Hold your weights properly. You don't have to grip the weight with a lot of tension in your hand, wrist, and forearm. Simply hold the weight lightly and let your back and arm muscles do the actual lifting. Also, be sure to keep the line of your hand, wrist, and forearm straight. Don't bend your wrist when you're holding the weight.

Maintain your flexibility. Stretching will help keep your muscles long and lean, and it will also help prevent injuries. After you've finished working each body part, do the stretching and flexibility exercises at the end of each series of exercises. Don't skip this very important step.

Your Weight-Loss Prescription

While exercise is one of the few weight-control aids that may help you get results, lots of people shy away from it. If you're among those who avoid it, one trick is to think of exercise as a prescription from your doctor. After all, it gives you many health benefits. You'll develop a healthier heart, lower your cholesterol, and help increase your bone density. And if you need another reason, some doctors think that exercise may help prevent some cancers. So here's your official prescription from a doctor:

To achieve or maintain weight loss, you need to create a program that incorporates aerobic exercise and resistance training. Aerobics will decrease the total amount of body fat that you carry, while strength training will change your metabolism.

—Dr. Blix

So just how do you go about filling that weight-loss prescription? Here are specifics from Victoria Johnson, a certified personal trainer and president of Victoria Johnson International, a fitness consulting group in Portland, Oregon. Try to apply yourself to exercise the same way you take other medications—on a regular schedule.

"You're going to do a cardiovascular activity five days a week and top that with two weight-training sessions each week," says Johnson. "The cardio activity can be walking, riding a bike, jogging, or any other aerobic exercise that you like."

Your goal is to eventually work up to 30 minutes of aerobic activity four or five times per week. But the intensity and duration of your cardiovascular workout will depend on the kind of shape you're in right now. During your workout, "you should be working hard but still be able to carry on a conversation," says Johnson.

If you're particularly overweight or out of shape, you need to work up to your 30 minutes of exercise. "Start with short intervals of light or no-impact activity," Johnson says. "You could swim, ride a stationary bicycle, or walk on a treadmill, for example. Your long-term goal, however, is to be able to sustain 30 minutes of exercise by the end of 12 months."

You don't have to do the same aerobic activity every time. If time is a problem, divide your 30 minutes into two segments. You could walk for 15 minutes every morning and then walk again for 15 minutes in the evening, for example.

Now You Need a Lift

The second half of your prescription is strength training. "Aerobic activity burns calories, but it's strength training that keeps weight off," says Dr. Cohen. Increasing the strength of your muscles doesn't necessarily mean that they'll grow in size. The fact is, muscle takes up less space than fat, and it helps your body burn more calories through the course of a day.

Do the following exercise routine on nonconsecutive days, such as Monday and Thursday or Tuesday and Friday. "This routine is short," says Johnson. "The whole series of exercises will take between 30 and 45 minutes, including the warmup and stretching. You're going to do one or two exercises per body part. Do two sets of each exercise, 12 repetitions per set."

For the exercises that use weights, you can choose among free weights—that is, dumbbells or hand weights, starting with three to five pounds. That's how Johnson designed this program. But if you have access to a gym or health club, you can use the exercise equipment that you find there, as long as you get some guidance.

"If you aren't familiar with these exercises, ask an instructor at your gym to help you," says Johnson, "or hire a personal trainer for a few sessions. She'll be able to teach you how to do the exercises and make sure that you're doing them correctly." Rest assured that you won't always have to use a trainer, but you may want to have an expert help get you started.

Ready to Go

All of these exercises can be done at home. Warm up first by doing five to eight minutes of a low-intensity aerobic activity such as walking, riding a stationary bike, or aerobic dance. Then follow with these exercises for specific areas of the body.

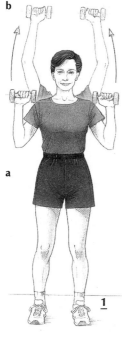

Shoulders

Overhead press. Stand with your feet hip-width apart, your knees slightly bent, and your weight evenly distributed on both feet. Holding a light weight in each hand, start with your hands just above your shoulders, with the palms facing forward. Your shoulders should be relaxed and your abdominal muscles pulled in (*1a*). Exhale and raise your hands past your head until your arms are straight, but do not lock your elbows (*1b*). Lower the weights slowly as you inhale.

Posture check: Don't stretch too far—you're trying to raise your shoulders without moving them up toward your ears. Keep your arms extended straight up and close to your head.

Bent-arm lateral raise. Hold a light dumbbell in each hand. Keeping your upper arms close to your sides, bend your elbows at a 90-degree angle. Your hands should be just in front of your body, facing each other (*2a*). Exhale and raise your arms to the sides (with your elbows bent like chicken wings) so that your hands are facing down (*2b*). Inhale and slowly lower your arms.

Posture check: Don't raise your arms and elbows too high. They should end up just at or below shoulder level.

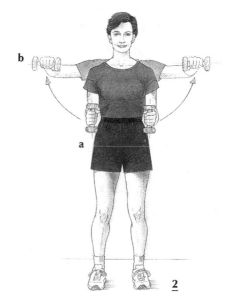

Fronts of the Arms

Bicep curls (palms facing up). Stand with your feet hip-width apart and your knees slightly bent. Hold a light dumbbell in each hand, grasping it so that your palms are facing

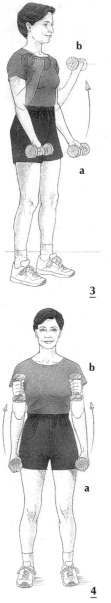

3

4

up (*3a*). Exhale, bend your elbows, and slowly raise your hands and forearms toward your shoulders (*3b*). Then inhale and slowly lower your hands to the starting position. Don't lock your elbows on the way down.

Posture check: Only your arms should move during this exercise. Your upper body should stay completely still to isolate your biceps.

Hammer curls (palms facing in). This exercise is the same as the bicep curl, except that your palms face in, toward your body (*4a and b*). Again, do the repetitions slowly.

Posture check: Keep your upper back relaxed, and try not to bounce your hands against your shoulders.

Backs of the Arms

Triceps kickback. Stand with your feet hip-width apart and your knees slightly bent. Hold a light dumbbell in each hand, with your palms facing in. Bend your elbows so that your hands are by your sides at chest level, pressing your elbows into your body (*5a*). Exhale and straighten your arms, extending them slightly behind your body. Don't lock your elbows. Instead, focus on feeling the tension in the

muscles in the backs of your arms (*5b*). Inhale and return to the starting position.

Posture check: Keep your back relaxed and try not tense your shoulders when you extend your arms.

French press. Stand with your feet hip-width apart and your knees slightly bent. Holding a light dumbbell with both hands, raise your arms over your head, making sure that your shoulders are relaxed. Bend your elbows and bring the weight down behind your head. Then exhale and extend your arms back over your head, pushing the weight up. Straighten your elbows, but don't lock them. (For an illustration of this exercise, see page 400.)

Posture check: Be sure that your elbows are close to your head rather than flying out to the sides. Also, don't rock your upper body—keep everything still except your forearms.

5

Chest

Chest press. Lie on your back with your knees bent and your feet flat on the floor. Holding a dumbbell in each hand, rest your upper arms on the floor with your elbows out and away from your sides. Bend your elbows and raise your hands, with the palms facing forward (*6a*). Exhale and push your hands up toward the ceiling, bringing the weights together end to end, which will squeeze your chest (*6b*). Inhale and lower your arms to the starting position.

6

Posture check: To take the strain off your back, press your stomach toward your back and your back into the floor. Your pelvis should be tilted slightly upward.

Flies. Lie on your back with your knees bent and your feet flat on the floor. With a dumbbell in each hand, stretch your arms out to the sides with your hands raised slightly off the floor and your palms facing up (*7a*). Your arms will form a very slight arc. Exhale and raise your arms toward each other above your head (*7b*). Keep the arc in your arms, as if you were hugging someone, but don't try to bring your elbows together. Inhale and lower your arms slowly.

Posture check: Again, press your stomach toward your back and your back into the floor to tilt your pelvis slightly upward.

7

Upper Back

Bent-over rows. Holding a dumbbell in each hand, stand with your feet hip-width apart.

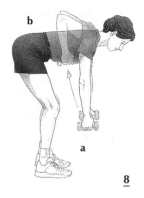

8

Bend your knees very slightly and bend forward at the hips until your upper body is parallel to the floor. Your arms should hang straight down from your torso (*8a*). Exhale and raise your hands toward your shoulders, bending your elbows upward (*8b*). Squeeze the muscles behind your shoulder blades together. Inhale and slowly lower your arms and hands.

Posture check: Keep your neck relaxed during this exercise. Don't lower your chin to your chest or look up at the ceiling. Instead, keep your neck in a neutral position, with your eyes looking at the floor a few feet away from your toes. Your lower back should be relaxed and as flat as possible.

Back extensions. Lie face-down on the floor with your arms straight above your head. Exhale and simultaneously raise your left arm and right leg off the floor about 6 to 12 inches (*9*). Keep your head and neck down. Don't hold your breath, but try to keep your arm and leg extended for about 10 seconds. Stretch your left fingers and right toes in opposite directions. Then return to the starting position and repeat with your right arm and left leg.

Posture check: Don't worry if you can't get your arms and legs very high. Instead, focus on the stretch and on keeping your back and neck relaxed.

9

Upper-Body Stretches

Curl-up. Lie on your back, then exhale and bring your knees up to your chest, holding them with both hands. Raise your shoulders and head toward your knees without putting your chin on your chest. Keep breathing and hold for 20 to 30 seconds (*10*). It lengthens your entire back.

10

11

Shoulder hug. Standing or sitting, put each hand on the opposite shoulder, wrapping your arms around your body without raising your shoulders (*11*). You should feel a stretch in your upper back.

Abdomen and Hips

Crunches. Lie on your back with your knees bent and your feet flat on the floor a comfortable distance from your buttocks. Cross your hands over your chest or keep them by your sides. Keeping your eyes focused above your knees or on the ceiling, exhale and raise your head, neck, and shoulder blades off the floor. Inhale and bring your upper body down,

stopping just before your shoulders touch the floor. Do these repetitions slowly. (For an illustration of this exercise, see page 337.)

Posture check: Be sure that your abdominal muscles are pulled in toward your back and your back is pressed against the floor.

Oblique twists. Lie on your back with your knees bent and your feet on the floor a comfortable distance from your buttocks. Put your right ankle on your left knee, bending your right knee out to the side. Cup your hands behind your ears and keep your elbows out to the side (*12*). Raise your head, neck, and shoulders (*13*). Exhale and twist your upper body as if you were going to touch your right knee with your left shoulder, keeping your elbows back (*14*). Inhale and come

12

13

14

down. Do one set on each side, then repeat.

Posture check: Be sure that your abdominal muscles are pulled in toward your back and your back is pressed against the floor.

Bicycles. This exercise is very similar to oblique twists. Lie on your back and bring both knees toward your chest. Cup your hands behind your ears and drop your elbows down to the floor (*15*). Straighten your left leg away from your torso and bring your left shoulder toward your right knee, keeping your elbows back (*16*). Repeat, alternating sides. The stronger your abdominal muscles, the closer to the floor you will be able to bring your legs when they are extended. Do the exercise slowly, inhaling when you have one knee up and exhaling when you have the other knee up.

Posture check: Be sure that your abdominal muscles are pulled in toward your back and your back is pressed against the floor.

Buttocks

Squats. Stand with your feet hip-width apart and a light dumbbell in each hand, then rest the ends of the dumbbells on your shoulders. Tuck your buttocks under slightly to make sure that your back isn't arched (*17*). Inhale, bend your knees, and lower your body as if you were sitting down. Go down as far as you can without dropping your buttocks lower than your knees; make sure that your knees are right over your toes (*18*). Exhale and stand up, squeezing your buttock muscles together as you do.

Posture check: Try to keep your back straight as you lower your body. Don't bend forward or backward at the waist. Keep your neck straight and your head up.

Lunges. Stand with your right leg straight ahead of you as if you were taking a larger-than average step forward. Your right foot should be flat and your left heel should be raised off the floor (*19a*). Bend both legs, bringing your left knee directly over but not beyond the toes of your left foot and lowering your right knee to about six inches above the floor (*19b*). Come up slowly. Do one set on each side, then repeat.

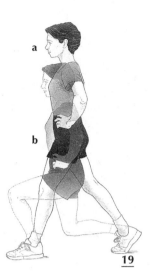

19

Posture check: Be sure your torso is centered over your legs. You shouldn't have to lean forward or back to maintain your balance. You may have to experiment to determine how far apart your feet should be.

Abdominal Stretch

Cobra pose. This yoga move is a great way to stretch the front of your torso while strengthening your back muscles. Lie on your stomach with your legs stretched out behind you, your hands on the floor on either side of your chest, and your face to the floor (*20a*). Exhale and raise your head, neck, shoulders, and chest. Your elbows will straighten a bit, but don't try to straighten them all the way. Look ahead, but try not to bend your neck back (*20b*). Inhale and hold. Keep breathing and hold for 20 seconds.

Posture check: Be sure that your shoulders are lowered and not squeezing up into your ears.

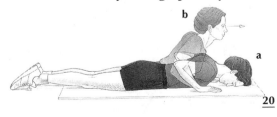

20

Backs of the Legs

Leg curls. These are also called hamstring curls, named for the hamstring muscles in the backs of your thighs. Stand facing the back of a chair and hold on to the chair with your hands at about waist height (*21a*). Extend your right leg back and lift it a few inches from the floor. Point your right foot out at a slight angle from your body (*21b*). Then flex your foot with the toes pointing down and bend your knee, slowly bringing your heel in toward your buttocks (*21c*). Lower your leg to the starting position, then switch legs and repeat.

21

Posture check: This move is not as big as it might seem. Don't try to kick your leg. Instead, focus on maintaining proper posture. Tuck your pelvis in and under your torso and don't arch your back. Only the working leg should move.

Fronts of the Legs

Leg extensions. Sit on a chair with your knees bent and your feet flat on the floor. Raise and straighten your left leg, keeping both thighs level but not locking your knee. Flex your foot and point your toes. Return to the starting position. Do one set with each leg, then repeat. (For an illustration of this exercise, see page 337.)

Posture check: Be sure that you're sitting up straight when you bring your leg up.

Buttocks Stretches

Knee squeeze. Lie on your back with your knees bent and your feet flat on the floor. Put your right ankle on your left knee, then grasp your left knee with both hands and pull it toward you (22). You should feel the stretch in your right buttock. To increase the stretch, press your right knee out away from your body. Hold for 20 to 30 seconds. Switch legs and repeat.

Posture check: Your head and neck should be relaxed against the floor as you look up.

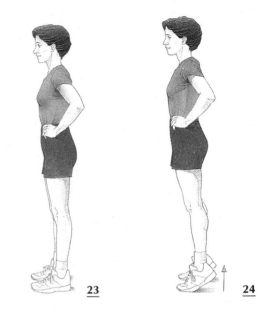

Calves

Toe raises. Stand with your feet about hip-width apart and your hands on your hips or your arms out to the sides for balance (23). Slowly rise up on your toes (24), then lower your heels slowly.

Posture check: Tuck your pelvis in and under your torso and don't let your back arch.

Leg Stretches

Quadriceps stretch. Stand with your feet about hip-width apart. If you need to, hold on to a chair for balance. Bend your left knee and raise your left foot toward your left buttock. Clasp your left ankle with your left hand and hold for 20 to 30 seconds. Keep your knee pointed down. Release the leg and lower it slowly. Switch legs and repeat. (For an illustration of this exercise, see page 336.)

Hamstring stretch. Stand with your right foot one stride ahead of your left and your knees slightly bent (25). Place both hands on

the front of your left thigh. With your right leg straight and your back flat, bend over from the waist. Bend your left leg without raising your heel and lift the toes of your right foot (26). You should feel the stretch in your right hamstring. Hold for 20 to 30 seconds, then straighten both legs and lower your foot. Repeat on the other side.

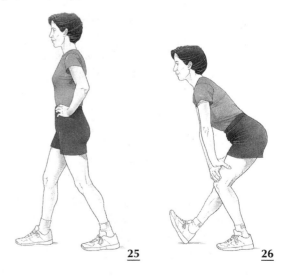

PANIC ATTACKS

Breathing in Relief

In his first inaugural address, when the United States was deep in the Depression, President Franklin Delano Roosevelt reassured Americans that "the only thing we have to fear is fear itself."

He attacked fear as a "nameless, unreasoning, unjustified terror which paralyzes." FDR's famous quote led to the creation of the New Deal, a successful plan that brought the country's economy back from the brink by pouring billions of dollars into relief and job programs to increase employment and consumer confidence.

In many ways, FDR's "we-can-beat-fear" approach can also help those who are paralyzed by panic attacks. People who have panic attacks sweat, feel dizzy and nauseated, experience chills or hot flushes, gasp for breath, and fear that they're losing control. (When new symptoms develop, such as chest pain or pressure, difficulty breathing, dizziness, loss of consciousness, or an intense sense of fear for the first time, it is important to see your doctor. Every so often, these symptoms can be early indications of a more serious problem.) Perfectionists, defeatists, and folks who are hypercritical of themselves seem especially susceptible to the unleashing of these physical and emotional symptoms, doctors say.

"People who have panic attacks feel like they're going to die," says Jane Sullivan-Durand, M.D., a behavioral medicine physician in Contoocook, New Hampshire. If your symptoms are indeed a panic attack, you need to realize that you're going to survive the panic, no matter how terrible it feels. "Do some self-talk to reassure yourself that you're not going to die. Then look at what triggered the attack," she says.

Fortunately, panic attacks can be prevented or reduced up to 90 percent of the time. Even in the grip of a feeling that makes you sweat, swoon, and feel terror in every part of your body, you can use some physical as well as emotional techniques to help stop the awful rush of feelings. You may be able to conquer fearful panic attacks with a health-minded New Deal that blends breathing techniques, some mental mantras, and acupressure.

Catching Your Breath

We do it about a dozen times a minute, every hour, every day, yet some of us still have not recognized the healing magic of healthy breathing. Here are some breathing methods to pull you out of a panic attack—or prevent one from occurring.

Brown-bag it. There are many changes happening in your body during a panic attack. Your sympathetic nervous system—which triggers the fight-or-flight response—is activated, releasing adrenaline to various organs. Your heart rate can increase to 150 beats per minute. You may feel as if your heart is thumping so loudly that it might jump out of your chest. Your respiratory system is also affected, increasing the rate of breathing and leading to hyperventilation.

The quickest and best remedy to halt hyperventilation is to breathe into a paper bag, says Dr. Sullivan-Durand. "A person who is hyperventilating is breathing out carbon dioxide at a faster rate than normal," she explains. "When this happens, the pH in the blood begins to rise, and the longer someone hyperventilates, the longer the pH change will continue to progress and eventually involve body organs and the nerves. That leads to symptoms like tingling lips, numb fingertips, and getting a lump in your throat. You actually feel short of breath, even though you're breathing too much."

If you're prone to panic attacks, Dr. Sullivan-Durand advises that you keep a folded paper bag handy. Open the bag, cover your mouth with the open end, and inhale and exhale slowly for five minutes. Don't worry about running out of oxygen, she adds, because with each exhalation into the bag, you breathe out oxygen, and the bag is porous and will let oxygen in.

"Most times, the bag does the trick," says Dr. Sullivan-Durand. "When you breathe into the paper bag, you are breathing in the carbon dioxide you let out and restoring your carbon dioxide level to normal." (If this doesn't improve your symptoms of panic within 5 to 10 minutes, or if the symptoms feel different from what you usually experience, call the doctor. It

---

Loving That Lobe

You'll need your ears and a mirror for this panic attack reliever, says David Nickel, O.M.D., a doctor of Oriental medicine and licensed acupuncturist in Santa Monica.

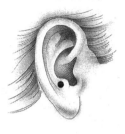

Using your right index finger and thumb, pinch the spot where your right earlobe meets the cartilage of your lower ear, just below the ear canal. Apply firm pressure for five seconds and then release for five seconds, repeating the sequence for a total of one minute. Try to exhale while pressing the point, and visualize your panic attack fleeing as you press. You should feel a sensation, usually hot and stinging, at the ear point during pressure. For added effectiveness, you can do the same steps on your left earlobe, says Dr. Nickel.

---

could be an emergency, and you could be suffering from something besides a panic attack.)

Expand your horizons. Taking deep, slow breaths from the bottom of your belly engages your body's natural relaxation response to curb a panic attack, says J. Crit Harley, M.D., a behavioral physician and director of Un-Limited Performance, a stress-management company in Hendersonville, North Carolina.

It is important to never try to inhale or exhale beyond your capacity, Dr. Harley adds. To make sure you're breathing correctly, he suggests placing a big rubber band around your waist, over your belly button. It should stretch each time you inhale to indicate that you're breathing from the belly and not from the chest. You can't find a rubber band that large? Just use your imagination. Pretend that you have a rubber band around your belly and make sure that your stomach moves out before your chest on each inhalation, says Dr. Harley.

"Diaphragmatic breathing is the healthiest and the most natural way to breathe," he says. "We breathed like that as babies. But it's a forgotten habit unless you are a singer or have been trained to project as a speaker."

Go ahead and guffaw. Laughter does more than put a smile on your face. It actually changes your breathing pattern, which is healthful for anyone prone to panic attacks, says Dr. Sullivan-Durand. The next time you feel a panic attack approaching, shift your focus to a funny scene from your favorite comedy and let out at least 30 seconds of laughter.

Sing your praises. Singing, like laughter, alters your breathing pattern, says Dr. Sullivan-Durand. Select a motivating song and sing it out loud to curb a panic attack. Whether your choice is "We Shall Overcome" or "The Battle Hymn of the Republic," sing it with conviction to help march past the panic.

Roar and purr. When panic sends your breathing into a shallow, swift staccato, try visualization to slow your pace, says Dr. Harley. In your mind, picture your heart as a caged lion. At first, the lion is roaring and pacing in quick circles. But as you take steady, slow breaths, you can imagine the lion becoming quiet and finally lying down.

Relax progressively. Call this the head-to-toe panic calmer. For this progressive relaxation exercise, Dr. Sullivan-Durand suggests that you find a quiet place in your home that has soft lighting. Unplug the phone, close the doors, and keep the pets out of the room so that you can focus on relaxing.

Sit in a comfortable chair with a firm back and place your feet flat on the floor. Rest your hands on your lap or on the arms of the chair. Close your eyes.

"Focus on your breathing, every detail of it. In this way, you will quiet your mind. Don't get caught up in any self-criticism if your mind wanders. Just bring your attention back to the breath," says Dr. Sullivan-Durand.

You're going to tense and then relax each part of your body from head to toe. Here are the steps.

▶Inhale from the diaphragm. As you exhale, mentally say the word *relax* or *peace*. Repeat this word with each exhalation.

▶Starting with your face, tense your eyes and facial muscles, squeezing tightly for 5 seconds and then relaxing them for 10 seconds.

▶Slowly work down your body, tensing and relaxing your neck, shoulders, back, upper and lower arms, hands, chest, abdomen, thighs, calves, ankles, feet, and toes.

▶Once you've completed the head-to-toe exercise, stay put for a few minutes and enjoy this relaxed state.

"The practice of progressive relaxation training activates the parasympathetic nervous system, which helps reduce the heart rate, blood pressure, and other body changes associated with panic or stress symptoms," says Dr. Sullivan-Durand, who has created several audiotapes that teach relaxation techniques. "Relaxation is a skill that can be learned. It's best to set aside the same time each day to make it a habit."

Rooting Out Panic at Its Source

Experts also recommend some mental exercises designed to develop our abilities to concentrate and silence irrational thoughts.

Turn to juggling. One way to tune down panic is to shift your attention elsewhere—to juggling, for example. "When you juggle, you can only juggle," explains Dr. Harley. "You can't be worried about this or that. It's a great way to stop a panic attack." To ensure success, Dr. Harley offers these two options.

One: Instead of tennis balls, use three long scarves. "Scarves are easier to catch, and knowing that helps to relax you," he explains.

Two: Juggle with only one ball, and purposely let it hit the ground. Watch it bounce and roll. "By allowing it to hit the ground, you realize that the world did not end and that everything is okay," Dr. Harley says.

Shift to the moment. One mental way to stop panic is to engage your observation powers, says Dr. Harley. He recommends selecting some scenery or an object—perhaps your front door—and describing it in as much detail as possible, as if you were speaking to someone who is blind.

How many hinges are there? Does it open in or out? Which side is the doorknob on? What type of wood is the door?

"This exercise forces you to deal with facts, not worrisome 'maybes' that might trigger a panic attack," says Dr. Harley. "Paying attention to observing what you see takes your mind off your panic, which is what you can't see."

Play the phrase game. You don't need a partner or a scorecard to play this game. The goal is to rid yourself of panic tendencies by replacing negative talk with positive talk, says Dr. Sullivan-Durand.

The next time you feel panicky, instead of declaring, "I'm going to have a heart attack," try substituting the words, "It's only uneasiness. It will pass." Instead of thinking, "I am going to die," try calmly saying, "It's just a little dizziness. I can handle it." This helpful self-talk can give you more control over the problem, she adds.

"Relaxation is one way to curb distorted beliefs and negative self-talk," adds Dr. Sullivan-Durand. "When you turn your attention away from the negative chatter that races through your mind, you are more able to focus on what is positive in your life. In this way, you will feel more energetic.

"This is not to imply that we can deal with all of life's difficulties by distracting ourselves from them. Tragic events and conflicts with other people are bound to happen. We must face them in order to resolve these problems once and for all," she says.

Get feedback. During a panic attack, the temperature drops in your hands and feet, your blood pressure rises, and your heart rate zooms up. A few sessions with a biofeedback therapist may help you prepare for panic attacks and reverse these symptoms, suggests Jo Anne Herman, R.N., Ph.D., a biofeedback expert and associate professor of nursing at the University of South Carolina in Columbia.

During a session, with your body hooked to a biofeedback monitor, you can receive information about your heart rate, blood pressure, muscle tension, and skin temperature. A therapist can teach you breathing and relaxation techniques to calm yourself down, and you can see—and feel—the results. The more relaxed you become, the higher the temperature in your hands and feet and the lower your blood pressure and heart rate.

PILATES

Strengthen Yourself to the Core

WHAT IT IS

A few years ago, the actress Jane Seymour appeared on TV soon after having twins—with a flat stomach. Her secret? Pilates.

The Pilates Method of physical training was developed by a German athlete named Joseph Pilates (pronounced Puh-LAH-teeze). During World War I, Pilates was in England, studying to become a nurse and caring for British patients who had been immobilized by wounds. To help these patients, he developed exercise equipment by using springs that he attached to hospital beds. Pilates later refined his training approach and labeled it contrology, thus creating a modern exercise program that he believed everyone could use. Today, contrology has been trademarked as a method that's simply called Pilates, after its creator.

The principle behind Pilates is that you have to strengthen the area of your body that trainers call your core—specifically, your abdominal muscles, lower back, and buttocks. Pilates uses carefully controlled, precise movements to improve strength and flexibility without bulking up your body. That's why dancers love it. Two of our most distinguished dancers—Martha Graham and George Balanchine—were among the earliest converts to the method.

There are more than 500 Pilates exercises to choose from, according to Julie Sorrentino, owner of Body Balance, a Pilates Method studio in New York City. "All Pilates exercises teach people how to find and strengthen their center," she says. "When most people begin their study, they can't even feel their transverse abdominal muscle, which is where their center is." This muscle, which wraps across the front of your torso and compresses the abdominal area, starts approximately at the center of your chest and continues down to the area of the pubic bone.

"If you have a strong core, you'll have maximum mobility of your limbs and be able to move your arms and legs more gracefully and with less effort," says Leslie Scheindel, a Pilates Method teacher and owner of Leslie's Total Fitness in Chicago and in Northbrook, Illinois.

Over the years, the Pilates Method has become a crucial part of some dancers' training as well as a rehabilitation aid for people who need physical therapy. "Exercise, like the Pilates Method, may also help prevent back problems," says Liz Henry, P.T., a physical therapist and certified orthopedic specialist with West Side Dance Physical Therapy in New York City.

There are two types of Pilates Method classes. One involves floor exercises, which are held in sessions called mat classes. The other makes use of the machines that Joseph Pilates designed. With names like the Reformer and the Cadillac, Pilates machines use pulleys, springs, and sliding platforms to create resistance. The equipment is expensive, and you have to go to a Pilates Method studio and work with a teacher one-on-one to learn how to use them. Although there's one type of machine for home use, teachers agree that if you're going to use the machines, the best investment is to take classes with an instructor.

Pilates exercises are done slowly and in a very controlled manner. The combination of slow pace with control means that you might only do 10 repetitions of an exercise, but you'll feel the intensity of the movement immediately, says Jennifer Kries, a dancer, choreographer, Pilates instructor, and founder of Jennifer Kries, The Method, in New York City.

Compare a standard abdominal crunch—the kind you might do in exercise class—to the Pilates mat class exercise called "dipping your toes in the pool." When doing abdominal crunches, you try to do as many repetitions as possible to build your abdominal muscles. With the Pilates exercise, the objective is to do it very slowly, with positions and motions that are exactly right. To dip your toes in the pool, for instance, you lie on your back with your arms by your sides and your knees pulled in toward your chest. Very slowly, you move one foot toward the floor, with your leg bent and your toes pointed, without dropping it all the way. Before your toe reaches the floor, you pull the same leg back toward your chest in a smooth motion. At most, you do five repetitions with each leg—very precisely and slowly. Then you do five repetitions of the same exercise with both legs together, says Kries.

WHAT IT DOES

If you've ever admired the proud and elegant way in which a dancer can cross a stage, then you've already noticed the benefits of Pilates. "Pilates shows you how to use the right muscles to keep your posture correct," Henry says. "Most people strengthen their rectus abdominus muscle, which runs down the front center of your torso. But it's really the obliques and transverse abdominus that support your trunk and spine and keep your gut in."

The other advantage of Pilates is that by working at the end ranges of motion and by stretching and strengthening, the method increases the range of motion of many of your joints, especially the hips, shoulders, and spine, explains Henry.

REAPING THE BENEFITS

One of the best ways to get a sense of what Pilates is like is to check it out by using a series of home exercise videos by Jennifer Kries. "I recommend either the "Method Toning" tape or the "Balanced Zones" tape for beginners," says Kries. "Just try to do about 15 minutes of the tapes three times a week when you're starting out, then work your way into the whole tape, also a few times a week." You can find her tapes at video stores, or ask to special-order them.

To get the full benefit of Pilates training, however, you at first need to have one-on-one sessions with an instructor. "At most studios, new students have their first 10 sessions alone with an instructor," explains Sorrentino. You can find a gym or studio in your area by writing to the Physical Mind Institute at 1807 Second Street, Suite 47, Santa Fe, NM 87505. Then you can join a group class, which might include only three other people. Sessions may cost anywhere from $40 to $75 per hour.

The one-on-one sessions guarantee that you'll learn the method as well as possible. "Even though I'm a physical therapist and have a lot of knowledge about the way the body moves, I can't jump out of my skin and look over myself to give myself corrections," says Henry, explaining why she takes classes in Pilates as well as using Pilates-based exercises to treat some of her patients. "What feels centered and balanced to me isn't actually centered and balanced. I need my teacher to watch me and give me cues."

TRIAL RUN

"If you want to try a Pilates exercise on your own, your first step should be to learn how to breathe properly through the exercise," says Kries. You need to inhale deeply through your nose, she says. And the trick is to breathe deeply while not letting your stomach get bigger. "Even though you're breathing in, you should still hold your abdominal muscles into your spine," she says. "Inhale for five counts, then exhale for five counts."

Now it's time to move with your breath. One exercise to try is called the roll-up. Here's how it's done.

▶Lie on your back with your knees bent and your feet on the floor just in front of your buttocks.

▶Stretch your arms above your head, palms up.

▶Pull your stomach muscles in toward your spine, flattening your lower back to the floor while still trying to breathe naturally.

▶While inhaling, lift your arms toward the ceiling until they're perpendicular. Press your chin to your chest at the same time. This is a slow, controlled movement.

▶Exhale and begin to "peel" your shoulders, head, and upper body off the floor, rolling up toward your feet.

▶Inhale as you stretch your arms past your bent legs, with your fingers reaching past your toes.

▶Roll back down slowly and with as much control as you used to get off the floor.

▶Try to do 10 repetitions, rolling up as far as possible each time. This should take about five minutes.

At first, you may not be able get very far when you're rolling up from the floor, but keep trying. Eventually, you'll be able to roll completely up. "The important thing is to keep your feet flat on the floor," says Kries. "Don't put something on your feet to keep them down. Just roll up as far as you can."

POSTURE PROBLEMS

How to Get Out of a Slump

Your mother was right to nag you about slouching. True, she was probably right for the wrong reason—she was more concerned about your dating life than your health—but there's no doubt that the good posture she bugged you about is a wellness plus.

If you stand tall and straight, with your head aligned properly and your shoulders back and down, your lungs get to stretch and you breathe better, experts say. You also get gut-level benefits. With good posture, you'll digest your food more efficiently because your stomach and intestines have more room. Most important, the upright among us often feel more energized because there is a clear path for the blood to flow to our brains. And by using the muscles of your back correctly, you'll avoid strains and be less likely to suffer chronic back pain.

Even if you ignored Mom and have, over the years, mastered the slouch, it's not too late to straighten up and stride right. Here are some tips to help you get vertical and do a favor for your lungs, tummy, back, and brain.

Posture 101: The Basics

Despite years of being crooked, a healthy stance is easy to achieve, says Peggy Anglin, P.T., a physical therapist who practices at Duke University Medical Center in Durham, North Carolina. Here's how.

First, get those ears over those shoulders. Most of us tip our heads forward. That's wrong. Hold your head high and keep your chin parallel to the ground. Next, align your shoulders over your hips. Don't pull your shoulders way back—just keep them slightly back and down. Then, once you have your earlobes and shoulders in place, lift your breastbone and maintain a little hollow at the lower back. Keep your knees unlocked (1).

1

Of course, a good stance is only step one. If you're going to be posture-perfect, you'll need a good sitting style, too. Jane Sullivan-Durand,

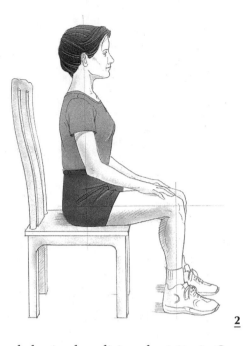

2

M.D., a behavioral medicine physician in Contoocook, New Hampshire, says that it's important to sit up straight on the front edge of your chair so that you're not touching the back. Whenever possible, make sure your head is sitting square on your shoulders and your shoulders are back and down (2). This military style will help straighten your spine. Sitting properly can prevent rounded shoulders and minimize the strain on the ligaments and muscles in your back. Try looking at yourself sideways in a mirror to check if you're doing it right.

Proper Posture: Part II

Okay, now that you've got the basics of standing and sitting up straight, here are some exercises to improve flexibility in your back and give you a healthy pose, says Meir Schneider, Ph.D, a licensed massage therapist, founder of the Center and School for Self-Healing in San Francisco,

and creator of the Meir Schneider Self-Healing Method. At all times, he adds, check your comfort level and stay within its limits. If you have back pain, consult a doctor before beginning these or any new exercises.

Go prone at home. This posture promoter allows you to finish that novel or plow through your junk mail while stretching your back muscles, says Anglin.

Lie flat on your stomach with your legs straight out. Prop yourself up on your elbows and forearms and allow your gaze to fall naturally downward while maintaining a neutral neck curve. Don't raise your head to look out at the room or to watch TV. Allow your body to stretch and relax (3).

"This is a passive stretch that helps you unwind from the modern-day life of sitting behind

3

a computer, inside a car, and on a couch," Anglin says. Do not attempt this exercise if you have spinal stenosis (the narrowing of passages through the vertebrae), she adds. Instead, try simply lying on your stomach and resting your forehead on a rolled towel to maintain a neutral neck alignment. Work up to resting on your forearms.

Stand tall like a tree. If you'd like to try something a little more exotic, Alice Christensen, founder and executive director of the American Yoga Association in Sarasota, Florida, offers this do-anywhere exercise that should enhance both your stance and your concentration.

<u>4</u>   <u>5</u>

►Stand with your legs together and your feet parallel to each other.

►Shift your weight to your right leg and place the heel of your left foot against your right ankle. You can hold onto a wall or countertop if you need support.

►Slowly raise your left foot until the heel rests on your right inner thigh. You can use your left hand to help get your foot into position (4).

►Stare at a spot on the opposite wall to help you balance. Let your arms drop loosely at your sides, then slowly raise them above your head as straight as possible until the palms meet (5).

►Relax your stomach and breathe deeply. Hold for several seconds, then slowly lower your left leg and repeat with your right leg.

Act like a cobra. Not much for trees? Dr. Schneider suggests this reptilian alternative. This exercise improves flexibility in the spine and strengthens those all-important abdominal muscles, says Dr. Schneider. Note that the arching in this exercise may be uncomfortable for people whose lower backs are too flat.

►Lie on your stomach on a mat or carpet and place your hands palms down on either side of your chest.

►Slowly raise your upper body from the floor as far as is comfortable by straightening your elbows, keeping your pelvis on the floor.

►Hold for two deep breaths and then slowly lower yourself. If it feels good, repeat 10 times.

►Still lying on your stomach, lift your legs off the floor as far as possible 10 times.

►If you're still feeling okay, place your hands on top of your buttocks and command your back muscles to lift your torso and head as far as possible. Pretend you're a cobra ready to attack. Hold that pose for two deep breaths and then slowly lower yourself to the floor. Repeat 10 times.

Play wall tennis. This exercise, recommended by Dr. Schneider, gets the tension out of tight back muscles so that you can stand taller. While it may not make you the next Pete Sampras or Steffi Graf, it should create more balance in the forces around the spine and help you to stand straight, he says. Skip this one, though, if you have torn back muscles, osteoporosis, or a compression fracture in the spine.

►Stand with your back against a wall, with your feet about 12 inches away from the wall and hip-width apart. Keep your knees straight.

►Place two tennis balls between the wall and your back, one on either side of your lower spine above your buttocks. (Make sure the balls are pressed against your back muscles, not against your spinal column.)

►Press yourself against the wall to keep the balls from falling.

▶Slowly bend your knees as far as you can, sliding your back downward against the wall so that the balls roll from your lower back up to your shoulders.

▶When the balls are between your shoulder blades, stretch your arms in front of you with your elbows straight and clasp your hands together.

▶Make circles with your hands, first in one direction and then in the other. You'll feel the tension in your upper back melt away.

▶Slowly straighten your knees, sliding your back upward against the wall and rolling the balls downward to your lower back.

▶When the balls are in the small of your back, bring one knee at a time to your chest, bending your head toward your knees.

▶Repeat five or six times, until you feel so refreshed that you're ready for a real tennis game.

Elevate your literary powers. Here's another use for that thick War and Peace–type novel that you've been promising yourself to read. Let that hefty tome improve your posture, says Dr. Schneider.

Kneel on all fours and place the book on your lower back. Move your lower back up and down without straining your abdominal muscles or dropping the book.

This exercise, says Dr. Schneider, activates your lower back, increasing its flexibility and mobility, and releases tension in your abdomen.

Two Straight-Up Tips

Even if you're tuned in to good posture most of the time, there may be situations in which you backslide. Here are Anglin's tactics for two of life's tough posture moments.

Get a leg up. Stuck with a sinkful of dishes? To get relief from standing for a long time and still maintain a healthy posture, Anglin suggests that you raise one foot. Either put a stool near the sink and put one foot on it or open a lower cabinet and place one foot on the ledge. If you have a lot to wash, switch feet midway through.

Be a posture-savvy spectator. Are your kids soccer phenoms? Do you spend a lot of time in the bleachers watching them get their kicks? Anglin offers this posture reminder.

"Most of us tend to sit on a bleacher and slump, straining our lower backs and rounding our shoulders," she says. "Sit more erect. When you do, you find your balance on your pelvis—that is, you shift your weight forward so your back doesn't take all the weight. It's definitely less of a strain," says Anglin.

PREMATURE EJACULATION

When Slower Is Better

In this wound-up world of ours, speed and efficiency usually win the biggest kudos. Meandering, moseying, dallying, and digressing are, more often than not, frowned upon as time wasters. But in the art of love, the man who takes his time is generally regarded as the more skillful paramour.

In most cases, timing trouble in bed—premature ejaculation, or climaxing sooner than desired—is little more than a troublesome response that can be changed, says Robert Birch, Ph.D., a psychologist and sex therapist in Columbus, Ohio, and author of *Male Sexual Endurance*.

Premature ejaculation can often be blamed on increased sensitivity in that already oh-so-sensitive body part, says Dr. Birch. In addition, some studies suggest that men who ejaculate rapidly are probably working in a faster gear than other men, because they also become sexually excited more quickly. The good news is that exercise can help you learn how to take your time and control your responses.

There are two good reasons for learning to hold your fire. First, a quick finish can compromise your partner's pleasure. Second, it can lessen your own. Lingering on the brink can be a sweet sensation.

Prolonging the Dance

"I think that most men would definitely like sex to last longer, for themselves as well as for their mates," says Dudley Danoff, M.D., senior attending urologic surgeon at Cedars-Sinai Medical Center in Los Angeles and author of *Superpotency*.

The best way to become master of your emissions requires a loving and patient partner. The method is called the stop/start technique, and it's been around since the 1950s. The goal is to teach a man to recognize when he is approaching the point of no return, otherwise known as the point of ejaculatory inevitability.

Here are the steps to successfully mastering the stop/start technique, according to Dr. Birch.

Begin by bestowing. Since working through the stop/start technique asks a lot of a partner's patience, be sure to give pleasure where it's due. Lovingly tending to your partner's needs first and bringing her to orgasm is a wonderful way to guarantee that the favor will be returned. If you find yourself ejaculating as you please your partner, don't worry about it. Have fun and try again the next night, because practicing frequently will benefit you both.

Ease into ecstasy. At this point, the genitals are definitely off-limits. Ask your partner to begin touching other parts of your body, like your chest and stomach, in a nonsexual way. This step should get you both relaxed and calm, but not necessarily aroused. If you do become erect, that's fine, too.

Get a little closer. Now you and your partner can let the passion warm up a bit. Once you are fully erect but still feeling in control, let your partner know that it's okay to begin touching your genitals, using only her dry hands.

While this is happening, use your imagination to picture yourself at the bottom of a ladder. As you become more excited, picture yourself climbing the ladder one rung at a time. Reaching the top of the ladder corresponds with reaching the point of no return, where you know you can't stop.

The trick at this stage is to keep your mental focus on just where you are on that ladder. If you keep track of where you are at all times, you'll be able to stop before "flying over the top."

Take a pit stop—or five. If you've been paying close attention to the sensation of your partner's touch, you'll know when you're nearing the point of no return. And, incidentally, it's better to stop too early than too late, says Dr. Birch.

When your arousal reaches a fever pitch, tell your partner to immediately stop what she's doing. Then just lie still and wait it out. Once you're positive that you're back at the bottom of the ladder—back in full control of your sensations—you can let your partner know it's okay to begin again.

Stop and start like this five or six times before climaxing. Continue being intimate this way—hands-on only—for at least two weeks, two or three times a week, if possible. That should be enough time for you to develop a real sense of control over how aroused you get before allowing yourself to release.

Get wet. The next step in the stop/start method brings you closer to the goal of penetration with control. For the next two weeks, after pleasing your partner, ask her to use a lubricant when touching your genitals. This will surely be more exciting, as it will feel more like real intercourse, but you must still remember to halt the action before you reach the top of the ladder. And again, stop and start five or six times, just as before.

Let your partner be tops. Finally! Now that you've developed a good sense of cruise control, it's time to get fully intimate with your partner. To help you continue your slow road to success, Dr. Birch recommends using the female-superior position to make love.

With your partner on top, your sexual sensations will be somewhat reduced, which is a bonus in this case. Plus, by being on the bottom, your movement will be limited, a good safeguard against becoming unwittingly carried away by your passion.

Consider condom control. Since premature ejaculation can often be blamed on increased sensitivity, try wearing a condom for a change if you don't normally use one, suggests Dr. Danoff. It may decrease sensation enough to prolong the pleasure for both of you.

Get a grip on yourself. If sexual encounters are few and far between, premature ejaculation is often a bigger problem. That's because the less frequently you ejaculate, the more quickly it's likely to happen when you finally do. Using the stop/start technique while you masturbate can help you find your own rhythm and increase your awareness of the stages along the ladder, explains Dr. Birch.

PROGRESSIVE MUSCLE RELAXATION

Relief for Tension, Pain, and Even Insomnia

WHAT IT IS

Tense? Angry? In pain? During a progressive muscle relaxation (PMR) session, you can unburden both your mind and your body simply by intensely contracting and then deeply releasing each muscle within your body.

Having trouble imagining how it works? Well, think about this: You're angry. You unconsciously clench your fists so hard that your knuckles turn white. When you become aware of the tension, you exhale completely and shake out your hands to try to let go of the intense feeling as well as the tight muscles.

Voilà, instant relaxation! Using PMR, you'll repeat the same contraction and relaxation process throughout your whole body.

One of the best things about PMR is that it can be done anywhere, at any time. All you need to do is find a quiet spot and lie down or sit comfortably. Most people like to close their eyes. Then you tighten a muscle or group of muscles in your body, starting either at the top, with your head, or at the bottom, with your feet. You hold the clenched muscles for 10 seconds or so, then release them, noticing the difference between the tension and the relaxation. After that set of muscles is relaxed, you go on to a new group of muscles, working your way down (or up) your body.

"It only takes about 15 minutes to go through all of the muscle groups in your body," says Alison Milburn, Ph.D., a licensed psychologist with Eastdale Psychology Group in Iowa City, Iowa. "And it's best to practice PMR at least four times a week."

WHAT IT DOES

Stress begets muscle tension, which begets pain, which begets more stress. Progressive muscle relaxation can help stop this cycle, says Dr. Milburn. "Progressive muscle relaxation is good for two reasons. First, it teaches you the difference between what a tense muscle feels like and what a relaxed muscle feels like, which is important so that you can recognize when your body needs to relax. Second, muscles are more likely to deeply relax if you tense them first."

According to a panel from the National Institutes of Health, relaxation techniques, including PMR, decrease oxygen consumption and lower breathing rate and blood pressure. Because of the physical action of tensing and then relaxing muscles, PMR actually creates a deeper relaxation in the body than stress-management techniques that utilize only the mind, such as imagery or visualization, says Dr. Milburn.

"Progressive relaxation is especially good for people who have a muscle tension component to their pain," says Dr. Milburn.

PMR can be particularly helpful for people who have either chronic pain or painful episodes, such as tension headaches or temporomandibular disorder (TMD), which causes pain and tension in the lower jaw and face, says Dan Hamner, M.D., a physiatrist and sports medicine specialist in New York City. You can also use it for a visit to the dentist or if you're having a medical procedure. If you know how to do PMR, you'll probably be able to reduce both the amount of tension and the amount of pain that you feel during and after the procedure, he says.

"Pain is really an attention-driven phenomenon, so relaxation techniques can serve as a distraction from pain or stress," says Dr. Milburn. "Relaxation can also be effective in treating insomnia because in addition to slowing down your heart and respiration rate, it also helps your mind slow down."

REAPING THE BENEFITS

"Progressive muscle relaxation is a learned technique," says Dr. Milburn. "One good way to start your practice is to make a recording of yourself reading out the instructions."

When you're making the tape and when you're practicing PMR, be sure to unplug the phone and get rid of any other distractions that might interrupt you. Also, feel free to use background music that you find relaxing. Or you can use a recording of a particular sound, such as the rain or waves, to help induce a state of relaxation.

"You can personalize your practice by con-

centrating on specific muscles that cause you tension," says Dr. Milburn. If you suffer from back spasms, for example, you might pay special attention to your upper or lower back during a relaxation session by doing more tensing and relaxing repetitions. But if you feel pain in your back while using full tension, stop tensing, ice the pain, and use a lower level of tension the next time you practice PMR, says Dr. Hamner.

Maybe you think that PMR sounds right for you, but you want some expert advice on the practice. "Find a psychologist who specializes in pain or stress management," says Dr. Milburn. "They will often teach their clients how to do PMR, and many of them will make tapes for their patients to practice with at home."

TRIAL RUN

To make a PMR tape for yourself, you can use the following script just as it is or adjust it to suit your own needs. You may want to start at your feet rather than at your head, for example, or you may pay special attention to specific parts of the body in which you carry stress or have some pain by repeating the tense-and-relax cycles in these areas more often.

Either way, read the script a few times before you tape it so you don't stumble over words as you record. Use the tape as your guide through a few sessions and then try the technique without it. "Eventually, you'll be able to relax very early in the monologue," says Dr. Milburn. "The very act of starting to relax yourself will begin to reduce your fight-or-flight response. In other words, the practice itself will become a signal that it's time to relax."

Practice your progressive muscle relaxation

four times a week in order to become proficient at it. And don't practice when your pain is severe. Wait until you're comfortable with the process before you use it as a solution for pain or insomnia, says Dr. Milburn.

When you're recording the tape, go slowly to give yourself enough time to hold the tension. Or if you prefer, use a stopwatch or time yourself with a watch that has a second hand so that you can time 10 seconds of tension alternating with 10 to 20 seconds of relaxation. Don't count out loud, though—use that time to let some quiet into the practice.
Here's a possible script for your recording.

1. Breathe in deeply through your nose. Raise your eyebrows and wrinkle up your forehead, making your whole head tense. Hold for a count of 10 and then release. Exhale completely through your mouth.

2. Breathe in deeply through your nose. Squeeze your eyebrows together and scrunch up your nose. Hold for a count of 10, then release. Exhale completely through your mouth.

3. Breathe in deeply through your nose. Pull your lips into a grimace. After a count of 10, exhale completely through your mouth.

4. Breathe in deeply through your nose. Squeeze your shoulders up to your ears. Hold for 10 seconds, then relax. Breathe out completely through your nose.

5. Breathe in deeply through your nose. Expand your lungs deeply to create tension in your chest. Hold for a count of 10, then relax. Exhale completely through your nose.

6. Breathe in deeply through your nose. Pull your arms in tight to the sides of your body, tensing your biceps and triceps. Hold for a count of 10, then relax. Exhale completely through your mouth.

7. Breathe in deeply through your nose. Tense your forearms by holding your arms straight out tightly. Hold for a count of 10, then let go. Exhale completely through your mouth.

8. Breathe in deeply through your nose. Squeeze your hands and fingers, making tight fists. Hold for a count of 10, then relax. Exhale completely through your mouth.

9. Breathe in deeply through your nose. Pull your stomach in hard, creating tension in the center of your torso. Hold for 10 seconds, then exhale completely through your mouth.

10. Breathe in deeply through your nose. Squeeze your buttocks muscles together. Hold tightly for 10 seconds, then let go. Exhale completely through your mouth.

11. Breathe in deeply through your nose. Squeeze your thighs together. If you can't feel these muscles tense enough, point or flex your toes enough to raise your legs off the floor or chair. Let go after 10 seconds. Exhale completely through your mouth.

12. Breathe in deeply through your nose. Tense your calves and lower legs by flexing or pointing your feet. Hold for 10 seconds, then relax. Exhale completely through your mouth.

13. Breathe in deeply through your nose. Scrunch your feet and toes into balls. Squeeze for 10 seconds, then relax. Exhale completely through your mouth. (If you get bad cramps in your feet, skip this step of tensing and relaxing.)

14. Breathe in deeply through your nose. Do a mental scan of your body to find those areas that are still tense. Return to these tension spots and repeat the tense/relax cycle. Exhale completely through your mouth.

Feel free to repeat this script a couple of times on the same tape, so you won't have to interrupt your relaxation to rewind it.

Raynaud's Disease

Delivering Warm Thoughts

Cam Vuksinich loves skiing down a fresh snow-fall on a Colorado mountainside. And she doesn't freeze up in the face of Raynaud's disease.

Although she was diagnosed with this condition at age 15, Vuksinich has learned a safe and effective way to control it and still pursue her favorite pastime. Instead of medications, she relies on a mental technique known as autogenics to restore warmth to her suddenly chilled fingers.

"All of a sudden, I feel a tingling sensation and my fingers go white in less than a minute and feel as if they are frozen and numb," says Vuksinich, now in her midforties, a certified personal trainer and a certified clinical hypnotherapist in Denver. "It's strange. It can happen with or without gloves, even in the summer. What triggers it seems to be a shock or change in temperature."

Vuksinich is among the 5 to 10 percent of Americans with Raynaud's disease, a phenomenon that causes a big chill and a big hurt to fingers, toes, or both. Doctors know that it is caused by a spasm of the blood vessels that triggers a sudden temperature drop in the hands or feet because of poor blood flow to these extremities. In severe cases, a spasm can cause blood flow to stop.

Although medical experts aren't sure why it happens and have yet to find a cure, they do offer ways to kill the chill.

Playing Mind Games

Thinking warm thoughts—using your mind as a powerful weapon—is one of the best ways to counter the icy nature of Raynaud's disease. Two techniques with great success records are biofeedback and autogenics, say doctors.

Warm up with biofeedback. "Biofeedback is the treatment of choice against Raynaud's," says Jo Anne Herman, R.N., Ph.D., a biofeedback expert and associate professor of nursing at the University of South Carolina in Columbia. When blood vessels constrict, your fingers may turn blue or white, depending on the severity of the spasm; blue indicates a severe spasm and white a moderate one. "The muscles around the artery just clamp out all the blood. When your fingers turn red, that means the blood is returning," she says.

To effectively use biofeedback, experts recommend that people with Raynaud's first learn the technique from a trained therapist. When you go to the therapist, you'll be hooked up to a machine that provides data on certain body functions, such as finger temperature, blood pressure, heart rate, and brain wave activity.

"The goal of biofeedback for treating Raynaud's disease is to learn how to mentally increase finger temperature," says Dr. Herman. "The more relaxed you are, the more your arteries dilate and bring more warm blood to the cold areas."

Turn up your body's furnace. Autogenics is another effective way to tap into your mental warming images. Vuksinich is among those who rely on autogenics to control their unexpected Raynaud episodes.

Autogenics literally means "self-regulation." It is an easily learned technique that uses your conscious mind to supply proper cues to teach your body to relax.

Within five minutes of engaging autogenics, Vuksinich's fingers are flexible and rosy and capable of maneuvering her ski poles down the slope. "People have watched me do this and seen the color come back to my fingers and are just amazed," she says.

Although autogenic cues vary with each individual, Vuksinich offers her step-by-step process for restoring warmth to her fingers.

▶Stand still or sit down and close your eyes.

▶Take a few deep breaths, expanding your diaphragm as you inhale slowly and deeply, then exhale slowly, effortlessly, and completely. These breaths awaken the body to its own natural state of deep relaxation.

▶Say "I am" out loud on the first inhalation. Then as you exhale, say "relaxed" out loud. These are key trigger words that open the door to relaxation.

▶With each subsequent inhalation, say to yourself, "I am." With each exhalation, silently say "relaxed." Say it slowly. Follow your natural breathing patterns.

▶Imagine a ball of energy or light flowing down from your head and spine, branching at your heart, and then flowing down both arms and into each finger. Also imagine it flowing down your legs to your feet. Allow it to go out of your feet into the ground to give you a feeling of stability.

▶Imagine with every inhalation that this ball of energy or light recharges to the perfect temperature for you. Imagine with every exhalation that it flows through your body and relaxes and warms you.

▶Stay relaxed and allow this image to happen. Don't try to rush or force it.

"The key is to be still and focus all your attention in what I call relaxed awareness," says Vuksinich. "For me, this mental imagery of directing warmth through my body works best. It takes some practice to learn, but once you do, you can address it immediately and return warmth to your fingers."

Moving Warmly

Experts say that the power of motion may also defrost Raynaud's disease in some people. Here are some of their top choices.

Calm the palms. Walking from the sunny outdoors into an air-conditioned living room can often send shivers down the fingers of people with Raynaud's disease, says Joan Merrill, M.D., assistant chief of rheumatology at St. Luke's–Roosevelt Hospital Center in New York City.

The next time you find yourself moving from a warm house to the cold outdoors or from the warm outdoors into an air-conditioned building, try this discreet, easy, hand-warming movement.

With your hands at your sides, bend your fingers and tuck them into your palms as if you were making loose fists. This allows your fingers to warm each other. Let a few minutes pass for your body to acclimate to the new, cooler temperature before releasing your fists, recommends Dr. Merrill.

Keep your fingers moving. Fingers that are chilled by Raynaud's disease can often stiffen to the point of immobility, cautions Dr. Merrill. To keep your fingers warm and limber, she suggests that you head to the nearest sink the next time you feel the temperature plummet in your fingers. "Plunge your hands into a sink or basin of moderately warm water and wiggle your fingers around until they move freely," she suggests.

Practice glove love. Simple tasks like grabbing a bag of broccoli out of the freezer or rinsing out your panty hose in cold water can trigger a spasm and constrict the circulation, says Dr. Merrill.

The solution? Keep a pair of oven mitts within reach of the freezer door and pull on a pair of oversize rubber gloves whem you wash panty hose, she recommends.

Lap up some heat. The next time your hands turn chilly, drink two cups of water to increase blood volume and then find a comfortable chair. Sit and place your hands on your lap in a relaxed position, says Robert Markison, M.D., a hand surgeon and associate clinical professor of surgery at the University of California, San Francisco.

"Your best thermometers are your cheeks," says Dr. Markison. If you place your hands against your cheeks and your hands don't feel warm, place them on your lap and take a few deep, relaxed breaths, he suggests. "Have the image of relaxing all the muscles in the area of the chest wall and the collarbone," he says. "This is the muscle web intertwined with blood vessels. Use this imagery to warm your hands and improve blood flow. You should find after a few minutes that you've warmed your hands by a few degrees."

REAR-END SAGGING

Make Your Own Happy Ending

Mark Twain was wrong about death and taxes. They're not the only two sure things in life. There's a third certainty: some downward drift in our derrieres. While there's nothing wrong with some southerly migration, and some consider it a sign of wisdom and experience, nobody wants their buns to bottom out before their time. Exercise can slow the slippage.

Here's some background on your background. The buttocks, or gluteal muscles, don't actually sag, says Robyn Stuhr, exercise physiologist at the Women's Sports Center at the Hospital for Special Surgery in New York City. Rather, they shrink from lack of exercise, so they can't hold up their end of the bargain. The fat and skin covering the glutes have nowhere to go but down.

Improving Your Bottom Line

To keep your buttocks muscles strong and supportive, Stuhr recommends doing these butt-boosting exercises three times a week.

Do the butt beautifier. This is a variation on the old "donkey kick" maneuver.

▶ Start on your hands and knees, with your hands directly under your shoulders and your knees directly under your hips. Don't round your back or let your stomach sag.

▶ Lift your right leg behind you while keeping the knee bent at around 90 degrees. Try to raise your right thigh so that it's lined up with your right buttock (1). Don't overarch your back or kick the leg up. Instead, bring it up slowly and smoothly.

1

▶ Hold for a count of three, then lower your leg almost to the starting position, but don't touch your knee to the floor.

▶ Raise your leg again and repeat 10 to 15 times if you can (you might need to start with a more modest number). Switch sides and repeat.

Get cute with a glute semi-squat. This exercise is what Stuhr calls a functional movement—one that mimics an everyday activity. "This seated squat variation strengthens muscles that we rely on all the time," she says. "Not only does it work the glutes, but the quads and hamstrings get involved as well."

▶Start by sitting on the edge of a sturdy chair (do not use a chair with wheels) with your feet flat on the floor about shoulder-width apart. Your knees should be directly over your ankles.

▶Lift your head to sit tall, keeping your shoulders back. Extend your arms in front of you for balance (2*a*).

▶Using your buttocks and leg muscles, stand up slowly, keeping your back as upright as possible. Be sure that your knees do not extend beyond your toes (2*b*).

▶Pause for a moment when you reach a standing position, then prepare to reverse directions (2*c*).

▶Begin to sit down slowly. Thrust your buttocks way out behind you as you aim for the chair. Don't let your weight sink until you feel the chair under you.

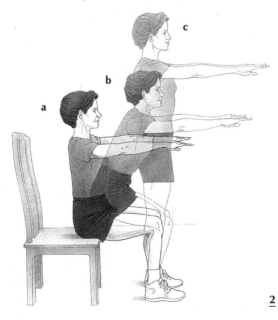

2

▶Let your buttocks tap the chair, sit tall, and begin moving your body to stand up again.

▶Repeat 10 to 15 times.

If your fanny isn't where it used to be, try these fun ways to boost your caboose.

It's Better on a Bike

It's a big modern-day fitness mistake—driving *everywhere*, even a quarter-mile down the road to Betty's house. For quick trips, get out of the car habit. The next time you have to buy stamps, scoot down to the post office on your Schwinn. Why? Cycling is a great tush tightener. Get spinning whenever you can.

Ride in place. When you visit the gym, keep your rear in gear by making a date with an exercise bike three times a week, says Stuhr.

Get the kids in the act. Start a new family tradition—make Saturday road-trip day with a bike outing. It doesn't have to be anything too ambitious. A ride into town for breakfast could be just great. Just pick a destination, strap on your helmets, and roll on. Keep your eyes open for bike paths: Many communities now have them, and they're perfect for easy, safe riding.

Climb to a Better Backside

Climbing stairs is a serious rear-reducer. But in a world of elevators and escalators, taking the stairs may take some ingenuity. Here are some ways to do it more often.

Make some machine magic. Stair-climbing machines are common at just about every gym. Experts say that climbing in place for 20 minutes three times a week will do the job.

Stick with stairs. Does it seem that there's always one too many flights of stairs to climb at the end of the day? Well, think of it this way: With every flight you climb, your butt rises with you. You're not wasting energy—you're working out!

Swear off elevators. If you work or live anywhere above the ground floor, you—and your rump—are in luck. You've got a built-in, no-excuses opportunity to take some stairs. But to make sure you don't overextend and give up, be reasonable. If you work on the 27th floor, walk up a few flights and take the elevator from there.

Join the Dance

Ballet dancers have bodies that seem immune to the effects of gravity and time. To get a taut "ballet butt" of your own, try these tips recommended by ballet-master-in-chief Peter Martins in his book *The New York City Ballet Workout.*

Plié, please. Stand with your back straight and your feet turned to either side as far as they'll go without straining. Make a circle with your arms and hold them in front of your hips. Then bend from your knees and slowly descend, raising your arms and leading with your elbows. You should lower yourself far enough to feel your backside muscles working. Hold the position for a beat or two at the bottom and then slowly come back up. Do 10 to 15 repetitions once a day.

Sweep and lift. Develop a nicely rounded derriere with this simplified version of the arabesque. Use your left hand to hold onto a chair or countertop for balance. Keeping your back and neck upright, shift your weight to your left leg and move your right leg out to the side. As your leg straightens, gracefully point your toes as much as you can.

Keeping your toes pointed and touching the floor, sweep your straight leg around to point behind you. Without arching your back, lift your leg back and off the ground no more than a few inches. Hold there for a beat or two, then lower your leg and lift it again. Do 10 to 15 repetitions, then switch sides and repeat.

REFLEXOLOGY

Fancy Footwork

WHAT IT IS

Reflexology is a method of pressure-point therapy used specifically for healing purposes. Its therapeutic theory is that certain reflex areas and points in the feet correspond to every organ and gland in the body and that there are zones in the feet that are recognized as energy pathways that correspond to energy zones in the body (1). According to practitioners, "working," or applying specific pressures to these points, has a direct and powerful effect on the parts of the body with which they are associated. Reflexology may also be done on the hands and ears.

If you've never had a foot reflexology treatment, be prepared for a sole-soothing experience. The feet, in case you haven't noticed, are exquisitely sensitive, so touching them is sheer sensual indulgence, says Laura Norman, a nationally certified reflexologist, reflexology trainer, and director of the Laura Norman Reflexology Center in New York City. "There are over 7,000 nerve endings in each foot," says Norman, "and most of the time, they're kept layered under socks and shoes." Reflexology, she says, helps bring attention to a touch-starved part of our bodies.

In some ways, foot reflexology is similar to massage therapy. The setting is usually a quiet room, with the reflexee resting comfortably on a padded table. But there's no need to undress any further than removing shoes and socks, which is a plus for those who may shy away from the disrobing that's common when receiving full-body massage.

Reflexology is also very user-friendly. With a little knowledge, you can experience its healing effects by working your own feet.

WHAT IT DOES

Reflexology offers some very clear health benefits. "One of the prime effects of reflexology is the relief of tension," says Ray Wunderlich, M.D., head of the Wunderlich Center for Nutritional Medicine in St. Petersburg, Florida. Having a reflexology session induces a noticeably deep state of relaxation, which he says can be very beneficial to ailments such as migraine and PMS.

But the effects of reflexology extend beyond simple tension relief. It can "tone up," and restore better function to some organs, according to Dr. Wunderlich. Some problems that experts believe are often helped by reflexology are high blood pressure, constipation, lung and sinus problems, hyperactivity in chil-

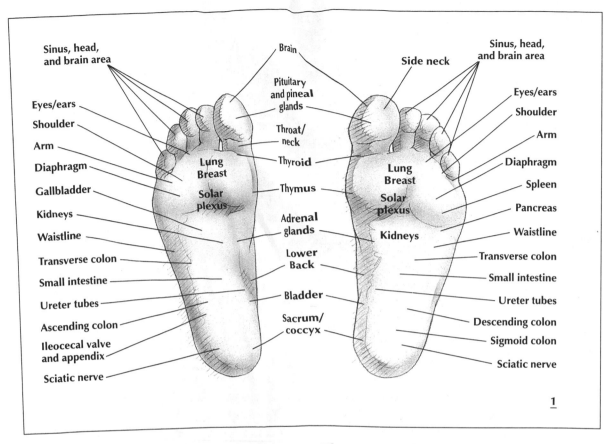

Sinus, head, and brain area
Eyes/ears
Shoulder
Arm
Diaphragm
Gallbladder
Kidneys
Waistline
Transverse colon
Small intestine
Ureter tubes
Ascending colon
Ileocecal valve and appendix
Sciatic nerve

Brain
Pituitary and pineal glands
Throat/neck
Thyroid
Thymus
Adrenal glands
Lower Back
Bladder
Sacrum/coccyx

Side neck
Sinus, head, and brain area
Eyes/ears
Shoulder
Arm
Diaphragm
Spleen
Pancreas
Waistline
Transverse colon
Small intestine
Ureter tubes
Descending colon
Sigmoid colon
Sciatic nerve

Lung Breast
Solar plexus

Lung Breast
Solar plexus
Kidneys

1

dren, poor digestion, obesity, anxiety, athletic injuries, and irritable bowel syndrome.

How can this kind of footwork have such far-ranging results? There's more than just reflexology's soothing side effects at work here, says Terry Oleson, Ph.D., chair of the department of psychology and director of behavioral medicine at the California Graduate Institute in Los Angeles. Rubbing reflex points on the feet has a circuit-breaker-like effect, says Dr. Oleson. "The 'reflex' in reflexology refers to the nerve pathways—or neurological loops—that deliver sensory information throughout the body," he says.

A study performed by Dr. Oleson and Bill Flocco, a certified reflexologist with the American Academy of Reflexology in Los Angeles, involved 35 women with symptoms of PMS and proved the power of these mapped-out reflexology points. Half of the women got the real thing—ailment-specific reflexology treatments given by trained practitioners. The others received pleasant but otherwise medically powerless foot rubs. If relaxation were enough to do the trick, all of the reflexees should have shown the same positive response. But in fact, the women who received true targeted reflexology found much more relief from their monthly woes than those who simply had a soothing foot massage.

When nerves in the feet are worked, says Dr. Oleson, a stimulating message is routed through the spine and out to far-flung parts of the body like the arms, hands, neck, head, and even the internal organs. Reflexing the big toe, for instance, increases blood flow to the head and brain. And pressing the arch of the foot helps relieve aching back muscles. The trick to reaching the right destination, he says, is to have a map—a foot-shaped directory of where to push for what problem.

REAPING THE BENEFITS

Here are some expert recommendations for getting the most from your reflexology session. (*Note:* These tips apply whether you work on yourself or have a professional handle your feet.) If you'd like to find a certified reflexologist in your area, contact the American Reflexology Certification Board, P. O. Box 620607, Littleton, CO 80162. Include your name, phone number (including area code), city, state, and ZIP code when you make your request.

Give it a month. Reflexology may help chronic ailments like digestive problems or high blood pressure, but it takes a little time. "Try having treatments once or twice a week for a month," says Dr. Wunderlich. "By then you should be able to determine whether it's having the desired effect."

Scoot around sore spots. If you should hit a tender spot when working your own feet, move on and come back to it later, says Norman. Then press the area gently until the soreness eases. Likewise, if you feel pain during treatment by someone else, let her know what you're feeling. Reflexology should relax you, not make you writhe. ·

Skip the pre-reflex repast. "You really shouldn't eat a big meal before a reflexology session," says Norman. "Digestion and relaxation are conflicting bodily processes."

TRIAL RUN

At the end of a tough day, there's nothing quite like having your feet worked by an attentive slave—uh, make that companion. But even when you're all alone, you can experience the sleep-inducing effects of reflexology, says Norman. Try her do-it-yourself anti-insomnia reflexology workout to get you ready for bed. (This routine can also be used on a partner.)

Getting started. Sit comfortably in a quiet room. Using a light, absorbent, greaseless lotion, massage both feet with squeezing, stroking, kneading, and wringing motions to prepare for the reflexology to come. Then sprinkle your feet with cornstarch body powder and massage until all traces of lotion are gone.

Working the feet. Hold the ankle, heel, or toes of one foot firmly in one hand. Place the thumb of your other hand on the sole of your foot near the inner heel. Then apply steady, even pressure with the edge or ball of your thumb, keeping the thumb slightly bent at the joint. Using a forward, caterpillar-like motion called thumb walking, press one spot, move forward a half-inch, press again, and so on. When you reach the toes, return to the heel area, pick a new starting point, and repeat the process.

Continue until the entire bottom of your foot has been worked. Then do the top of your foot, walking with your fingers instead of with your thumb, in the same way. Switch to the other foot and complete your treatment—if you can stay awake that long.

RESISTANCE TRAINING

More Important the Older You Get

It's Sunday morning, and a dozen sweaty weight lifters are putting a variety of exercise machines through their clanging paces in the gym at Carleton University in Ottawa. Another dozen or so are hefting barbells and dumbbells, up and down and from side to side.

It's pretty much what you'd expect at a gym, except for a few particulars. All of the weight lifters here are 50 or older, and the music being piped through the sound system is Schubert.

"On Sunday mornings, they play waltzes and other music that people in the program like," explains Diane Withnall, a 56-year-old grandmother enrolled in the university's Weight Lifting for Older Adults program, a twice-weekly course that teaches 50-and-up adults the ins and outs of resistance training.

"It's really been very good for me," she says. "Since I started lifting weights, I can do all sorts of things—go up the stairs, carry groceries— much more easily."

Resistance training isn't just for the 20- something Olympian brutes that you see on TV, notes Greg Poole, an exercise physiologist and assistant director of programs at the university. "Older adults have a great deal to gain from weight training."

WHAT IT DOES

Also known as weight training and strength training, resistance training actually includes a number of related activities—workouts with weight-training machines, with stretchy elastic exercise bands, with dumbbells, and with "equipment" that you can find around the house. Two plastic milk jugs filled with water, for instance, make a perfectly respectable set of weights for resistance training.

All resistance-training programs require you to move your muscles against some sort of resistance, explains Poole. That might be the resistance supplied by the elastic exercise band that you're going all-out to stretch or by gravity as it pulls unflaggingly on the barbell you're trying to lift.

Resistance training offers a wide range of benefits. For starters, it's your very best bet if you want to build muscle and get stronger. Resistance training won't build big, bulky muscles, unless you stick to it for hours and hours a day. But it does build what are known as type two muscle fibers, explains Wade Lillegard, M.D., co-director of the sports medicine division of the department of orthopedics at St. Mary's Duluth Clinic in Minnesota. Type two fibers provide short-term, high-intensity power—just the kind you need to run up a flight of stairs to a train platform or heft a Thanksgiving's worth of groceries. The greater the resistance, the greater the muscle gain.

Because it builds muscle, weight training can also help you lose weight and keep it off. Pound for pound, muscle burns considerably more calories than body fat does, notes Susan Puhl, Ph.D., associate professor of physical education and coordinator of adult fitness at the State University of New York in Cortland. So the more muscle you have, the easier it is to slim down.

Weight training can also build your endurance, says Philip Ades, M.D., professor of medicine and director of cardiac rehabilitation and preventive cardiology at the University of Vermont College of Medicine in Burlington. In a study headed by Dr. Ades, elderly adults who worked out with weights for three months were able to walk longer without feeling fatigued. Stronger muscles simply don't become fatigued as quickly as weak ones.

Since stronger muscles do a better job of holding joints in their proper places, resistance training can lessen the joint wear and tear associated with osteoarthritis, the type of arthritis that most often afflicts older adults. What's more, studies find, weight training can strengthen your bones, offering added insurance against osteoporosis. That's because your bones and muscles are intimately connected. When you work your muscles against resistance, they pull on the bones they're attached to. In medical lingo, your muscles exert stress on your bones, and your bones, under stress, respond by laying down more calcium to reinforce themselves, explains Dr. Ades.

Resistance training can even improve your posture and coordination, says Poole. That's because out-of-balance musculature—say, strong chest muscles but weak back muscles—often contributes to rounded shoulders, a rounded back, and other posture problems. "Weight training helps with posture because it helps build muscle groups so that they're properly balanced," he says.

REAPING THE BENEFITS

If you'd like to start throwing some weight around, you have some options. You can join a gym somewhere. Better yet, get a few items and set up an area at home. All you need are a couple of milk jugs, a chair, a blanket or two, and some inexpensive ankle weights. You'll want to start with a half-gallon milk jug and move to a gallon jug as your strength increases. Use water to fill the jugs to the appropriate weight, keeping in mind that two cups of water equals a pound. The ankle weights you use should be no more than two pounds.

Before you get started on any program, the experts suggest you commit to memory these 10 Commandments of Resistance Training.

1. Ask Marcus Welby. Have your doctor okay any weight-training program that you plan to follow. This is essential, particularly if you have high blood pressure or heart disease or an inflammatory disorder like rheumatoid arthritis, says Dr. Ades. Your blood pressure can rise temporarily while you're lifting weights, he notes. And training can hurt your joints if you do it when they're inflamed. "When the inflammation settles down, though, it's okay," he says.

2. Favor your feet. Wear sturdy footwear—tennis shoes or walking shoes are fine—that will give you good traction and help prevent slips and falls, suggests Dr. Puhl.

3. Turn up your thermostat. Do the warmup exercises on page 334 before moving on to the resistance exercises that follow. "Warming up warms your muscles and loosens your joints a little, so you're able go through a full range of motion with decreased risk of injury," explains Dr. Puhl.

4. Remember to breathe. Remind yourself to inhale *and* exhale while you exercise. Lots of people hold their breath when they bear up under the resistance of a pugnacious exercise band or the unrelenting pull of gravity. "Holding your breath under a strain raises your blood pressure in and of itself," says Poole.

5. Think "slow motion." Move slowly and steadily through every exercise. The longer you work against the resistance, the greater the benefits, says Dr. Ades.

6. Chill and stretch. Once you're finished with your resistance exercises, cool down (you can follow the cooldown and stretching exercises on page 337). "Resistance exercise raises your heart rate somewhat, and cooling down will safely bring your heart rate back down to normal," says Dr. Puhl. The stretches? They'll keep you from turning into a stiff.

7. Make it semiweekly. Fifteen minutes of resistance training for two or three workouts a week should do the trick, says Dr. Puhl. There's no need to go overboard.

8. Rest for a day in between. Take at least one day off between resistance training workouts, says Dr. Lillegard. Muscles need at least 48 hours to rebuild after training. If you want to work out on consecutive days, alternate the muscle groups that you exercise by working your legs one day and your arms the next, for example.

9. Stop if it hurts. Resistance exercises should leave your muscles feeling a bit tingly and warm but not painful. If an exercise hurts, stop and check the instructions or consult an instructor. If your muscle pain is more than aches related to the exercise and persists for more than a week or worsens, it should be evaluated by a doctor, advises Dr. Lillegard.

10. Mount more resistance. Every time you increase the resistance your muscles are working against, you build more strength. To keep getting stronger, you can either repeat each exercise more frequently or add resistance. If you're following the program below, you can do that by using small arm and ankle weights during your workout, says Dr. Puhl. Start with one-pound weights if you haven't been working out. If you've already been exercising, begin with two-pound weights. These weights are sold at most sporting goods stores.

You don't have to keep increasing the resistance indefinitely, though. "Once you've reached the point where you're satisfied with your strength, you can maintain it by sticking with the number of reps and the resistance you're used to," Dr. Puhl says.

TRIAL RUN

Always take 5 to 10 minutes to warm up and stretch before you do resistance exercises and 5 to 10 minutes to cool down and stretch afterward, says Dr. Puhl.

That routine will help prevent injuries and keep you from stiffening up. And it's easy. Simply do the following exercises from the American College of Sports Medicine's *ACSM Fitness Book*, of which Dr. Puhl was co-author.

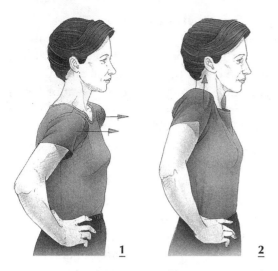

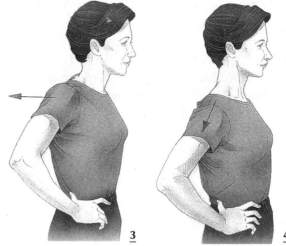

<u>1</u>　　　　<u>2</u>　　　　<u>3</u>　　　　<u>4</u>

The Warmup

Hold each of the following positions for 10 to 20 seconds—without bouncing—then relax for a moment or so before moving on to the next.

Roll your shoulders. While standing or sitting, put your hands on your hips and rotate both shoulders together, tracing a circle with each shoulder (*1, 2, 3, and 4*). Circle forward three times, then backward three times. You should feel the stretch in both shoulders.

Tilt your head. Stand or sit straight with your head upright (*5*). Slowly tilt your head forward, bending until you feel the stretch in

the back of your neck (*6*). Hold, then return to the starting position (*7*). Then slowly tilt your head back (*8*).

Walk in place. Step up and down at a moderate pace for two minutes.

Cherry-pick. Stand straight with your arms at your sides (*9*). Reach straight up with your left arm as if you were plucking a cherry from a tree directly overhead (*10*). Feel the stretch in your left side. Hold, then do five repetitions. Repeat with your right arm.

Imitate your cat. Begin by standing with your feet shoulder-width apart and your knees

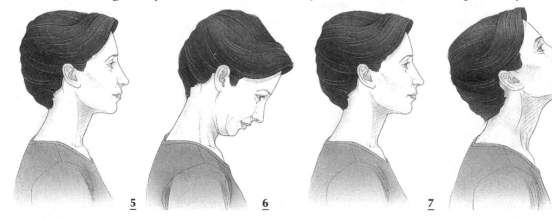

<u>5</u>　　　　<u>6</u>　　　　<u>7</u>　　　　<u>8</u>

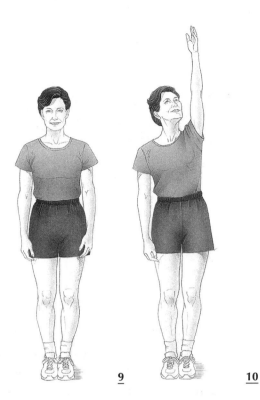

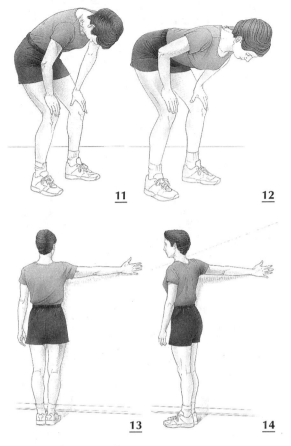

slightly bent, then bend at the waist until you can touch your fingertips to your knees. Rest the heels of your hands on your lower thighs and bend a bit farther so that your elbows jut out, keeping your hands on your thighs. Round your back (*11*) and then straighten it, transferring your weight through your hands to your lower thighs (*12*). You should feel the stretch in your chest and back. Hold, then repeat twice.

Stretch your chest. Stand facing a wall, about six inches away. Reach out to the side and place your right hand on the wall at shoulder level (*13*). Turn away from your outstretched arm, feeling the stretch in the side of your chest nearest the wall, and hold (*14*). Then face the wall again and repeat the stretch to the opposite side.

Lean for your legs. Stand facing a wall, about six inches away. Place the palms of both hands on the wall about shoulder-width apart, then bend your right knee slightly and step back with your left foot (*15a*). Lean into the wall until you feel a stretch in your left calf (*15b*). Hold, then switch positions and repeat with the other leg.

Stretch your quads. Stand next to a table and rest your right hand on the tabletop to steady yourself. Bend your left leg at the knee, grab your left foot with your left hand, and gently pull your foot straight up behind your left buttock. Push your foot against your left hand, away from your buttock (*16*). You should feel the stretch in your upper thigh. Hold, then change positions and repeat with the other leg.

16

17

The Workout

The following at-home resistance exercises are also from the *ACSM Fitness Book*. For starters, you can use the household items described, but after you're more accustomed to the routine, you might want to get "official" dumbbells, a bench, and an exercise pad at a sporting goods store. The first week, do each of the following exercises twice. Each week after that, increase the number by two, until by the end of six weeks, you're doing each exercise 14 times.

Shore up the walls. To build your arm, chest, and shoulder muscles, stand facing a wall, one or two steps away. Reach out and place your palms on the wall at shoulder height (*17a*). Slowly bend your elbows outward, lowering your body toward the wall (*17b*). Then slowly push away.

Lift a jug. To exercise your arms and shoulders, fill a milk jug with enough sand or water so that you feel some tension in your arm and shoulder when you hold it. Start with one to two pounds and add more if you need it.

Place the filled milk jug near your right foot. Stand with your knees bent and your feet shoulder-width apart. Bend at the waist, put your left hand on your left knee for support, and pick up the jug in your right hand (*18a*). Bending your right arm at the elbow, pull the jug up toward your shoulder (*18b*). Then slowly ease it toward the floor, stopping at your knee. Continue lifting and lowering to complete a set, then repeat with the other arm.

18

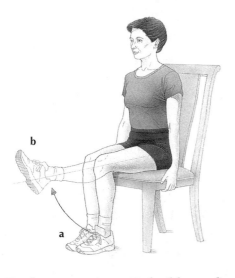

19

Try leg extensions. To build your leg and hip muscles, sit in a sturdy, straight-backed chair with your feet flat on the floor and your hands holding on to the sides of the chair (*19a*). Slowly straighten your left leg, keeping your thigh on the chair (*19b*). Slowly lower your leg, then repeat with your right leg. Eventually, as your strength increases, you can lift your thigh off the chair.

Crunch. To tighten your abdominal muscles, fold a blanket in thirds the long way and lay it down as a pad. Lie on your back on the blanket with your knees bent and your feet flat

20

21

on the floor. Cross your arms over your chest (*20*). Keeping your arms crossed and your shoulders and neck as straight as possible, slowly lift your head and shoulders off the floor (*21*). Slowly return to the starting position.

The Cooldown

Repeat the shoulder roll, chest stretch, leg lean, and thigh exercises that you did in the warmup. Then do the following stretches.

Touch your toes. Sit on a folded blanket or exercise mat with your legs straight in front of you and your knees slightly bent (*22a*). Reach out with both arms and try to touch your toes (*22b*). You should feel the stretch in your legs and hips. Hold. then repeat three to five times.

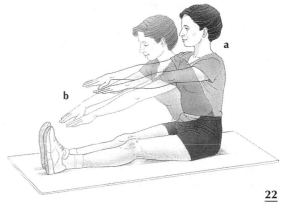

22

Stretch out. Lie on your back on a folded blanket or exercise pad. With your legs straight and your arms stretched back over your head, press your lower back to the floor as you reach overhead (*23*). Hold, feeling the stretch in your back and chest. Repeat three to five times.

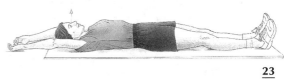

23

337

RETIREMENT BLUES

Make Getting in Shape Your Job Description

It wasn't until Agnes McGlone retired that she realized how central work was to her social life.

"I really missed the daily contact with people," says McGlone, who retired from a job as a school activities coordinator at 63. She was lonely and a bit blue. A widow, she lived alone in a condo complex where the neighbors said hello, but not much more. But retirement got considerably more enjoyable when McGlone started exercising.

"I take exercise classes three times a week, I walk every day, and I have square dancing every Sunday night," she says. "The exercise classes are very social, which is one of the reasons I look forward to them."

Social isolation is just one of what might be called the nonoccupational hazards of retirement, according to Ross Goldstein, Ph.D., a psychologist and president of Generation Insights, a San Francisco consulting firm. The other common laments of retired people? Their postjob lives lack structure, and they have no arena for achievement.

Exercise can fill each of these voids, so now that you've finally got the time to perfect your backhand or your backstroke, cash in on all of your retirement benefits. Make exercise a part of your daily routine.

Wanted: The Structure of 9 to 5

After retirement, some people report that their days have no structure, that one day just drifts into the next. Exercise can add the routine that's missing if you make working out the skeleton of your daily schedule.

Design a plan. Schedule time for exercise, says Carol Goldberg, Ph.D., a clinical psychologist and president of Getting Ahead Programs, a New York–based corporation that offers workshops on stress management, health, and wellness. Commit yourself. Make a standing appointment to go running at the same time every Monday, Wednesday, and Friday. On the other days, you could take adult education classes in low-impact aerobics or swimming. Variety is best for your body and adds interest for your mind.

"What you want to avoid is having one day after another, totally unplanned and with nothing to look forward to," Dr. Goldberg says.

Write down your reps. You make notes about all your other commitments. If you write down your exercise obligation, too, you're more likely to stick with it. Pencil in "jogging time" every afternoon. No, write it in ink. Once you've come up with a schedule, stick to it. Don't skip exercise appointments with your body.

Wanted: A Goal Line to Run Toward

Even if you were delighted to quit your job, working probably gave you other satisfactions besides a paycheck. Now that you no longer have them, you're likely to be seeking other achievements. "You can use exercise to set goals and create challenges for yourself," says Dr. Goldberg. Invent exercise ambitions and start jogging toward them.

Begin modestly. Let's say that you've taken up golf. Give yourself an assignment to play nine holes three times a week for three months. If you've taken up yoga, commit to three half-hour sessions a week. Try to achieve these goals before setting larger ones.

Set a long-term goal, too. To help yourself stay motivated, pick a goal that's six months or a year down the road. If you're a beginning golfer, you might aspire to play at least one charity golf tournament by the end of your first year. If you're a neophyte yogi, pick a slightly advanced set of poses and aim to be comfortable doing them by your next birthday. "Exercise gives you a sense of accomplishment, both short-term, each time you exercise, and long-term, such as when you participate in some kind of competition or event," says Dr. Goldstein.

Wanted: The Schmoozing at the Water Cooler

If you miss the camaraderie of the workplace, exercise can fill that void.

Team up. Recruit a partner for ballroom dancing lessons. Make a date to walk in the park every morning with a friend. Join the Masters swimming team at your local pool.

Miss Being the Boss?

If you were lucky enough to call the shots at Scrooge and Marley's, once you retire, you may miss the authority that came with your job. In that case, look for a leadership role in some exercise-related endeavor, suggests Ross Goldstein, Ph.D., a psychologist and president of Generation Insights, a San Francisco consulting firm. "Retirement is a good time to explore new roles and interests," he says.

Lead the pack. Call your local United Way volunteer center and offer to lead a Boy Scout or Girl Scout troop. Share your hiking, knot-tying, campfire-building, and mosquito-swatting skills.

Orchestrate meets. Offer your organizational talents gratis to the local swim team. Keep track of everyone's best time in the butterfly.

Play mentor. Talk to the head of the town recreation department about starting a much-needed kids' sports program in your neighborhood.

Sign up. Join an exercise class—one you have to pay for. Some people find that if they make a financial investment in getting fit, they're more likely to stick with the program.

One added benefit is that classes can be great venues for making new friends, says Dr. Goldstein. "Exercise classes can be enjoyable," he explains. "And when you're doing something you enjoy, you're most likely to be at your most pleasant." Thus, you're most likely to make a good impression and win friends.

RUNNING

Taking It All in Stride

WHAT IT IS

Cavemen did it to catch a meal. The speedy Road Runner did it to avoid Wile E. Coyote. No one is quite sure why Forrest Gump did it. The first marathoner, Pheidippides, ran 26 miles to ancient Athens to announce victory over the Persians. Centuries later, on April 24, 1994, in Toledo, Ohio, his mirror image, American Bud Badyna, ran the world's fastest backward marathon in 3 hours, 53 minutes, 17 seconds.

Thus, the reliability of running is a given—as a means of getting from here to there, escaping threats, crossing finish lines, and catching buses. And the simplest definition? Running is a series of fast steps. Here's how scampering can enhance your health.

WHAT IT DOES

Running is a medicine cabinet, family doctor, mental health therapist, and weight-loss center all rolled into one. With each stride, runners are working to lower their blood pressures, lose unwanted pounds, bolster their hearts, fight stress, and lessen the dangers of diabetes. On top of all that, the sweat that's generated when you run helps your body get rid of toxins such as carbon dioxide. Here are a few of the health blessings that running bestows.

Circulation stimulation. "Moderate daily running is good for every organ of the body because it improves circulation," says long-time runner Dan Hamner, M.D., a physiatrist and sports medicine specialist in New York City. When your blood flows strongly, all of your organs and tissues benefit from the oxygen the blood carries.

Help with weight control. A steady diet of running—20 miles or more a week—can help you keep your weight where it belongs. And running helps you lose weight by burning calories and modifying your appetite, too.

"Running increases your metabolism," says Dr. Hamner. "The longer you rev your engine, the more running acts like a mild appetite suppressant. For some people, running can change their food cravings. They find themselves craving carbohydrates like pasta instead of red meat."

Of course, the first few days of a running program may make you have a Fred Flintstone appetite. But be patient, says Dr. Hamner. By week three, running's mild appetite-suppressant effect will kick in.

The farther and the longer you run each time, the higher the rate at which you will burn body fat. "If you run for 15 minutes, you'll get a calorie burn-off for about two hours afterward," says Dr. Hamner. "If you run for a half-hour, you'll burn calories for four hours."

Running can be a fabulous weight-loss workout. Of course, the final tally of how much you'll lose depends on that time-honored struggle between calories burned and calories consumed.

Heart help. Running's coronary benefits go far beyond weight control. "Running improves your general cardiovascular fitness," says John Cianca, M.D., assistant professor of physical medicine and rehabilitation and director of the sports and human performance medicine program at Baylor College of Medicine in Houston. "Being in better shape allows you to perform daily activities with more ease and leaves you less vulnerable to illness and injury," he adds.

Running helps to dilate the capillaries you do have and also helps produce a lot of extra capillary beds, according to Dr. Hamner. New capillaries help prevent blockages anywhere in the body by developing new channels of circulation. "Due to the increased profusion of blood through the arteries, the whole system is better off," says Dr. Hamner.

Cholesterol containment. The farther a woman runs, the better her cholesterol levels, according to a study reported in the *New England Journal of Medicine*. Researchers at Lawrence Berkeley National Laboratory in Berkeley, California, evaluated the health of almost 2,000 female runners and found that the greater their mileage, the higher their blood levels of the "good" cholesterol (high-density lipoprotein, or HDL).

Stress management. Is the boss giving you a splitting headache? Are your kids testing your patience? Did you pull in to the supermarket just as the manager was locking the door for the night? Welcome to the hotbed known as s-t-r-e-s-s. Take a time-out and go for a run.

"Running is a good stress release. You're physically outputting energy that's built up inside you," says Dr. Cianca.

Dr. Hamner says that he usually feels better and is able to think more clearly after he jogs. A mile or two into his run, he is a bit fatigued but feels more able to focus on solving a problem. "My hypothesis is that your brain neurons reorganize into a different way of thinking about the problem," he says. "It hasn't been proven scientifically, but I'm convinced that running cuts down the extraneous types of thinking and helps you focus your mental energy."

REAPING THE BENEFITS

One of the allures of running is that it's easy to get started. It's not equipment-intensive or complicated to learn. Here are some tips from doctors and top runners on how to avoid the common pitfalls of running.

Don't give your feet fits. Invest in good running shoes. On average, your feet strike the ground 8,000 to 10,000 times per day. A 150-pound jogger who runs three miles feels the cumulative impact of 150 tons on each foot. This is no time for those bargain sneakers with cardboard insoles.

"You need a pair of running shoes that fit properly for your feet and your body type," says Dr. Cianca. "A heavy runner will need more support in his shoes than someone who is light. A runner with flat, flexible feet needs supportive shoes, and a runner with high-arched feet needs more flexible shoes. If you're unsure, your podiatrist or doctor can write a prescription for the kind of shoe you need," says Dr. Cianca.

If you run 20 miles per week, you should probably replace your shoes every four to eight months. The shock-absorbing capabilities can wear out after 350 to 500 miles, says Stephen Pribut, D.P.M., a podiatrist and surgeon in Washington, D.C.

Choose your landing pad. The beauty of running is that you can do it outdoors or indoors, on asphalt roads or a wood-chipped trail in the woods, on a treadmill or an indoor track. Be aware, however, that each running surface comes with its advantages and handicaps.

The road in front of your house is probably the most convenient, but you may have to dodge cars and bicycles. And there are those unforgiving hills on the horizon.

"Don't run downhill any more than you have to, because that's where the highest rate of injuries can occur," says Terry Nelson, M.D., an orthopedist at K Valley Orthopedics in Kalamazoo, Michigan, and a former marathon runner. "Running downhill increases the stress on your knees, legs, and feet, which can lead to shin splints."

Whenever possible, Dr. Hamner prefers running outdoors, for two reasons. First, "it is important for the mind to be outside in a quiet place next to nature," he says. "I feel that running outside helps in reducing stress." Second, he likes asphalt surfaces in the summer because they are softened by the sun. Trails offer

soft landing surfaces, too, but they can be uneven and cause ankle turns or sprains. On isolated paved roads, Dr. Hamner favors running in the center of the road, where the surface is the most level.

Treadmills offer more shock-absorbing cushion than roads and take away the excuse of canceling because of nasty weather. They also enable you to set a precise pace.

Give yourself a good stretch. Warm up your muscles with a slow first half-mile to a mile, then spend five minutes stretching. Continue your run at your pace. After a run, finish up with five minutes of slow, controlled stretches. This will reduce your risk of injury, doctors say.

Focus on stretching one set of leg muscles at a time, says Dr. Pribut. Make sure that you stretch your hamstrings, quadriceps, knee muscles, calf muscles, and Achilles tendons. Do a back stretch as well.

Stretch gently, and be sure to hold each stretch for at least 20 seconds and avoid bouncy movements, adds Dr. Hamner. Here are some recommendations.

Hamstring stretch. Lying on your back, bend your left leg so that you can comfortably rest your left foot flat on the floor. Keep your right leg straight and slowly raise it as high as you can. Try to reach the point where it makes a right angle with your body. You should feel a stretch in your hamstring. Hold, then change positions and stretch the opposite leg.

Quadricep and knee stretch. Keeping your quadriceps flexible can help stabilize your knees. Stand with your back straight and your feet shoulder-width apart and place your right hand on a tabletop or chair for support. Pull your left foot straight up behind you, grabbing

Scentsational Weight Loss. Why? Because exercise prompts a temporary drop in your blood sugar, which appears to prompt your nose to sniff out food. When you raise your heart rate just a bit, your sense of smell becomes more sensitive—to all scents. And with the extra help your nose provides, your sense of taste gets a boost, too. As a bonus, you don't even have to get all sweaty: "Walking for 20 minutes or so should do it," says Dr. Hirsch.

Avoid the pool. If tonight's your fancy anniversary dinner at Chez Expensive, take a brisk walk instead of a swim. The chlorine in swimming pools temporarily dulls your ability to smell and, by extension, your sense of taste, says Dr. Hirsch.

Pay attention. To get more from your senses of smell and taste, concentrate on what you're experiencing. Tomorrow morning, when you cut open your breakfast grapefruit, take a moment to enjoy its fruity, acidic aroma. Ask yourself, "What does this smell like?" Try to describe it. Then take a bite and chew slowly, breathing through your nose while you do. That gives food odors more of a chance to get up to the olfactory receptors, the structures that tell your brain, "Ummm! Grapefruit!" explains Dr. Levin-Pelchat.

Make a habit of tuning in to odors and flavors whenever you're in the vicinity of good ones. Doing yardwork this weekend? Focus on and try to describe the green, wet, yeasty smell of cut grass. Use your own words. At the Petersons' dinner party tomorrow night, pay attention to the subtle flavors in Mrs. Peterson's special "mystery casserole." Try to figure out what's inside.

Practice, practice, practice. "Smelling is like learning the piano; you have to practice,"

says Klumpp, who honed his sense of smell by spending countless hours sniffing compounds in the Gillette lab. Try doing the same at home. Take jars of several different herbs and spices out of the kitchen cabinet, uncap them, and arrange them on a table. Sit down, close your eyes, reach for a jar, and sniff deeply. Try to identify the contents of each jar by scent. Think hard: Was that the sweet and slightly peppery scent of cinnamon or the green-sour aroma of dill? Repeat this scent test regularly.

Become a taste tourist. Practice tasting different foods and you'll hone your sense of taste as well, says Joe Spellman, wine director at Charlie Trotter's restaurant in Chicago and one of only two Americans to win France's prestigious Grand Prix SOPEXA du Sommelier, an award much coveted by wine stewards. "Try different foods and drinks—try lots of different flavors," Spellman advises.

Each time you take a bite or sip of something new, try to identify the ingredients and flavors. Was that lemongrass in the Thai noodle dish? A hint of something with nutmeg in the burgundy? To stay fresh while you practice, cleanse your palate between courses. Try a few sips of water or some bland crackers or bread.

Drills for Better Listening

The world gives voice to all manner of beautiful sounds: the pebbly chatter of surf on beaches, the chants of kids in schoolyards, the phrase, "You've won the Connecticut state lottery." If you're finding it harder to discern the sounds you love, exercise may help, says Helaine Alessio, Ph.D., associate professor of physical education and sport studies at Miami Univer-

your ankle in your left hand. Keep your knee pointed toward the floor. As you pull your foot back and up, you should feel a stretch in the front of your thigh. Hold the stretch, then switch legs and repeat. (For an illustration of this stretch, see page 336.)

Calf and Achilles stretch. From a standing position, lean forward against a table with your hands palms down on the table edge. Stand far enough away so that you feel a moderate stretch in your calf muscles. Your arms should be straight. Lock your left leg at the knee, then slide your right leg forward and bend it so that your knee comes directly over your toes. Keep your heels on the floor. Press your upper body forward (don't bend at the waist) until you feel a moderate stretch in the calf muscles of your left leg. Hold the stretch for 15 to 20 seconds. Then, keeping both heels on the floor, bend your left leg until you feel a moderate stretch in your Achilles tendon. Hold for 15 to 20 seconds, then change positions and repeat the steps on the other side. (For an illustration of this stretch, see page 19.)

Lower back stretch. Lie on your back with your legs out straight. Use both hands to bring your left knee to your chest. Hold the stretch, then repeat with the right knee. Repeat once more, bringing both knees up at the same time.

TRIAL RUN

Grete Waitz gained her international fame by running. She drew headlines for winning nine New York City Marathon titles. *Runner's World* magazine selected her as the world's best female distance runner of the past quarter-century.

It wasn't the medals, awards, or honors, though, that persuaded Waitz's mother, Reidun

Andersen, to join her in a road race. It was simple prodding by her daughter. Waitz convinced her mother to join her in the annual Grete Waitz Run in their home country of Norway. This 5-K (3.1-mile) run draws more than 45,000 women annually.

"I remember my mother watching the race one year and remarking about the number of women she saw who were in their fifties and participating," says Waitz, who lives in Gainesville, Florida. "I got this feeling that she wanted to be part of it but needed to be pushed."

The push came in the form of a pair of running shoes and a running outfit that Waitz gave to her for Christmas. At 65, Reidun Andersen was getting her first pair of sneakers.

"I told her that by May, she will be ready to run with me in my race and that she can't be my mother without being in that race," says Waitz.

The daughter-mother training began in February. For the first couple of weeks, they walked one mile three times a week. Slowly, the walking distance was increased to three miles. Once Andersen was comfortable with that distance, Waitz introduced jogging. At first, they ran from one light pole to the next. Gradually, they built their jogging distance.

"I read her reactions to the exercise, like her breathing and level of fatigue," says Waitz. "One week before the race, I could see how motivated she was. She was telling her friends, 'I'm jogging with Grete.'"

In her first race, Andersen crossed the finish line with her daughter in 38 minutes. She participated in the race several times, until a hip problem from the wear and tear of old age developed. Now, in her midseventies, she still walks regularly and swims for exercise.

"I see that my mom now has a social life that so many people her age can't have because they are not in the same shape," says Waitz. "She has more energy and she is healthier, and a lot of it is due to regular exercise."

For anyone age 50 or older who wants to start jogging, Waitz offers the following specific training tips to make the transition from walking to running a 5-K race. Depending on your condition and determination, expect to take 10 to 15 weeks to build up to this distance.

Get a thorough physical exam. It's important to get the go-ahead from your doctor, especially if you've led a fairly sedentary life.

Make the decision to run and set a goal. Maybe you have your heart set on completing the annual local road race next summer or losing some weight before that 30th class reunion.

Schedule the time. You pencil in appointments to see the doctor and the car mechanic and attend numerous business meetings. Give your daily jogs the same treatment. "Exercise is most likely to happen if it is a definite appointment you keep with yourself," says Waitz.

Prioritize your daily runs. Don't allow anything to interfere with your running. Instead of dwelling on why you can't run on an upcoming busy day, find a creative way to ensure that you do it.

Add value to your running. By jogging with your family or friends, you can strengthen your relationships. A lot of Waitz's most meaningful talks with her mother came while they were running together.

Form a group. If you have a pack of friends who like to mall shop, try motivating them to join you in a running group. You can give yourself a name like Wichita's Women on the Run and wear it on T-shirts during your jogs.

Be a foul-weather exerciser. Prepare for those rainy or snowy days or for being away from your favorite jogging path with a backup plan. Try swimming, riding a stationary bike, stair-climbing, or jumping rope.

Have reasonable expectations. Don't attempt to be the next Grete Waitz in a month. If you can comfortably run 30 to 40 minutes at a slow pace several times a week, you're doing your body a great favor. "Don't pay any attention to that old adage 'No pain, no gain,'" adds Waitz. "Exercise doesn't have to be painful, long, or disruptive of your life to be beneficial."

SENSORY LOSS

Exercises That Sharpen Your Sensitivity

The chief perfumier at the Gillette corporation, Carl Klumpp, has one of the world's most influential noses.

Klumpp is the man who decides how products like Soft & Dri and Right Guard should smell. He spends his workdays sniffing. Usually, Klumpp's duties involve sniffing chemical compounds in the search for new scents for deodorants, soaps, and other potions that mask human smells. But sometimes he samples people, too.

"The other day, I smelled 60 armpits," says Klumpp, chuckling. "We were testing deodorant, and there's no better way to find out if it works."

Although our sense of smell—as well as our other four senses—tend to become less acute as we get older, Klumpp's is still going strong after 64 years. A few key exercises can help you keep your sense of smell as keen as ever, too, he says. In fact, experts agree that a variety of exercises can help keep all of your senses in shape.

So read on, then go ahead and start exercising. Why take a chance on missing out on the spicy, fruity scent of Thanksgiving's pumpkin pie, the sound of sweet nothings whispered in your ear, or the satiny softness of your granddaughter's cheek?

Strategies to Help You Smell (and Taste) Better

A keen sense of smell helps you enjoy life's rich fragrances *and* flavors. That's because smell and taste work in tandem. On their own, your tastebuds detect just four simple tastes: sweet, salty, bitter, and sour. To make more subtle distinctions, such as discerning that the sweet stuff you're eating is rocky road, not butter pecan, they need help from your sense of smell.

"The brain takes information about odor from the nose and about taste from the mouth and combines it to make a single perceptual experience that we call flavor," explains Marcia Levin-Pelchat, Ph.D., a sensory psychologist and associate member at the Monell Chemical Senses Center, a nonprofit research institute in Philadelphia that studies taste and smell.

Try the following exercises—both physical and mental—and odds are, you'll find life's flowers smell more flowery and Aunt Pearl's pecan pie more piquant.

Make tracks. Aerobic exercise can temporarily sharpen your sense of smell and, consequently, your sense of taste, says Alan Hirsch, M.D., neurological director at the Smell and Taste Treatment and Research Foundation in Chicago and author of *Dr. Hirsch's Guide*

sity of Ohio in Oxford. Here are some tricks to sharpen your hearing.

Sweat for at least an hour a week. In a series of studies, Dr. Alessio and her colleagues found that people who weren't in shape were able to hear better once they got in shape. What's the connection? Boosting your cardiovascular fitness level boosts blood flow to all parts of your body, including your inner ear, she explains. The better the blood flow to your inner ear, the better it seems to work. "We've been led to believe that with age, your hearing level will decrease," she says. "But it can also increase."

You don't have to relocate to the gym to improve your hearing. Thirty minutes of moderate aerobic exercise—walking, running, cycling, swimming, jumping rope, and the like—twice a week should do it, says Dr. Alessio.

If you're already in good cardiovascular shape, additional aerobic exercise won't improve your actual hearing ability, says Dr. Alessio. But the following focusing exercises will improve your listening skills, to similar effect.

Read my lips and eyebrows. When it's time for the nightly news, tune in to your favorite peppy news team and turn down the sound so you can barely hear it. Then focus on the lip movements and facial expressions of the anchors and use the clues they provide to decipher the just-audible soundtrack. You'll be practicing "speech reading," which is slightly different from lip reading, since you read both lips and facial expressions.

Regular practice can improve your listening skills significantly, says Stephen W. Painton, Ph.D., an audiologist and clinic director of the John W. Keys Speech and Hearing Center at the University of Oklahoma Health Sciences Center in Oklahoma City. Speech reading will train you

to be a more attentive listener and help you pick up all sorts of visual information that will supplement the sounds. You can practice speech reading while watching just about any show—excluding, of course, cartoons. But news programs are a good bet since they usually provide an unobstructed view of the anchors' faces.

Find the sitar. The next time you give the Beatles' *White Album* a whirl on the CD player, try picking out the sitar. Or keep an ear out for the trumpet in the overture to Handel's "Water Music." Picking out the sounds of a particular instrument when you listen to music can help you sharpen your auditory focusing skills, says Dr. Painton. If necessary, close your eyes to lessen distraction.

Exercise for Your Eyes

Do-it-yourself exercises can help with a number of common vision dysfunctions, including the headaches, discomfort, and fatigue that result from staring at a computer screen for more than four hours a day, says Arthur S. Seiderman, O.D., an optometrist in suburban Philadelphia. Try these, from Dr. Seiderman's book, *The Athletic Eye.*

Step up to bat. Eye-hand coordination tends to go a bit awry as we get older, raising golf scores and lowering rankings in tennis tournaments. To help your eyes and appendages get back in sync again, find a Whiffle ball and bat (check the garage) and a piece of string (try the tool chest). Then tie several feet of string to the Whiffle ball and tape the free end to the ceiling; the ball should fall between your shoulders and knees. Practice hitting the ball for three minutes. Repeat regularly.

Follow your thumb. Your "eye tracking ability" is your talent for following targets (fluorescent yellow tennis balls or lines of print in a book) with your eyes. Your tracking ability can fade over time, too. To keep it keen, try this exercise.

Raise your arm in front of you so that your hand is at eye level. Make a fist, point your thumb at the ceiling, and focus your eyes on your thumbnail. Move your hand from side to side, up and down, diagonally, and in circles. All the while, keep your head still and your eyes fixed resolutely on your nail. Practice for two minutes.

A Soft Touch to Sharpen Sensitivity

Your partner's caress. A bear hug from your Uncle Earl. A child's small, only slightly sticky, ice-cream-coated hand in yours. Touch feels good. And, research finds, it's good for you. Touch not only boosts the levels of all sorts of health-promoting hormones and brain chemicals, it also depresses the levels of bad ones like the nasty stress hormone adrenaline. With all touch has going for it, wouldn't it be nice if you could feel *better*? You can, with practice.

If your partner's willing, try this "20-minute sensate focus" exercise suggested by Barbara Bartlik, M.D., a researcher with the human sexuality program at the New York Hospital–Cornell Medical Center in New York City.

►Find a warm room where you both feel comfortable and turn down the lights (the better to focus on touch).

►Decide who gets the massage and who gives it. (Next time, switch roles.)

►If you're the one receiving the massage, undress, sit or lie in a comfy position on a bed or

Smell Better, Weigh Less

You already know that exercise can help you lose weight by shifting your body into calorie-burning overdrive. Well, there's another way that it will help you get slimmer.

Research suggests that exercise can also help you drop extra pounds by sharpening your sense of smell, says Alan Hirsch, M.D., neurological director at the Smell and Taste Treatment and Research Foundation in Chicago and author of *Dr. Hirsch's Guide to Scentsational Weight Loss*. Studies have shown that people with sensitive smellers eat less before feeling full than those with less well-honed noses.

How come? Whenever you put a bite of cheesecake or lasagna into your mouth, odors from the food filter up to your olfactory bulb, a small organ inside your nose. Your olfactory bulb has a direct line to part of your brain called the satiety center, the place that decides when you've eaten enough. Since exercise sharpens your sense of smell, your brain gets more powerful messages of satisfaction, explains Dr. Hirsch.

couch, and do nothing more. "Simply focus on what you feel," says Dr. Bartlik.

►If you're giving the massage, get some baby oil or massage oil. "With a lubricant, you get the most pleasant sensation of touch," she says.

►Then start softly and slowly. Use soft strokes, not heavy, kneading ones. Stroke with your hands and your lips.

►Forget sex for the moment. Limit your stroking to nonsexual areas like the neck, the backs of the knees, and the earlobes. Avoid the breasts and genitals, the traditional erogenous zones. The idea here isn't to get your partner sexually excited; it's to give your partner, and yourself, an opportunity to focus on the sensation of touch.

►After you're finished, the two of you should rest a bit. Take some time to talk about what each of you liked most, and least, about the exercise. Wait at least a half-hour before changing places. "Back-to-back massages is an overload, " says Dr. Bartlik. "Either take a break before switching or schedule two massages at two different times." The better your tactile skills, the more you'll enjoy the feel of everything—from fabrics to faces.

SEX

Shape Up in the Sack

WHAT IT IS

Although it may feel more like playtime than a workout, healthy sexual activity is heartily endorsed by many experts. And while they're not saying that you can get fit with l'amour alone (the exertion involved in a single round of romance is no more than you'd use to climb two flights of stairs), they do believe that every little bit provides a tonic. As a nice complement to your workout plans, sex works just fine.

WHAT IT DOES

For starters, vigorous encounters of the amorous variety can help heart hardiness. That's because sex, like other moderate-intensity activities, increases blood circulation and improves your respiration. When your heart and lungs work harder, they tend to get stronger.

If you need a little more inspiration, consider this: Science seems to indicate that frequent sexual activity may actually play a role in prolonging your life span as well as adding extra zest to the years that you span. A 10-year study of more than 900 men between the ages of 45 and 59 was performed at the University of Bristol in England. The mortality from all causes—including heart disease—among the men who enjoyed the most frequent orgasms was half that of their less frisky brethren.

The reasons for this protective effect aren't exactly clear, but it's likely that people who have more sex also have less stress, says Barbara Bartlik, M.D., a researcher with the human sexuality program at the New York Hospital–Cornell Medical Center in New York City. "More sex may make us more relaxed," she says. And that could be the factor that helps lower your risk of many medical problems.

Whether you're a man or a woman, sexual arousal increases blood flow to and from the genitals, while orgasm creates internal muscular movements.

For women, lovemaking can help squelch menstrual cramps. The reason lies in the nature of orgasm. When a woman climaxes, a muscle spasm spreads throughout the pelvic region, including the uterus. This helps relieve the accumulation of fluid in the area, explains Dr. Bartlik, and that in turn helps ease the cramping and abdominal pain that some women experience with a monthly vengeance.

Postmenopausal women also have an excellent excuse to indulge in intimacy as often as possible. Vaginal dryness is a symptom often associated with menopause, and if you increase sexual activity during menopause, you'll help improve moistness. With sexual arousal, the genital tissues become swollen with blood, which helps keep them healthy, says Dr. Bartlik. Maintaining regular sexual activity, including masturbation, may help ensure that dryness never becomes a problem.

Frequent sexual activity offers special help to men as well, says Joel Block, Ph.D., a sex therapist and psychologist in private practice in Huntington, New York, and author of *Secrets of Better Sex*. Having erections is a boon to mental and physical health, according to Dr. Block. When a man is aroused, oxygenated blood surges into the penis, keeping those crucial blood vessels in good working order.

Among both men and women, researchers have found that regular sex with the one you love can help keep sexual self-esteem at a healthy high, says Dr. Block.

And there's a skin-deep bonus as well. While sex can leave a smile on your face, it also soothes the skin, says Dr. Block. The increased blood circulation, combined with the stress relief that lovemaking brings, may improve your complexion and add a lovely glow.

REAPING THE BENEFITS

It's obvious that making love feels good. But doctors say that you get a special reward when you achieve orgasm.

Orgasm—both his and hers—triggers the release of pleasant painkillers called endorphins, according to Barbara Keesling, Ph.D., in her book *Sexual Healing*. That's why you feel extra-good at that wave-crashing moment. What you may not realize, however, is that an orgasm can help provide pain relief that can last for a couple of hours or more. Hormones released in

the heat of sex leave a kind of afterglow that actually reduces pain, she explains. Sex can even help alleviate headaches and chronic pain that's due to problems like arthritis.

TRIAL RUN

The "trial runs" are up to you, of course, and it's likely that you don't need much specific guidance. But for many partners, the hardest part is finding time for some all-round, good-for-you sexual contact. Here are some suggestions from Dr. Block for making time for lovemaking.

Get some daily hugs in. Every day, make time for at least one hug, one kiss, and the words "I love you." Sometime during the day, find at least 30 minutes to be with each other, free of distractions.

Find an hour weekly. You might need to take a long walk together, sneak off to the bedroom, or get to a spicy movie, but somehow, find an hour or more once a week when you can be alone together.

Once a month, check out for an evening. It may mean hiring a baby-sitter, but you and your partner need at least one full evening to go on a date together.

Get a quarterly escape. Once every three months, get away for a weekend or at least an overnight excursion. The more luxuries, the better: Check out the "package deals" at hotels where you can spend the weekend pampering yourselves and focusing on one another.

Give yourselves an annual bonus. Once a year, take a week-long vacation together. "You need it," says Dr. Block. "Remember that there's no way that you would have gotten together in the first place if you hadn't spent some time alone."

SHINSPLINTS

Soothing Your Lower Legs

Margaret Gutgesell, M.D., Ph.D., started running when she was 45. She enjoyed it instantly, and in no time, she was logging 50 miles a week. But when she started to get pains in her lower legs, she knew she had the problem that runners dread—shinsplints.

Shinsplints are usually caused by exercise or other kinds of repetitive physical activity. Constant pounding—from running, playing tennis, or sometimes even walking in high heels—can irritate the tendons that link the leg muscles to the lower leg bones. The result can be shin pain and often swelling.

If exercise is causing the shinsplints, varying your exercise routine can be the cure, says Melvin Williams, Ph.D., exercise physiologist at Old Dominion University in Norfolk, Virginia. Just for starters, he notes, you should stop all high-impact, weight-bearing exercise. Tennis and jogging are definitely out for a while. You might even limit your walking if it aggravates the shinsplints.

The inflamed fibers in your legs need a chance to recuperate. If you apply ice, you can cut the pain and perhaps facilitate the healing process, says Dr. Williams. It's also important to see a doctor to find the correct treatment for your condition, he adds, because a diagnosis of shinsplints might also involve a more serious condi-

---

Retire Those Running Shoes Periodically

Aging running shoes may be a cause of shinsplints, says Margaret Gutgesell, M.D., Ph.D., associate professor of pediatrics at the University of Virginia Health Science Center in Charlottesville. An avid runner, she now keeps a log of the miles she jogs and replaces her shoes every 300 to 500 miles. You should select shoes with soles that flex sufficiently all over and have a good "roller action" from heel to toe.

---

tion such as tendinitis or even a stress fracture.

If you want to continue to exercise, you need to switch to no-impact exercises and then take some measures to speed the healing.

Get Wet or Start Spinning

Dr. Williams has also felt the pain of shinsplints. When his shins start barking, he turns to water workouts as a low-impact alternative to running.

Take some strides on the wet side. "A lot of people tend to lay off exercise altogether when they get shinsplints," says Dr. Williams. "But I rest for a couple of days and then do alternate training like running in water." Water has a nice cushioning effect. Try running in your local pool or lake. It's probably best to run in place, but if the pool is big enough, jog slowly back and forth in the shallow end; waist-to chest-deep is best. If the lake is shallow enough and the bottom isn't too mucky, run up and down along the shore.

Wear flotation fashion. Dr. Williams suggests wearing a buoyancy vest while you take your wet run. This minimal life preserver helps keep you upright and further minimizes impact to your shins.

Cycle for your shins. Bicycling is also a good exercise alternative to ease your way back from shinsplints, says Dr. Williams. It involves far less trauma to your legs than the constant banging of running or other weight-bearing exercises.

"Bicycling is non-weight-bearing," says Dr. Gutgesell, associate professor of pediatrics at the University of Virginia Health Science Center in Charlottesville. "Just be sure to stay off hills and set the gears at the easiest level so there's no stress on your knees or ankles, and don't forget to set the bicycle seat at the correct height."

You might also check other options, such as some of the newer stair-climbing machines that may reduce impact, says Dr. Williams. If you experience no pain when exercising, these may be good alternatives as well.

Stretch for Your Shins

Shinsplints can sometimes hang on for a while. Stretching before and after exercise is one good way to speed your recovery, says Dan Hamner, M.D., a physiatrist and sports medicine specialist in New York City.

As the pain recedes, try this calf muscle stretch suggested by Carl Fried, P.T., a physical therapist at K Valley Orthopedics in Kalamazoo,

---

Skate Away the Pain

To bring immediate relief from shinsplints, take a cup of ice and skate it across the shin bone for five minutes, suggests Carl Fried, P.T., a physical therapist at K Valley Orthopedics in Kalamazoo, Michigan.

"Ice will not only numb the area so you don't feel pain, it will also decrease the blood flow to that area to help control the inflammation," he explains.

Simply fill a few bathroom paper cups with water and store them in the freezer. The next time your shin aches, Fried suggests that you take one of the frozen cups and peel back the top to expose about ½ inch of ice. Hold the cup by the bottom and rub the ice in a circular motion directly on your shin for about five minutes.

To keep pain at bay, try this ice massage a few times during the day as well as after a strenuous activity such as walking up hills or climbing flights of stairs, he adds.

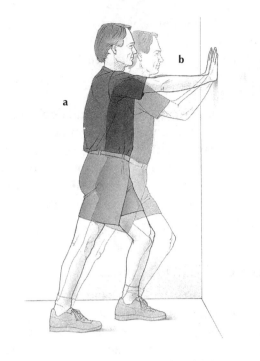

1

Michigan. "Every stretch should be a mild stretch," he recommends. "You should hold each one for at least 15 seconds to give your muscles time to get warm and loose."

▶Stand facing a wall with your left foot 12 inches from the wall and your right foot about 12 inches behind the left. Keep your arms straight and place your palms flat on the wall for support.

▶Bend your left leg at the knee, keeping your right leg straight (*1a*).

▶Keeping both heels on the floor, bend your arms and lean your upper body forward into the wall, letting your full weight press against your hands (*1b*). You should feel the stretch in the calf muscle of your right leg.

▶Hold the stretch for 15 to 20 seconds, then relax. Repeat two more times.

▶Change the position of your legs and stretch the other calf.

Here's another playful exercise that may help ease the pain of shinsplints and prevent them from coming back, according to Steven Lawrence, M.D., an orthopedic surgeon at the Lancaster Orthopedic Group in Pennsylvania.

▶Take off your shoes and socks and sit in a comfortable chair.

▶Drop a pencil at your feet and try to pick it up with all the toes of one foot (*2*).

▶Hold it for at least 10 seconds.

▶Drop the pencil and repeat the process five times with each foot.

▶Repeat twice a day.

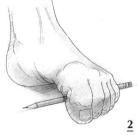

2

Shoulder Pain

Relief Is within Reach

Will everyone with shoulder pain please raise their hands? *Ouch*, right? If you have shoulder trouble, even simple everyday activities like reaching up and putting the groceries in the cupboard can be tough.

Shoulder pain comes in lots of shapes and sizes. Some folks ache in their upper backs and some in the front of their shoulders, along the line from the top of the breastbone to the armpit. Still others hurt along the top, from their necks to the curve of their shoulders. The kind of pain varies, too, from a dull ache to a sharp, stabbing pain. But all shoulder pain has two things in common—one bad and one good.

On the negative side, it's extremely debilitating. Since you use your shoulders so frequently, if one of them hurts, you may wince as often as you smile. But on the positive side, you can fight shoulder pain with exercise by building up the joints.

"Human body joints, like cars, need maintenance," says Thomas Meade, M.D., orthopedic surgeon and medical director of the Allentown Sports Medicine and Human Performance Center in Pennsylvania. "There is no body version of 10W-40 (motor oil), so we must think of these exercises as preventive maintenance for our bodies."

When to See a Doctor

If pain, tightness, or limited range of motion in your shoulder interferes with everyday tasks such as combing your hair or fastening bra hooks and persists for more than five to seven days, you should seek medical attention, advises Dan Hamner, M.D., a physiatrist and sports medicine specialist in New York City.

Stretch for Relief

Try these simple stretches to alleviate any aches you have and help prevent future pain.

Tug your towel. After you step out of your morning shower and dry off, turn your towel into an exercise tool for your achy shoulders, suggests Dr. Meade.

▶Standing straight, hold one end of the towel with your right hand and let hang it down behind your back. Hold it so that your right fist is a few inches above your right shoulder

▶Reach behind your back with your left hand and grasp the other end of the towel.

▶With your right hand, try to raise the towel as

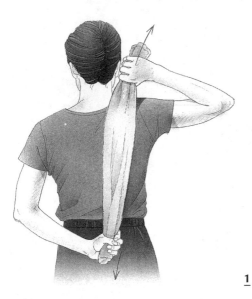

<u>1</u>

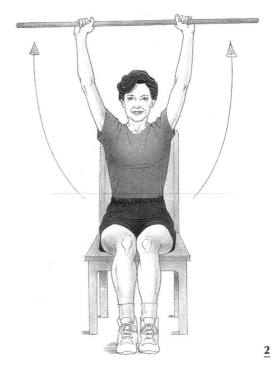

<u>2</u>

ders moving in all directions. Even a broom handle can help keep your shoulders supple.

Sit straight in a chair with both feet flat on the floor. Hold the broom handle in both hands, palms down, in front of your waist with your arms straight. Slowly lift your arms and raise the handle as high as you can—ideally, over your head (2). (If you have tendinitis, don't go higher than your shoulders.) Count to 5 or 10 and then lower the broom handle back to waist level. Try at least 15 repetitions twice a day.

high as you can while resisting the upward pull by bearing down with your left hand (1).

▶Hold for 10 seconds, slowly lower your right hand, and relax.

▶Try this stretch four more times, then repeat on the other side.

Hug yourself. You can warm your shoulder muscles with this body hug offered by Dr. Meade. He recommends doing this stretch twice a day to keep your shoulder muscles limber.

Stand straight and place your left forearm across your waist, with your left elbow bent at a 90-degree angle. Place your right forearm on top of your left so that your right hand cups your left elbow. Use your right hand to pull slowly and firmly on your left elbow for 20 to 30 seconds. Don't try to resist with your left arm. You should feel the muscles stretching in your shoulder and elbow joints. Switch arms and repeat.

Bring out the broom. There's no need for expensive gym equipment to keep your shoul-

"You should go to the point of feeling a little discomfort, but do not push it to the point of pain," advises Kim Fagan, M.D., a sports medicine physician at the Alabama Sports Medicine and Orthopedic Center in Birmingham.

Strengthen for Prevention

Shoulders need more strength training than most joints. "The shoulder's design is great for flexibility and range of motion, but it is lacking in stability," says Dr. Fagan. "For it to function properly, the surrounding muscles have to do their jobs well." She offers these stabilizing exercises to bolster the deltoid muscles, rotator cuff tendons, and shoulder blade stabilizers.

Soup up your shoulder. Grab a couple of cans of soup from the pantry and give yourself a workout before dinner.

Stand straight with your shoulders back. Your arms should be hanging at your sides with each hand holding a soup can.

Keeping your elbows straight and your palms facing the floor, slowly raise your arms to shoulder level. Hold for five seconds, then relax and slowly lower your arms. Repeat this lift four more times. Build up a little each day until you can hoist the cans 15 times per session.

Take cans in hand again. Before you put the soup away, try these moves, suggests Dr. Fagan. Hold one soup can with your arm hanging straight down at your side. Slowly and with controlled movements, swing the can back and forth like a pendulum. Be sure to keep your arm straight. For starters, try 25 swings with each arm three times a day. For variety, try making circular motions with the can. "The can-holding is a little resistance training for the rotator cuff," explains Dr. Fagan.

Shrug and strengthen. Don't expect Arnold Schwarzenegger deltoids, but here is a series of easy exercises to ensure healthier shoulders, recommended by Michael Ciccotti, M.D., an orthopedic surgeon and director of sports medicine at the Rothman Institute at Thomas Jefferson University in Philadelphia.

Stand straight with your arms at your sides. Lift your shoulders up for a count of two, then let them down on a count of two. Do 20 repetitions. Then try alternate left and right shoulder shrugs, repeating 20 times with each shoulder. Finish by moving both shoulders forward and then backward 20 times.

SHYNESS

Four Steps to Social Fitness

When you see Carol Burnett on television—clowning, singing, and yodeling like Tarzan—it's hard to believe that she was a shy kid. But she was. It wasn't until she landed a role in a comedy at the University of California at Los Angeles that she was able to shed her shy persona and connect with her audience.

"Bliss. I never wanted to lose it. And I knew then that for the rest of my life, I would keep sticking my chin out no matter what, to see if I could ever feel that good again," Burnett recalls in her autobiography, *One More Time*. "I got a rave in the *Daily Bruin*," she writes. "Some total strangers stopped me in the hall and said they didn't remember the last time they'd had such a good belly laugh. I couldn't get over it."

Most of us are shy to some degree, afraid of rejection or of making a bad impression. But acutely shy people fear flopping so keenly that their anxiety keeps them from fully enjoying their lives, says Lynne Henderson, Ph.D., director of the Palo Alto Shyness Clinic, co-director of the Shyness Institute, a nonprofit research and education organization in Portola Valley, California, and visiting scholar at Stanford University. Dr. Henderson believes that for the super-shy, social exercise can help reduce reticence.

Use Your Head

There's no on-the-money research to prove that physical exercise fights shyness, but consider the other effects of exercise. An aerobic workout habit enhances your circulation and pumps oxygen throughout your body—your brain included. Some of the neurotransmitters released during exercise are believed to create a feeling of well-being. The adrenaline and norepinephrine that spike with exertion create feelings of excitation. When you're working out, you're hitting on all cylinders. Doesn't sound exactly shy, does it?

On top of the chemical confidence that may come with exercise, physical activity may have a psychological anti-shyness effect. Not surprisingly, shyness often goes hand in hand with low self-esteem, research suggests. If you have the discipline to stick to an exercise plan—to go for your walk even when it's so cold out there and so warm in your bed—it may enhance your self-esteem, says Robert Motta, Ph.D., a psychologist and director of the doctoral program in psychology at Hofstra University in Hempstead, New York. That good feeling about yourself may translate into social situations.

Another reason that exercise can fight shyness is that it can function as an icebreaker.

Group sports or fitness classes, which combine both physical and social exercise, may be perfect starting places if you want to overcome shyness. First of all, soccer games and aerobic dance classes are to some extent scripted—you kick the ball in soccer and mount the plastic step in aerobics. Anxiety about doing the wrong thing may be minimized. If you think your two left feet might embarrass you, you can choose beginner classes to build up your skill or classes in which enjoyment rather than perfection is the goal, suggests Dr. Henderson.

Second, these settings suggest all sorts of topics of conversation—reliving a great save or an outrageous foul or discussing whether the workout would go better to the music of John Lennon or John Philip Sousa. Exercise also releases endorphins, the hormones that can help calm social anxiety.

Party Survival: Managing the First Three Minutes

Every shy person knows that the first few minutes of a social situation are the toughest to deal with. Here are some shyness stoppers from a woman who's an expert on parties, Letitia Baldrige, who was social secretary at the Kennedy White House during the glamorous days of Camelot and is author of *Letitia Baldrige's Complete Guide to the New Manners for the '90s*.

Like a Boy Scout, be prepared. Before you go to a party, make a mental note of a few emergency conversation starters or silence fillers. That way, if awkward moments arise, you'll be ready with, "How do you feel about the controversy on cloning," or "Have you seen the new museum? Isn't it great?"

Knowing that you have a few rehearsed conversation openers at the ready if you need them will cut your anxiety level and let you enjoy yourself more.

Enter breathing. Before you actually go in to the party, take a deep breath or two or three, says Baldrige. Deep breathing is a good tranquilizer.

Find a few friendly-looking types. Don't know a soul at the party? Look around for friendly faces and head their way. If the friendly folk are conversing, wait until they finish. When they turn to you with a quizzical look, introduce yourself. Try: "I'm Mary Jones from Schenectady, and I don't know one person in this room, but you look like the nicest, most intelligent people here, so will you talk to me?"

"Say this laughingly," says Baldrige. "They will laugh, put their arms around you, get you a drink, and show you around."

Stay in circulation. If the first group of natives you approached weren't friendly and didn't show you around, say, "It's been really nice to meet you all. Goodbye," walk off, and try another group, says Baldrige. If you haven't made any conversational conquests in half an hour, "you're in an unfriendly group situation," she says. So go home and watch a good movie or read a book. You made a heroic effort. Reward yourself.

Exercise Your Sociability

Although physical exercise may help, the most effective way to overcome shyness is to work out socially, building your social muscles until you're in top social shape. If you find yourself too timid to dive into social situations, remember—your social muscles are just like the muscles in your tummy and back. They need exercise, too. "Overcoming shyness is like getting physically fit," says Dr. Henderson. "You've got to get social exercise every day, or at least several times a week."

If you can overcome your initial fear, your shyness will be less inhibiting. And while you may not become a television star like Carol Burnett, you'll enjoy your life more. Here are a few social "exercises" to help you break through reticence.

Show off those eyes, that smile. Get in the habit of making eye contact and smile contact as well, says Dr. Henderson. Sometimes our social anxiety makes us avert our eyes or tense our expressions. A steady gaze and friendly smile are body language for "I'm approachable and amiable." They'll put others at ease and help break down the shyness walls.

Be careful not to stare. Hold someone's gaze for three seconds and no more, says Letitia Baldrige, White House social secretary during the Kennedy Administration and author of *Letitia Baldrige's Complete Guide to the New Manners for the '90s.* A longer stare may make the other person nervous.

Talk light. Strike up short, trivial conversations with people you meet in the airport lounge or in line at the bank, Dr. Henderson suggests. In situations where it's clear that your

---

Conversational ABCs

Many people are shy because they think they're bad conversationalists. But that's often because they worry too much. Conversation isn't as complex as you may think. Remember these two rules.

Neither a blatherer nor a cipher be. Some shy people get so nervous in social situations that they talk nonstop. Others clam up. If you have a tendency to chatter, pretend you're a reporter, and ask the other person questions now and then. If you never talk about yourself, remember to share. "If you don't do that, others may feel uncomfortable around you because they've shared a lot, but they don't know anything about you," explains Lynne Henderson, Ph.D., director of the Palo Alto Shyness Clinic, co-director of the Shyness Institute, a nonprofit research and education organization in Portola Valley, California, and visiting scholar at Stanford University.

Talk like Charles Kuralt. Keep your conversational style simple. Don't stretch to be interesting. Just be warm and open. "Never talk over another person's head or do anything to make another person feel inadequate," says Letitia Baldrige, author of *Letitia Baldrige's Complete Guide to the New Manners for the '90s.* Make a concerted effort to shift your focus away from yourself and onto the other person. When you realize that conversation is easy, you'll be on the way to beating shyness.

interaction will be brief, nothing deep or meaningful is required. The weather is a perfectly fine subject. Just exchanging sounds with another human helps exercise your social muscles. If you practice openness in situations where the stakes are not very high, you'll find yourself more at ease in more pressure-packed social situations.

Many of us are shy because we have self-defeating thoughts—"Nobody will like me" or "I'll say something dumb"—echoing over and over in our heads. Experts agree that if we carefully assess those pessimistic scripts, we'll see that they're overstated.

Cut catastrophe from your forecast. Shy people have a tendency to exaggerate the consequences of taking social risks and to see potential failure everywhere. Dr. Henderson advises dialing down the disaster meter. She suggests asking yourself, "Is it really a disaster whenever I go out? Maybe I had one bad experience, and I've forgotten the other 98 times when nothing terrible happened." She also points out that nobody else is as critical of your performance as you are.

Pretend you're Jackie O. There's no need to wear a pillbox hat, but sometimes shy people can overcome timidity by imagining that they're someone with great social skills. If you're lurking in a corner at a party, picture Jackie O working the room. Slip out of your shy self, take on the role of a social whiz, and start circulating.

"Think of someone whose social skills you admire and use that person as a model," says Dr. Henderson. The First Lady role doesn't lure you? Try one of the Hepburns—Audrey or Katharine. You're a man? Imagine that you're Cary Grant or Fred Astaire, the paragons of cool.

SINUSITIS

Kick That Block!

Our sinuses—eight little caves behind the eyes and nose—are defenders of our lungs. When everything's okay in there, the mucus that lines the sinuses acts like flypaper, catching airborne enemies such as viruses, allergens, and dust before they can infiltrate our lungs. In addition to sifting out debris, these remarkable little recesses prepare air for our lungs in other ways, warming the cool air, cooling the hot air, and moisturizing air that's too dry.

Sinus problems arise when the sinus membranes swell, preventing the mucus from draining properly. Mucus buildup leads to blockage and pressure. The result? Sinusitis— headaches, facial pain, difficulty breathing, and the need for multiple boxes of tissues.

Go with the Flow

The key to dealing with sinusitis is to break up the mucus clog. Here are some movement tips to help destroy the Hoover Dam behind your honker.

Hang down your head. An easy way to open clogged sinuses is to lie on your stomach on your bed with your arms stretched overhead and dangle your upper body over the side until the top of your head rests lightly on the floor, says Charles Gross, M.D., professor of otolaryngology–head and neck surgery at the University of Virginia in Charlottesville. In this position, gravity can help drain mucus from the maxillary sinuses, located behind your cheekbones. Hold this position for no more than five minutes.

For the most effective sinus relief when you are in this position, Dr. Gross recommends first spraying your nose with phenylephrine (Neo-Synephrine or Alconefrin), an over-the-counter decongestant that will help open your nose, and waiting five minutes before giving yourself to gravity. You can practice this move every six hours as needed. If you try these decongestants, however, you shouldn't use them for more than three days to avoid the rebound effect, advises Dr. Gross.

Sleep right. If only one nostril is clogged, you can open it if you sleep in the right position. If your left nostril is open, for example, lie on that side, says Alexander C. Chester, M.D., clinical professor of medicine at Georgetown University Medical Center in Washington, D.C. There are nerves in the chest wall that influence the blood flow to your nose, he says. By assuming this side-sleeping position, he says, you trigger the nerve reflexes that can open up the clogged nostril. By morning, there could be a new, well-rested, easier-breathing you.

Resist being Rip Van Winkle. Sinusitis can make you tired, but you should try not to sleep, or even lie down, more than usual. In fact, one study of patients with chronic sinusitis showed that lying down increased nasal congestion and made it 20 percent more difficult to breathe. Limit the time you spend lying in bed, says Dr. Chester. If you're feeling poorly, rest upright in a chair, he adds.

Work it Out

Even if you're tired and achy from sinusitis, a brisk walk or a light bike ride is wonderful for relieving sinus pain, says Dr. Chester. "In about three-quarters of people with sinus problems, aerobic exercise is a very effective decongestant."

Exercise produces adrenaline (epinephrine), which shrinks blood vessels in your nose. When the swelling in your sinuses goes down, trapped mucus can flow freely. About 20 to 30 minutes of a light workout should help clear your cavities.

Put Your Sinuses on the Steam Setting

Breathing in warm, steamy air is a great way to open up sinuses and get mucus flowing again, says Alexander C. Chester, M.D., clinical professor of medicine at Georgetown University Medical Center in Washington, D.C. "The incidence of sinusitis is greater in climates that have dry, cold winter months."

You would think that swimming a few laps in your local YMCA pool would be a great sinus opener because you're exercising and breathing humid air at the same time. But as it turns out, even if water is good in the form of steam, it's bad in liquid form if you have sinusitis. Activities such as swimming, diving, and scuba diving can actually drive mucus farther back into your sinuses.

While aquatic sports are a poor choice for loosening mucus, here are a few of Dr. Chester's recommendations for getting a snoot full of steam.

Hit the showers. If possible, leave the vent fan off while you take a warm shower. Spending some time in a steamy bathroom is an easy way to help your sinuses drain.

Get steamed. Anytime you're feeling clogged during the day, get a steaming beverage and practice this trick for instant relief. Put your nose close to your cup of steaming coffee, tea, or soup, and cup your hand over it to direct the vapor up your nose. Just be careful not to burn yourself.

Be a spectator. Go watch a water polo or swim team compete. Just sitting in the humidified pool environment can make your nose run.

Tent yourself. Fill your bathroom sink with hot water, place your head close enough to be able to breathe the vapor from the steam (but not so close that you burn yourself), and stretch a towel over your head and the sink bowl. The towel will trap the vapor and encase your head in a warm, mucus-loosening haven.

Although a little exercise is good, however, it doesn't necessarily follow that more is better. If you're really feeling lousy, don't exercise strenuously or for too long, says Dr. Chester. And don't try an exercise cure if you have a fever to go with your sinusitis, he adds.

If you're going outside to exercise, try to avoid the things that trigger your sinusitis, says Dr. Chester. If you know that ragweed is your nemesis, for example, try exercising in the afternoon, when pollen counts are lower. Try fitting in your workout during your lunch break, for instance.

Pace it off. Relief may be as close as your front door. Whenever you feel especially stuffed up, put on your sneakers and head out for a leisurely stroll. "Exercising in the fresh air is great for relieving sinusitis pain," says Dr. Chester. Choose a fairly level course that will take you about 20 minutes to complete when you're going at a relaxed pace.

Scarf up for moisture. If it's chilly, use a scarf to cover your mouth and nose, Dr. Chester advises. This will help humidify the air and start the mucus running. Cram some tissues into your coat pocket—you'll need them.

Cycle your sinuses clear. Dig your kid's bike out of the garage and take a spin around the neighborhood. Put it into a low gear and pedal at a comfortable pace. The point here is to unblock your sinuses, not begin training for the Tour de France. "Your sinuses should open up shortly after beginning exercise," says Dr. Chester, so take some tissues along.

Ride inside. If you have a stationary bike and a humidifier, you can really unclog the works. Exercising in moist air will help loosen the mucus even more, says Dr. Chester.

Dodge the dry chill. During any season of the year, the best indoor environment for exercise is both humidified and well-ventilated with outside air. "It's best to avoid exercising in air-conditioned health clubs because the air is too dry and can make sinusitis worse," says Dr. Chester.

SMOKING

Take a Run at Kicking the Habit

A few years ago, Trish Beichner was like most smokers—she'd tried to quit many times but relapsed again and again. Then she found the secret that helped her quit for good. It wasn't a magic drug or a secret technique. It was exercise.

Beichner, who is in her thirties and a manager at a magazine fulfillment company in Pennsylvania, took up running. A few months later, when she tried to kick the smoking habit again, she succeeded. She's been smoke-free ever since.

Quitting cigarettes is never easy, because nicotine is so powerfully addictive. But there is evidence that exercise can help you kick it. If you're a smoker, it's not a good idea to dive into a demanding exercise regimen, but little by little, you can just start becoming more and more active. People who lead more active lives have greater success in bidding cigarettes a final farewell.

Why? For two reasons. First, exercise reduces stress and boosts the feeling of self-control that is crucial to kicking. Second, exercise picks up the pace of the purging. It helps get rid of the toxins, such as carbon monoxide and tar, that have poisoned your body, and it helps your lungs get better faster.

Don't Touch That!

For some ex-smokers, keeping their hands busy without a cigarette can be a challenge. If this sounds like you, it may be helpful to try some ways to get a handful of something other than a Salem.

Try these tips from Raymond Kowalcyk, Ph.D., a clinical psychologist and hypnotherapist in private practice in Allentown, Pennsylvania.

► Do some knitting or crocheting.
► Teach yourself to juggle.
► Keep a doodle pad next to the phone.
► Work in the garden.
► Give yourself a manicure.
► Finger-paint with a child.
► Spend the day baking bread.
► Write a batch of overdue letters.

Dealing with Stress and Claiming Control

Feeling out of control is a big issue for many smokers who quit but relapse. That's understandable, says Raymond Kowalcyk, Ph.D., a clinical psychologist in private practice in Allentown, Pennsylvania. "When a smoker quits," he says, "she is giving up her chosen method of stress reduction." Without the familiar crutch, even everyday anxieties seem overwhelming.

Exercise eases this stress. An animal study at the School of Biomedical Engineering at Banaras Hindu University in India showed that rats that were subjected to stress and given a treadmill workout had brain wave activity that showed less stress than rats that got the stress without the exercise.

Having the self-discipline to go cycling or swimming every day may give people the confidence that they have the self-control required to stop smoking as well, says Dr. Kowalcyk.

Healing the Lungs

Exercise is especially important to the lungs of current and former smokers, says nurse practitioner Barbara Forbes, director of the Smoking Cessation Institute at Vanderbilt University in Nashville. "While the lungs will eventually recover from smoke damage on their own," she says, "exercise can speed up the process somewhat."

Specifically, Forbes says, regular activity aids the alveoli, the delicate, grapelike clusters of lung tissue where oxygen is infused into the bloodstream. "Since the body views cigarette smoke as dirt," says Forbes, "it reacts by producing excess mucus to trap it." Unfortunately, too much mucus will clog the alveoli,

Hypnosis: This Mental Exercise Might Help You

"You are getting sleepy. . . your eyelids are getting heavy. . .when I snap my fingers, you will cluck like a chicken. . . . " If you still think hypnosis is nothing but a magician's stage trick, think again. Hypnosis is a unique tool recommended by health-care professionals to help people quit smoking.

Using relaxation and the power of suggestion, hypnosis deals directly with the mind, says Raymond Kowalcyk, Ph.D., a clinical psychologist and hypnotherapist in private practice in Allentown, Pennsylvania. Hypnosis actually erases the habitual urge to smoke, he says.

Hypnosis for smoking cessation is usually given in individual sessions that take about two hours. After an introduction, the lights are dimmed and you are told to get comfortable. Then, speaking quietly and slowly, the therapist guides you into a deeply relaxed state. Once relaxation is reached, the therapist begins to mention the positive rewards of not smoking. He'll also explain very gently that after the session, all cravings for cigarettes will simply disappear. Amazingly, this simple, nonaggressive technique gets good results.

Success rates from just one session of hypnosis-induced suggestion can be very high, according to Dr. Kowalcyk. "In my experience, 60 to 80 percent of people will actually be able to stop smoking using hypnosis after one individual session," he says. "But after a year, that drops to 20 to 25 percent who continue to not smoke. People start smoking again for a variety of reasons."

causing smoker's cough and other breathing difficulties. Even moderate exercise, performed while breathing deeply, can frequently clear blocked alveoli and speed the lungs to a healthy state.

Any regular low-key aerobic exercise regimen will help you quit smoking. You can ride your bike for a couple of miles a day or take an easy morning jog. You might try a simple swimming strategy to help your body get over nicotine. But perhaps the best aerobic exercise for smokers who want to be ex-smokers is the easiest of all—plain old walking. It's the main activity that Forbes recommends at the Smoking Cessation Institute.

As you get into the walking habit, make it a point to gradually walk farther and faster. When you first start out, meandering may be okay. But according to Forbes, you should try to gradually increase your speed and distance, especially if you want to improve your cardiovascular health or to lose weight. For cardiovascular health, walk for 20 to 60 minutes three to five days a week. If your goal is to lose some weight, walk for 45 minutes six days a week, says Forbes.

As with any new activity, start slowly, but do it regularly. Start by walking comfortably for 20 minutes a day. Gradually increase the distance you walk. After about two weeks, gradually increase your speed for a few minutes at a time, so that by the end of a month you can walk for 45 minutes, alternating between a comfortable pace and a brisk pace, she recommends. After two months, you should be able to walk at a brisk pace for the entire 45 minutes.

Start looking for opportunities to walk wherever you can. Consider these options.

Explore your city. Add some history to your walking program with a walking tour. Most

Exercise to Ward Off Weight Gain

Unfortunately, roughly two out of three people who quit smoking will gain some weight, and among these people, the average gain is between 10 and 12 pounds, says Martin Katahn, Ph.D., professor emeritus of psychology at Vanderbilt University in Nashville and author of *How to Quit Smoking without Gaining Weight*.

Although cigarettes are bad for your health, they do suppress appetite. Nicotine raises levels of serotonin and dopamine, which are two appetite-moderating neurotransmitters in the brain. Less nicotine often means more hunger. Nicotine also increases the secretion of catecholamines, which rev up metabolism and make the body burn more calories.

Exercise not only makes quitting easier, it also helps prevent weight gain, says Dr. Katahn. "Regular physical activity can counter the metabolic slowdown that comes with quitting," he says. In addition, appetite is dampened immediately after vigorous exercise. In the following few hours, appetite increases gradually, but not enough to fully compensate for all the calories burned during exercise.

cities have them. Pick self-guided tours rather than joining a group, so you can really step up the pace. There is usually a variety of tours to choose from—representing different neighborhoods or other themes—so you can make a hobby out of touring by foot. Call your local tourism or visitors' bureau, chamber of commerce, or public library for schedules and meeting points.

Make a habit of hiking. Who says walking has to be done on level ground? Hiking uses the same skills as walking, with the bonus of burning many more calories. The Sierra Club has chapters nationwide, and you can check local bulletin boards or the Internet for an informal hiking group near you.

Befriend a beagle. Walking a dog adds a whole new dimension of fun to your daily constitutional. If you're dogless, borrow your sister's schnauzer. Before long, reaching for the leash may seem more natural than lighting up.

Stair all you want. You don't need to sweat your afternoon away on a stair-climbing machine, says Forbes. Just head for the nearest stairwell or set of steps. If you have a report to take down to the sales department, don't take the elevator, take the stairs. Make it a point to add some ups and downs to your day.

SNORING

Active Days Make for Silent Nights

If your snoring's gotten so clamorous lately that your partner refuses to share your bed, your best path back to cuddling may be running around.

The block, that is. Several times. And six or seven days a week.

Why? You may need to shed a few pounds, and exercise is the best way to do that.

One of the main contributors to snoring is being overweight, according to Alex A. Clerk, M.D., director of the Stanford University sleep clinic in Palo Alto, California. If you're only slightly overweight, as little as 15 extra pounds can make the difference between loud and silent nights, according to Dr. Clerk. Exercise may help you trim down—and pipe down—come sundown.

Anatomy of a Snore

Snoring is the sound caused when the fleshy parts of your mouth and airway vibrate in the gusts of air that you exhale. The vibrating parts include the back of your tongue, the walls of your throat, and your uvula, the little "punching bag" that dangles over the top of your throat. When you sleep, the muscles in your throat and mouth relax a bit, so these tissues droop a little, thereby narrowing your airway. The narrower the passageway, the faster the rush of air through it. The faster the air, the more those fleshy parts flap, like semaphore flags in a tropical storm.

But you didn't snore five years ago, so why now? Simple. As we age, we tend to gain weight. When we gain, we add fat not only to our thighs but also to our throats, and the fat deposits narrow the airway. Further, the muscles in our throats tend to lose their tone over time, so tissues in our throats sag into our airways. This can happen even without weight gain, Dr. Clerk explains.

For years, people have tried to come up with anti-snoring exercises to strengthen the muscles inside the throat. But there aren't any targeted exercises that get at those specific muscles, Dr. Clerk says. So the best approach is to try to get down to fighting weight.

Thirty Minutes a Day Makes the Rumble Go Away

To lose weight with exercise alone, you'll have to put in at least 30 minutes just about every day of the week, says Wayne Miller, Ph.D., professor of exercise science and nutrition at the George Washington University Medical Center in Washington, D.C. But there's no need to push yourself to the brink of exhaustion in the weight room. While doing your normal activities, make a point of pushing yourself beyond your normal exertion level. Try these aerobic upgrades.

Go natural in the garden. If you usually use a power mower to do the lawn, take the push mower out for a whirl, says Dr. Miller. Or use a good-old fashioned hoe instead of that 4,500-horsepower Roto-tiller. If you usually garden at a leisurely pace—sipping iced tea and chatting with your neighbor while you weed—

put down the tea, shun your neighbor, and pick up the weeding pace.

Cancel the cart. If you're a golfer, forgo the electric cart and walk the fairways, suggests Dr. Miller. (You won't need to carry your clubs; the walking will do the trick.)

Find a few hills to climb. Change the route on your evening walk so that you go up a couple of hills. Nothing alpine, though—just a few gentle inclines will crank up the calories burned.

Then, if you also limit your calories during meals and between meals, that 30 minutes of exercise every day should help you get rid of

---

The Serious Side of Snoring

Although snoring is the stuff of sitcom plots, it can be a serious business, says Michel Babajanian, M.D., a head and neck surgeon at Century City Hospital in California. Snoring can be a symptom of a dangerous disorder called sleep apnea. If you have apnea, your airway narrows so much that you more or less stop breathing—dozens of times a night. Starved of oxygen, you awaken briefly. You may not even be aware of each arousal, but you don't get the good, solid rest that you need.

Some studies suggest that apnea may contribute to high blood pressure, heart disease, and stroke as well as fatigue. No one knows exactly why it might do this, but research points a finger at adrenaline and other stimulating chemicals. When a lack of oxygen jolts you into wakefulness, your body releases lots of this stimulating hormone, which boosts

your blood pressure, says Alex A. Clerk, M.D., director of the Stanford University sleep clinic in Palo Alto, California.

Daytime sleepiness is a telltale symptom of apnea. So if you snore at night and nod through the day, see your doctor. Even if you don't snore, it's worth seeing your doctor if you're chronically tired during the day, says Dr. Clerk. Some folks with apnea don't snore.

Odds are that your doctor will invite you for a sleepover at a sleep laboratory. While you doze there, he'll check various vital signs, including your blood oxygen levels. If the diagnosis is apnea, he may suggest that you wear a special mask to bed that lets you breathe normally as it pressurizes the air to keep your airway open. Or he may suggest surgery to remove some of the drooping tissue that's blocking your airway and your access to a good night's sleep.

---

those extra dozen pounds in about two months, explains Dr. Miller.

Exercise early to stifle snoring. Exercise is stimulating, so it can keep you up at night if you work out too close to bedtime. That's bad, since a night without sufficient sleep will leave you exhausted the next day, and that can boost the odds that you'll snore.

What's more, when exhaustion knocks you out cold, your muscles relax more than usual, so all those fluttery tissues in your mouth and throat droop even further into your airway, amplifying your snore. The solution to exercise-induced insomnia? "Exercise as far from bedtime as possible, never any later than early evening," says Dr. Clerk.

The Anti-Snoring Triathlon

If you can't lose weight or if you're already at your ideal weight but snore anyway, it's clear that conventional aerobic exercise won't help you be invited back into bed with your partner. But there is hope for snoring cessation. Try these three unconventional "exercises" that frequently offer snoring relief by keeping airways open.

Practice the anti-snoring flip. Once you're in bed, flip from your back onto your side. "You're more likely to snore if you sleep on your back because in that position, your tongue falls back and obstructs your airway," says Dr. Clerk. To keep yourself from listing onto your back in the middle of the night, try this tip. Take a T-shirt with a pocket and sew a tennis or golf ball inside the pocket. Wear the shirt to bed with the pocket in back. If you start to flop backward, the ball should prod you to your side again.

Booze and Other Bad Guys

Sometimes, alcohol or prescription drugs can be the cause of lumber-cutting in the night because they can cause swelling of the esophageal tissues and excessive relaxation of the important muscles. So say "no, thank you" to the offer of a nightcap, and ask your doctor if medications might be making you a nocturnal nuisance. If so, adjusting your medications may be your ticket to a silent night.

An arid clime can dry out and irritate the linings of your airways and may also play a role in snoring, says Alex Clerk, M.D., director of the Stanford University sleep clinic in Palo Alto, California.

Perfect your mattress lift. Firmly grasp the top of your mattress, lift it, and insert a couple of hefty phone books or tedious novels, like *The Last of the Mohicans* (you'll never miss them), underneath. Slowly lower the mattress. This adjustment will raise your head a half-dozen inches when you're in bed. And when your head's raised, your tongue is less likely to flop back and block your airway, says Dr. Clerk.

Master the Marlboro heave. Throw your cigarettes—away. Go for distance. Aim for the trash can. Smoking makes the membranes in your throat swell, narrowing your airway.

Stage Fright

Turning the Tension Around

What strikes the most fear into your heart? Listening to the wail of a dentist's drill as he excavates one of your cavern-sized cavities? Looking down from the observation deck atop the 102nd floor of the Empire State Building? Contemplating your own inevitable death?

Now consider stepping up to a spotlighted microphone in front of an eager audience and launching into a speech. If you're like nearly 60 percent of Americans, you'll pick public speaking as the "scenario most likely to cause extreme anxiety."

The secret to doing away with stage fright isn't banishing your bouncing nerves. Instead, the answer is harnessing all of that natural energy for your own purposes. You just need to get moving, do a little meditation, and stand tall before that big meeting.

Getting the Butterflies in Formation

They don't call that jittery feeling "having butterflies" for nothing. The feather-winged creatures doing aerial dances down there beneath your diaphragm move with complete and utter randomness. And all you want is for that belly-ballet to end. But your performance, speech, story-telling, or plea bargain can actually benefit from your current state of excitement.

The trick is to take the nervousness of stage fright and turn it into ordered energy, says Dale Anderson, M.D., an urgent-care physician in Minneapolis, author of health humor books such as *The Orchestra Conductor's Secret to Health and Long Life*, and co-ordinator of the Act Now! Project, a health coalition of theatrical and medical professionals in St. Louis Park, Minnesota. Instead of letting them stir up unproductive anxiety that can hinder your performance, get your nerves into military formation. Turn the tension you're feeling into positive energy. Your prospective audience will catch your enthusiasm instead of noting your nervousness. Here's a step-by-step plan to help you prepare for your big event.

Imagine no performance anxiety. Starting for a week or so before the event, take some time every day to vividly picture the scenario from beginning to end, suggests Dr. Anderson. Begin by sitting or lying down in a quiet place. Take a few minutes to breathe slowly and become completely relaxed. Then, step by step, mentally rehearse your event. "See yourself having the very best performance you can imagine," says Dr. Anderson. Picture yourself being successful and you'll boost your confidence so you'll be ready for the real thing.

Get moving. You know that elated feeling you get after a brisk walk? That's just the attitude—and energy—that you need to exude during your performance. If you can't take the time for a workout an hour or so before "curtains up," you can still run up and down a few flights of stairs backstage, says Dr. Anderson.

Stretch to de-stress. You're next at the speaker's lectern. You're cool and calm and ready to go. Except for your shoulders—they're so knotted up they could double as earmuffs. A brief bout of stretching will counteract that intuitive tendency to tighten up in times of stress. "I like to use my wing-waiting time to do a few deep knee bends followed by some side bends," says Dr. Anderson. Finish with an upper-body stretch: Reach both arms overhead and then arch gently backward to relieve the stress in your shoulders, arms, and spine.

Take an Alexander minute. Take a tip from the Alexander Technique, a posture practice de-veloped by an actor. The technique strives to keep the head, neck, and spine aligned in a flowing, upward direction. Adjust your posture so your spine is lengthened, then tuck your chin in slightly so that the back and top of your head ease up toward the ceiling. Standing tall like this makes a big difference in how you feel—and in how you look. "You'll be ready to go," says Dr. Anderson, "and it will show."

Breathe your best. Now that your posture's perfect, take a few slow, deep breaths. A deep breath begins with a deep exhalation. Be sure to exhale completely before beginning each deep inhalation. Let your belly expand with each inhalation and relax your shoulders as you exhale. Use focused breathing to release any remaining tension in your body while sending a memory-jogging jetstream of oxygen to your brain.

Flex your face, too. Just before you step up to the mike, while you're still out of sight, flex your face muscles for all they're worth. "Making a really big, fake smile will turn your nervous chemistry into positive performance energy," says Dr. Anderson.

STRESS

Finding Serenity in Chaos

Feeling stressed? Wonder why? The complexity of modern life, that's why. And, if you don't believe it, just consider for a moment the evolution of a coffee shop.

Twenty years ago, coffee shops served up simplicity and contentment. You plunked a couple of quarters on the counter and received a cup of java. Today, a coffee shop is an anxiety-producing gauntlet of choices: Caffeinated or decaf? Cappuccino or latte? Whole milk or skim? French vanilla or mocha java? Iced or piping hot? Grande or a short cup? Care for a sprinkle of cinnamon on top?

Granted, you may be able to deal with the very minor stress involved in those choices. But it just goes to show how once-simple pleasures like a cup of coffee now come with a very thin but discernible extra layer of stress. And too much stress compromises our health. If we can't control it, we're at greater risk for everything from accidents to cancer.

Taming stress and reclaiming a peaceful mind requires learning how to pace yourself, simplify your life, and ignore the oodles of needless distractions. Regular exercise, a sense of playfulness and humor, and relaxation techniques can help you find serenity in our chaotic world.

Moving toward Relief

Physical exercise is a superb stress reducer. "Exercise releases endorphins in our brains that lead to an improved sense of well-being," says Jane Sullivan-Durand, M.D., a behavioral medicine physician in Contoocook, New Hampshire. Endorphins are chemicals that are responsible for relieving pain and stimulating relaxation. Countless studies have demonstrated that exercise helps defuse the threat of stress. A little laughter goes a long way toward helping us be laid-back as well. Here are some tips for getting exercise into your life and looking on the lighter side while you do it.

Try a bit-by-bit pace. Not a regular at Gold's Gym? That's okay—you can start slowly. "Increase your physical activity by bits and pieces during the day," suggests Dr. Sullivan-Durand. "Walking briskly for three 10-minute segments a day is almost as effective as a 30-minute walk."

Jog like Groucho. Lighten your step and your attitude by wearing a pair of Groucho Marx glasses the next time you head out for a jog after a stressful day at work. It's hard to maintain feelings of stress when you feel silly, says Steven Sultanoff, Ph.D., a clinical psychologist in Irvine, California, and president of the American Association for Therapeutic Humor.

De-stress through dance. Be your body's disc jockey and dance to a tune that puts you in a happy or mellow mood. Dr. Sullivan-Durand's mood-uplifting favorites are "Twist and Shout" and "Wipeout."

"Pick music that stirs your soul. When I'm stressed, nothing discharges bad energy out of me like those songs," says Dr. Sullivan-Durand. "I start dancing and drumming, and I feel my tension go away."

Stretch away stress. Experts agree that stretching your muscles can soften the effects of stress. Here are two ways to put a little stretch in your daily routine.

Easy workday stretch. Every few hours throughout the day, slip away to a private place—a stairwell will do—for a couple of minutes, says Dr. Sullivan-Durand.

Stand up straight and raise both hands above your head. Try reaching up as far as you can with your right hand, then do the same with your left hand, as if you were crawling upward. Alternate stretching right and left for a total of five times, then slowly bring your arms back down.

Then stretch the muscles in your legs. Stand straight and hold onto a chair or railing with your left hand for support, then raise your right leg straight out to the side as far as you comfortably can. You may also want to raise your right arm for balance. Hold your leg in that position for a count of five and slowly bring it back to the floor. Switch sides and repeat with your left leg. This stretch will ease tension and stress.

Neck stretch. Stress often shows itself in our necks. Here's a way to release the tension that

Freedom from Phone Stress

The almighty telephone seems to shackle us at times. Here are some strategies to give phone stress the busy signal, recommended by Jane Sullivan-Durand, M.D., a behavioral medicine physician in Contoocook, New Hampshire.

Stand up to chat. Don't contort yourself in a stressful position. If you cradle the phone between your ear and shoulder, all the muscles in your neck, shoulders, and back will start to knot up. It's time to take a stand—literally. Get off the couch, stand straight, and hold the receiver in your hand. Every five minutes or so, do a full-body stretch to keep you limber, suggests Dr. Sullivan-Durand.

Avoid phone lunging. Where is it written

that in the land of chat, we must lunge and pick up the telephone as soon as it rings, and never mind the rice bubbling over on the stovetop or that delightful game of Scrabble you're in the middle of playing with close friends. Let your answering machine be your messenger, and then you can return the calls when you're ready, says Dr. Sullivan-Durand.

"In life, there are few true emergencies, yet our society conditions us to believe that everything—including phone calls—must be done yesterday," she says. "The next time the phone rings, make a conscious decision as to whether this is a good time for you to talk. If it's not, either don't pick it up or let an answering machine take the message."

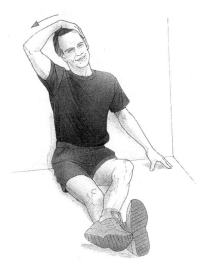

1

gathers there, says Dan Hamner, M.D., a phsyi-atrist and sports medicine specialist in New York City.

▶Sit on the floor against a wall with your back straight.

▶Extend your legs and cross the left ankle over the right.

▶To maintain your balance, extend your left arm to the side and let your fingertips touch the floor.

▶Place your right hand lightly on top of your head so that your fingers touch your left ear.

▶Keeping your shoulders down, gently pull your head toward your right shoulder and hold for 30 seconds (*1*). Return to the starting position.

▶Repeat this tilt three times, then do these steps using your opposite hand and shoulder.

Do the car seat wiggle. Sitting still can cause stress. If you're trapped in an office chair for long periods or stuck in a car during a traffic jam, tense and release your buttocks muscles. Lift your left buttock, tense it for a few seconds, relax, and let it drop back down. Then switch and do the same for your right buttock.

And don't forget those knotted shoulder muscles. Try rolling your shoulders forward and backward for at least 30 seconds every half-hour and you should feel tension melt away, says Dr. Sullivan-Durand.

Stress Humor and Humor Stress

You can fight stress with smiles and laughter, say doctors and humor experts. "We don't fully understand the biology of attitude, but muscle tension is actually reduced when we smile or laugh," says Dr. Sullivan-Durand. It's harder to remain stressed when you find a way to laugh about your situation.

Loretta LaRoche, a national stress expert, pres-ident of The Humor Potential in Plymouth, Mass-achusetts, and author of *Relax—You May Only Have a Few Minutes Left*, drives the point home with this question: "Did you ever try to lift a couch and laugh at the same time? You can't. By the same token, you can't feel stress and laugh at the same time."

Humor may also help to fend off emotional stress. Biochemically, humor offers these added health bonuses: It increases immunoglobulin A, an antibody designed to protect against colds, flu, and other respiratory conditions, and it de-creases the level of the stress hormone cortisol, says Dr. Sultanoff.

Try a smiling meditation. Here's a good way to start each day with a healthier outlook. After you've finished brushing your teeth, spend about 30 seconds grinning widely at your reflection in the mirror. Open your eyes wide and raise your eyebrows. Don't worry if you feel a little foolish. Then for the next 30 seconds, grab your belly with both hands and start laughing.

"Fake it till you make it," advises LaRoche. "After a while, you should start laughing naturally, and it can definitely improve your mood and relax you."

Laugh and love. The Eskimos call sex "laughing time," says Shirley Zussman, Ed.D., a sex and marital therapist in private practice in New York City. She recommends using sex as a tool to tame stress. "Sex can be fun, frivolous, and relaxing," she says.

Maintain a giggle library. Keep a joke book within arm's reach at work and in the car. When stresses mount, flip to a page on Dilbert's latest office escapade or The Far Side's twist on life, suggests Dr. Sultanoff. One good laugh may help drop your stress to a manageable level.

More Ha-Ha's from a Humor Master

Dr. Sultanoff is a promoter of humor. His strategies may seem a bit off-the-wall, but the next time you find yourself steaming in stress, remember these tips from the merry mirthologist.

Dish up a plate of chuckles. At your favorite restaurant, you're about to dig in to the boneless chicken breast and lightly buttered asparagus spears when your four-year-old erupts into ear-piercing screeches.

There's no need to gnash your teeth, says Dr. Sultanoff. Just whip a red clown nose out of your coat pocket and put it on. Then turn to the child, make eye contact, arch your eyebrows, offer a friendly smile, and wait for the tension around you to melt like butter on a warmed plate.

Acupressure Angles

A well-placed thumb or fingertip can deliver immediate relief from stressed muscles. Here are a couple of sample maneuvers from David Nickel, O.M.D., a doctor of Oriental medicine and a licensed acupuncturist in Santa Monica.

Head for the neck. One stress-relieving pressure point is at the back of your head. Slide your right thumb down the back of your head on the right side until you find the spot to the right of where your skull meets your neck. Apply steady thumb pressure on that spot, pressing upward toward the left for five seconds, then rest for five seconds (*right, a*). Alternate the on-and-off pressure for a total of one minute, says Dr. Nickel.

Shrug it off. This acupressure point is

located on the upper shoulder muscle where the neck meets the shoulder (*left, b*). Use

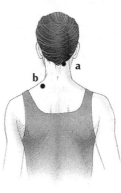

your index finger or thumb to alternate applying steady pressure for five seconds, then releasing for five seconds, for a total of about one minute, says Dr. Nickel. "This point affects the gallbladder and liver, which relate to the emotions of frustration, irritability, and anger," he explains.

"The child usually stops crying, starts gazing at you, and becomes quiet. It's a distraction that usually helps break the tension for both the child and everyone else in the restaurant," says Dr. Sultanoff.

This clown nose technique also works well in other tantrum-prone places like grocery stores, churches, and theaters, he adds.

Clear your head. The on-ramp to the highway is backed up to a standstill. Instead of pounding your dashboard, Dr. Sultanoff has a better way to blow off steam—with bubbles.

"I always carry a plastic bottle of bubbles, roll down my window, and blow bubbles into the air," he says. "There's nothing I can do to get anywhere faster. I'm stuck, but at least the bubbles help pass the time more quickly."

Chirp your presence. It's Saturday morning, and the line at the supermarket winds around like a boa. You're stuck behind a frazzled young mom with two toddlers tugging on her as she struggles to get her coupons in order. Now what? Dr. Sultanoff offers this stress fighter.

"I can chirp like a bird without moving my lips," he says. "So I start chirping and look for the 'bird.' When I get the kids' attention, I 'capture' this imaginary bird in my hands, pet it, and the give the bird to the kids to pet."

This playful act entertains those around him and helps them stop fuming about the long line. If chirping like a bird isn't in your repertoire, try using a noise-making toy hidden in your pocket.

Unbinding the Mind

Sources of stress are everywhere. But so are shortcuts to serenity. Try these.

Develop new self-talk. For a stress-free you, have a little pep chat with yourself. Sweep out the "should haves," "what ifs," and "have tos" in your vocabulary and replace them with "I can," "I accept," and "I'm looking forward to it" talk, says Dr. Sullivan-Durand.

"We sometimes distort the truth by overgeneralizing or by negative talk," she says. "We did a great job on a project, but one sentence did not seem right, and suddenly we view it as not being good enough. We need to remove our expectations of how we think things should be and accept how things are."

Be a daydream believer. When your mind wanders, see if you can let it wander into wonder. Take a five-minute break during your hectic schedule, stare out the window, and daydream about winning the lottery. Picture yourself picking the winning combination of numbers and collecting millions of dollars. This is a five-minute peaceful distraction from the hectic demands of the day, says Dr. Sullivan-Durand.

Notice beauty. Anything lovely—a person, a flower, a flashy sports car, the light sifting through trees—can give you tranquillity.

Work the day after Thanksgiving. Pick any day that is usually a day off for your officemates. You can get a lot more done without interruptions. Then reward yourself a week or so later by taking a day off when everyone else is working, says J. Crit Harley, M.D., a behavioral medicine physician and director of Un-Limited Performance, a stress-management company in Hendersonville, North Carolina.

Work the 800 lines. "I fully support catalog shopping as a stress solution," says Dr. Sullivan-Durand. After all, why stand in line during the Christmas rush when you can stay in your pajamas and order that sweater for Aunt Bess by phone? You won't have to jostle for a parking spot or grow old in the checkout line.

Skip rush hour. Everybody's a nine-to-fiver, right? Ask your boss if you can work 10:00 to 6:00 instead. Missing traffic may ease stress significantly.

Write a bad-to-glad list. Maybe you're fretting about your daughter's outdoor wedding that's scheduled for the rainy season in South Florida. Okay, write down the worst thing that could happen, suggests Dr. Harley. The entire wedding party could end up swimming toward the wedding cake. Now flip the page over and write down the best that could happen. The sun will shine, the trees will offer breezy shade, and everyone will look *mah-ve-lous*.

Then flip back over and write down what you could do to minimize the possible monsoons. Maybe you could rent a large tent to act as an umbrella for the wedding crowd.

"What this technique does is help you focus on concrete issues instead of free-floating worry," says Dr. Harley. "You learn to realize that perhaps the worry is not so bad or that you can do something to minimize it."

Find a happy buddy. Hang out with an upbeat pal who doesn't dwell in the days of whine and supposes, says Dr. Harley. Avoid complainers; they can increase your stress level.

Relaxing in Hectic Times

A key ingredient in the anti-stress lifestyle is learning how to truly relax. Grabbing five-minute relaxation breaks during the day can reduce the toll that stress takes on your body and mind, say doctors.

"These relaxation/meditation techniques activate the parasympathetic nervous system," explains Dr. Sullivan-Durand. "They can help reduce your heart rate, blood pressure, and other physical changes associated with stress."

Shift to a lower gear. Stuck behind a slow-moving train of cars on the way to work—again? When you start to fret, try focusing on taking slower, deeper breaths from your belly, recommends Dr. Sullivan-Durand.

"Diaphragmatic breathing can be used any time—at a meeting, in your car, or at your desk," says Dr. Sullivan-Durand, who has created several audiotapes that teach relaxation techniques. "You can restore peace by taking 5 to 10 seconds and shifting your breathing down to your belly. It lets you slow yourself down."

How do you know you're doing it right? Here's an easy clue: Place your right hand on your belly and concentrate on taking deeper, slower breaths. When you see your hand rise and fall, you'll know you are practicing diaphragmatic breathing.

"Diaphragmatic breathing works better if you breathe out twice as long as you breathe in, says Jo Anne Herman, R.N., Ph.D., a biofeedback expert and associate professor of nursing at the University of South Carolina in Columbia. "So inhale for a count of two and exhale for a count of four."

Head for the fresh air. A few hearty gulps of air on a mountain or even in your backyard can help oxygenate your tissues, improve circulation, reduce muscle tension, and lower your stress quotient, says Dr. Sullivan-Durand.

"I believe that people who spend more time outdoors are more aware of the cycles of nature and end up feeling happier," she says. "We are natural beings. When we isolate ourselves and get too far away from the cycles of nature, we feel out of sorts."

STRETCHING

Cats Know Where It's At

WHAT IT IS

Ever watch a cat wake up? Tabby usually starts with a slow, purposeful stretch. First one furry leg extends forward, then the other. After the classic cat back arch, she finishes with an all-out, tension-taming y-a-w-n. Well, there's wisdom in feline ways. Stretching—systematically elongating your muscles before and after exercise—is a gentle and effective way to enhance your fitness and stay spry.

WHAT IT DOES

The fundamental way that stretching betters health is by increasing flexibility, says Carol Espel, an exercise physiologist and general manager of a health and fitness club, Equinox of New York City in Westchester, New York.

"Maintaining flexibility means that your muscles keep their range of motion as you age," says Helen Schilling, M.D., medical director of HealthSouth–Houston Rehabilitation Institute in Texas. Flexibility is a fountain of many health blessings.

Protection from injuries. Stretching can prevent strains, sprains, and other injuries. "Limber people don't seem to get as many aches and pains," says Dr. Schilling.

A spring in your step. Sustaining a supple way of life may help you stay youthful in other ways. "Flexible people appear much younger than they really are," notes Dr. Schilling. "Just look at the easy way they move."

Relief from joint-related problems. If you stay flexible, age-related ailments like arthritis and back trouble may be less painful and inhibiting, says Dr. Schilling.

Better posture and balance. Stretching also promotes coordination, says John H. Bland, M.D., professor emeritus at the University of Vermont College of Medicine in Burlington and author of *Live Long, Die Fast: Playing the Aging Game to Win*. "Stretching affects the body at the cellular level," he says, "stimulating communication between the muscles and the brain." This can foster a healthy sense of balance and promote good posture.

REAPING THE BENEFITS

Stretching, unlike some other physical activities, is easy enough to learn, says Dr. Bland. But there is definitely a right way and a wrong way to do it. These tips will ensure that you get the most from your new stretching regimen, while avoiding injury.

Put a ban on bouncing. Bouncing—reaching farther than is comfortable and then quickly rebounding back—can actually do more harm than good. "Bouncing can lead to tears in the ligaments and tendons," says Dr. Bland. Instead, he suggests stretching gently until you reach a point of mild tension. Hold this easy degree of stretch until the tension subsides. Then, if it feels right, you can increase the range of the stretch.

Hold it for 20. Nerve endings near the "stretch tissues" are fired off in about 20 seconds, according to Dr. Bland. "So keep track of the time you hold each stretch, being certain that you hold the proper tension for at least 20 seconds," he advises.

Breathe like a natural. Normal breathing increases the benefits of stretching. "No tissue or body part can function at its best without oxygen," says Dr. Bland. If your breathing feels difficult in a stretch, ease up a bit until you can inhale and exhale naturally.

Be consistent. Becoming bendable is really a matter of consistency. "Contrary to what many believe," says Dr. Schilling, "age actually has little bearing on who can and can't increase their flexibility." If you stretch on a regular basis, she says, you will find yourself reaching farther and moving with greater ease.

TRIAL RUN

Stretching prepares specific muscles for the exertion to come—and that applies to more than strictly athletic endeavors. Try these stretching suggestions from Espel.

Prepare to promenade. Easy walking depends on flexible hamstrings. To loosen your legs before a stroll, try these hamstring stretches.

► Begin by sitting on the floor with your legs extended straight in front of you (*1*).

► Bend your right knee and place the sole of your right foot against the calf of your left leg so your right knee drops out to the side. Keep your left leg straight, with your toes pointing toward the ceiling (*2*).

► With your torso straight, bend from the hip joint slightly forward over your left leg. Go only as far as you can while keeping your spine straight; don't round your back (*3*).

► Breathe into the stretch and, if you can, hold for up to 30 seconds.

► On an inhalation, come up slowly, keeping your back straight.

► Return your right leg to the starting position and repeat the stretch to the left side.

Ready, set, weed. Yardwork often calls on the lower back muscles for strength and sup-

4   5   6

port. Try this easy, relaxing stretch before you dig, hoe, or mow, says Espel.

►Start by lying on your back on a mat, folded blanket, or other cushioned surface (thick carpeting is good) with your knees bent and your feet flat on the floor.

►Grasp both legs behind your thighs and pull your knees in toward your chest (*4*).

►Holding your legs, slowly roll your body from side to side, going from hip to hip as far as is comfortable (*5*). "You should feel a subtle massage of the lower back area," says Espel.

►Try rocking gently forward and back. Move for about a minute in each direction (*6*).

Get into the swing. A clean golf swing involves flexible shoulders and a supple upper back. Here's a stretch to put you in winning shape for next weekend's tourney.

►While either standing or sitting, raise your right arm and cross it in front of your body at about shoulder height.

►Bend your left arm so that your elbow is at 90 degrees and place your right forearm in the bend of your left arm just above the elbow (*7a*).

►Use your left forearm to gently pull your right arm further across your chest (*7b*). "You should feel a nice stretch in the back of your right shoulder and into the shoulder blade," says Espel.

►Try to relax your right shoulder down as you do this move, keeping it away from your right ear. Hold for up to 30 seconds.

►Repeat, stretching your left arm to the right.

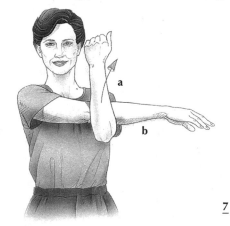

a

b

7

STUTTERING

Vocal Exercises to the Rescue

Here's a short list of things that won't cure stuttering: exorcism, bloodletting, singing (although many stutterers are fluent when they sing, it is not a cure), eating onions or horseradish, burning your tongue, vomiting, cutting the frenum (a membrane attached to the bottom of the tongue), and hearing someone tell you, "just relax and spit it out." Unfortunately, those are just a few of the "cures" prescribed for stuttering over the millennia.

Speaking fluidly requires remarkable choreography between parts of your brain and many muscles in your chest, throat, and mouth. Stuttering seems to stem from some subtle mistake in that choreography somewhere between mind and mouth, a problem that for some people may be inherited, says Edward G. Conture, Ph.D., professor of hearing and speech sciences at Vanderbilt University in Nashville.

Although the verbal faux pas isn't exactly *caused* by stress, tension can make it happen more frequently. "You're probably born with what causes stuttering, but you set it off more when you're under stress," says Peter Ramig, Ph.D., professor of speech and language pathology in the department of speech, language, and hearing sciences at the University of Colorado in Boulder. Because stuttering is often

---

A Sound Visualization

If you envision your vocal cords, jaw, tongue, and lips relaxing, you may find that you can relax them more easily, says Peter R. Ramig, Ph.D., professor of speech and language pathology in the department of speech, language, and hearing sciences at the University of Colorado in Boulder. Relaxed vocal cords look like a pair of tiny pink butterfly wings.

---

embarrassing and frustrating, speaking can be stressful, which tends to tense the muscles involved in speech—the ones controlling your lips, tongue, and vocal cords. And this excess tension triggers even more stuttering.

In Search of Smooth Talk

Stuttering is a tenacious problem that is best dealt with under the guidance of a certified speech therapist with proven expertise in the diagnosis and treatment of stuttering. But if professional help just isn't in the cards, Dr. Ramig suggests contacting the Stuttering Foundation of America at P. O. Box 11749, Memphis, TN 38111 for publications and videos that can help you work on your own. Experts also recommend the following vocal exercises that may help you coordinate your nerves and muscles if you sometimes find yourself stuttering.

Talk to yourself slowly. Every day, spend 5 to 10 minutes talking to yourself or reading aloud very slowly. Wade slowly into *Moby Dick*. Take a full second to say a single syllable. Why so slow? At this pace, the speech muscles in your throat, jaw, tongue, and lips will stay relatively relaxed and cooperative. They won't get overly tense. Practicing this technique will give you an opportunity to notice what your throat, jaw, tongue, and lip muscles feel like when they're sufficiently relaxed and you're speaking without a hitch, Dr. Ramig says.

Talk to yourself a bit faster. To get into the habit of keeping your tongue and lips more relaxed while you speak, spend another 5 to 10 minutes talking to yourself or reading aloud, but this time at your normal pace. Make a point of practicing words that start you stuttering. Often words that start with "p," "b," and "th" are problems. Try "peg leg " and "barnacle," and perhaps "Thar she blows!" Each time you start stuttering, make a conscious effort to relax the muscles in your tongue and lips, Dr. Ramig suggests. If you're squeezing your lips together, relax them so they touch lightly. If you're pressing your tongue hard against your gums or the roof of your mouth, ease up.

Pump your vocal cord muscles. To gain greater control over the muscles that work your vocal cords, set aside 5 to 10 minutes a day to rehearse words that often make you tense these muscles. Try repeating the word "Adam," suggests Dr. Ramig. To say it, you'll have to use the muscles around your Adam's apple, the ones that work your vocal cords. First, say "Adam" very slowly, so you don't tense and don't stutter. Pay attention to how the area around your Adam's apple feels. Then say "Adam" very quickly, so that you do tense up and begin to stutter. Again, pay close attention to how your

Some Like It Breathy

As sexy as it sounded, Marilyn Monroe's breathy voice wasn't a ploy to sound alluring but a technique to cover up her stutter. According to historians of stuttering, Monroe stuttered often as a child and occasionally as an adult.

Adopting a slightly breathy voice can help you speak fluently if you stutter, says Peter R. Ramig, Ph.D., professor of speech and language pathology in the department of speech, language, and hearing sciences at the University of Colorado in Boulder.

"When you use a slightly breathy voice, there's very little vocal cord vibration. It's a less complex way of speaking," Dr. Ramig explains. Since the verbal choreography is simpler, the opportunities for linguistic pitfalls are fewer, and speech is sometimes more fluent. Dr. Ramig points out that although it's rarely a long-term solution, whispering may at least convince you that you can speak smoothly.

Adam's apple and its environs feel. "By comparing and contrasting the way it feels when you speak fluidly and when you stutter, you can learn to relax those muscles," says Dr. Ramig.

Troubleshoot. Over time, you should be able to identify and relax the speech muscles that you're tensing, says Dr. Ramig. You're having trouble with "Pequod"? Get analytical. Where's the excess tension? Are you squeezing your lips together so tightly that no word stands a chance of escape? Let up the pressure and let the sound flow.

Glide into difficult words. When you're speaking in public, make things easier by sliding into words that usually make you stutter, Dr. Ramig suggests. Say the first syllable of a troublesome word very slowly, spending a full second or more as you ease through the word so you're less likely to tense up and stutter severely.

Jaw a lot. According to legend, the great Greek orator Demosthenes overcame stuttering by practicing orations with his mouth filled with pebbles. You can do without the pebbles, but not without the practice. As part of a much larger complete treatment program for stuttering, speak whenever possible, says C. Woodruff Starkweather, Ph.D., professor of communication sciences at Temple University in Philadelphia and co-author of *Stuttering*.

The more you speak to others, the less stressful speaking will become. The less stressful speaking becomes, the less often you'll stutter while speaking.

"Go into situations in which you need to talk," says Dr. Starkweather. "The practice will kill your fear of speaking. If you approach rather than avoid the situation, the fear dies."

Show and tell. If there's Career Day at the elementary school, volunteer to tell the class

Getting Help

To find a speech pathologist, contact the American Speech and Language Hearing Association's Resource Center for Consumers at 10801 Rockville Pike, Rockville, MD 20852. Ask for a referral to a speech therapist or speech-language pathologist with expertise in treating stuttering.

For more information on stuttering, contact the Stuttering Foundation of America at P. O. Box 11749 Memphis, TN 38111. The foundation publishes a 200-page self-help manual, *Self-Therapy for the Stutterer*, and will send a free copy to your local library at your request.

To find a support group, contact the National Stuttering Project at 5100 East LaPalma Avenue, #208, Anaheim Hills, CA 92807.

about your job. After all, this is a topic you know well, and you'll be speaking to a captive younger audience, so the setting is ideal.

Exercise honesty. Trying to hide your stuttering only makes speaking more stressful, which in turn makes stuttering more likely, says Dr. Starkweather. Be up-front. If you're striking up a conversation with someone new, say "I stutter sometimes," then go on with the conversation. Remember, stuttering is nothing to be ashamed of. Indeed, many folks who stutter score higher on IQ tests than nonstutterers. Famous stutterers include the talented likes of Moses, Lewis Carroll, and Winston Churchill.

TAI CHI

Slow-Motion Meditation

WHAT IT IS

Tai chi is an ancient Chinese martial art. Unlike its aggressive cousin, karate, tai chi works without violent chops or flying kicks. On the contrary, it's a gentle, slow-motion practice, a discipline that combines a series of fluid, dancelike movements with meditative attention to the breath and the body.

In San Francisco, tai chi classes congregate in the rectangular oasis of landmark Golden Gate Park, which stretches from midtown to the seashore. Since they were immortalized in Armistead Maupin's book series (and TV miniseries) *Tales of the City*, the great flocks of tai chi enthusiasts, moving together, have actually become a tourist attraction.

The graceful pattern of tai chi seems right at home in an outdoor setting. Preparing for class, students stand with their knees relaxed, as silent as the shadows moving over the grass. Beginning the individual movements traditionally called forms, arms and legs lift and lower in slow synchronicity, mimicking tree branches swaying in the breeze.

Even the names of the tai chi exercises echo nature: "Grasp the bird's tail," "wind blows lotus leaves," and "wave hands like clouds" are only a few examples. Like their names, the exercises themselves are poetically gentle. Knees are bent and unlocked, wrists are loose, and hands are held relaxed and open.

Because the ending position of one form is also the starting position for the form that follows, tai chi practice is like a smooth, slow, dance routine, but without the music. Tai chi also requires full attention to breathing and body movements.

WHAT IT DOES

In China, tai chi is considered medical therapy. "This is real medicine," says Martin Inn, O.M.D., a doctor of Oriental medicine and founder of and teacher at the Inner Research Institute of Tai Ji Quan in San Francisco. In the East, tai chi has been "prescribed" for everything from heart trouble and arthritis to depression and insomnia.

The health benefits of tai chi rival those of high-intensity Western exercise. Surprisingly, though, says Dr. Inn, tai chi provides its punch through mechanisms that are completely the opposite of sweaty, heart-pounding gym classes. "Standard aerobic exercise stresses the body, raising the heart rate, constricting blood vessels, and making you more tense," he says. "Its effect is similar to the classic fight-or-flight response."

Tai chi puts a stop to the stress cycle, says Dr. Inn. It quiets the body's emergency response system, allowing nutrients like blood and oxygen to flow freely. And this tonic effect is only part of tai chi's powerful portfolio. Here are some other tai chi rewards.

Muscle strength. Regular tai chi practice is a gentle way to build musculoskeletal strength. "The movements are performed very slowly," says Lixing Lao, Ph.D., assistant professor in the complementary medicine program at the University of Maryland School of Medicine in Baltimore and a tai chi instructor. "But slow work is actually more challenging to bones and muscles." And because tai chi involves the arms, legs, and torso equally, the muscular plusses add up throughout the body.

Better balance. In the elderly, loss of balance can result in a higher risk of falls. For older folks with osteoporosis, even one tumble could cause debilitating damage. Research shows that tai chi may help seniors significantly lower the risk of losing their balance.

A study conducted in Atlanta compared tai chi with computerized balance training. The study involved 200 people, all at least 70 years of age. At the end of the 15-week trial, those doing tai chi had a 47.5 percent lower risk of frequent falls than those who learned on the computer.

Mood mastery. Psychologically speaking, regular tai chi practice has been shown to lift the spirits, leaving tension, depression, and anger in the lurch. A study at the University of Massachusetts Medical School in Worcester showed that 16 weeks of tai chi beat the blues better than the same amount of walking. "Tai

chi seems to help uncover hidden emotions, releasing psychological tension," says Nikki Winston, a tai chi instructor at the Golden Door Spa in Escondido, California, and Win International in Delmar, California.

Arthritis relief. The stiffness and pain associated with arthritis are worsened by poor circulation, says Dr. Inn. Tai chi increases blood flow to the joints, he says, easing inflammation and possibly making movement less painful. In addition, the wide range of movement involved in tai chi increases flexibility around the joints, possibly preventing arthritis from developing in the first place.

REAPING THE BENEFITS

You can learn tai chi from an instructor, from books, or even from videos. If you've never done tai chi before, take note of these expert tips.

Feel the flow. Doing tai chi is like slow solo dancing—it's a series of steps that are linked together to form a flowing routine. "It's important to keep moving as you practice," says Winston. "If you try to stand still in a pose, you may get tense, which isn't at all what tai chi is about."

Tune in to comfort. Becoming comfortable is both a goal and a side effect of tai chi practice. "Tai chi will teach you to listen to yourself," says Winston. If you feel that something's wrong or if any pain flares up in some part of your body, stop the movement immediately. Try to identify the pain and seek immediate medical attention if it is severe. "You should never have any moments of pain," Winston says. "Tai chi should make you feel better all over, not worse."

Honor your playful self. An attitude of playfulness can help banish self-consciousness. "In China, people often say, 'Let's play tai chi,'" says Winston. Practicing in town squares and public parks, they let tai chi be fun and expressive. You can, too, she says, by giving yourself permission to play.

Make a daily date. To see effects quickly, a daily tai chi break is best, says Dr. Lao. "If that's not possible," he says, "try for at least two or three times a week." And you don't need to set aside a big block of time. Even 20 minutes a day can do the trick.

TRIAL RUN

If you'd like to try a quick tai chi–style mood booster, get ready for the "tai chi cheer." Winston adapted this simple move from traditional Asian exercises. The tai chi cheer "opens the body and the heart," she says, helping to ease the closed-in state so often associated with depression.

▶First, stand in a relaxed position with your knees slightly bent.

▶Place the palm of one hand flat over the other one and hold them just below your belly button (*1*).

▶Take a normal step forward, put your weight on your right foot, and rise onto the toes of your left foot.

▶As your legs move, open your arms and hands downward toward the Earth in an expansive gesture (*2*). Remain in this open position for one full inhalation and exhalation.

▶Next, step back, returning your arms to the original position (*3*).

▶Step forward again. This time, open your arms straight out to the sides (*4*). "Really open your

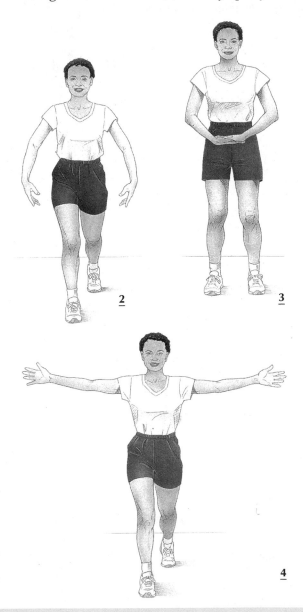

arms wide," says Winston. "Imagine that you are opening your heart."

▶Hold this outstretched position for a full breath.

▶Step back to the original position, with your hands together (5).

▶Finally, step forward again onto your right foot and open your arms high overhead. As your arms reach for the sky, let out a loud cheer, such as "Yay!" or even "Aahhh!" (6). Then step back.

▶Repeat the entire exercise, this time stepping forward with your left foot.

"Everyone loves this one," says Winston. "You simply can't help but feel more light-hearted after doing this move."

<u>5</u>   <u>6</u>

TEMPOROMANDIBULAR DISORDER (TMD)

Giving Your Jaw a Little TLC

Wake up! Shower. Forget your briefcase as you rush out the door. Merge onto that virtual parking lot called a freeway. Arrive late. Stain your shirt while guzzling coffee. Make some copies. Attend a meeting. Head home. Cook supper. Go to bed.

The stress producers in our daily lives are more prolific and intricate than the programming on cable TV. And stress is nasty stuff. Like excessive fat in our diets, it leads to a whole host of related problems, ranging from heart disease and insomnia to colds and allergies.

Well, now you can add pain in the joints and muscles of your face—called temporomandibular disorder, or TMD—to the list of stress-related problems. When you have TMD, your jaw may lock when you try to open and close your mouth, and you may find that it's painful to chew. The pain and chewing difficulty may also be associated with a clicking sound. (See your dentist if you experience these symptoms or find that your ability to function normally is limited—if you cannot open your mouth wide enough to eat, for instance.)

It was dentists who named this painful condition TMD, after the temporomandibular joints, which are little jaw-operating hinges right in front of each of your earlobes. It might be that you've had a defect in the joint since you were born and some lifestyle factor has prompted the pain. Other things can cause it, too, such as being socked in the jaw, for instance, or a head injury, possibly from a fall or accident. And although stress often contributes to TMD, it can also be triggered if you spend long periods with your mouth open (during dental work, for instance), if you have poor posture, or if you grind your teeth.

"TMD can be tough to diagnose because so many of its symptoms are shared by other medical conditions," says Charles McNeill, D.D.S., professor in the department of restorative dentistry at the University of California, San Francisco, and director of the Center for Orofacial Pain in San Francisco. It's not unusual for TMD patients' complaints to include earaches and headaches as well as the more common jaw and facial pain.

Dentists often initially prescribe a self-care regimen that includes jaw exercises for TMD. In fact, if your dentist immediately suggests jaw surgery or wants to grind down your teeth, you should seek other opinions before you agree, says Dr. McNeill. "In all of the patients I have, only about 2 percent ever need surgery for TMD," he observes.

Among your alternatives—and certainly the first to try—is healing with motion. Exercise can relieve the pain caused by TMD in a number of ways. First, depending on the cause of your TMD, posture improvement might be a total cure. Or you might need to start doing jaw exercises that can help rebuild connective tissues and retrain uncoordinated jaw muscles. When you do these exercises regularly, they can rehabilitate tight, shortened muscles, enabling them to handle the daily rigors of chewing, yawning, and talking.

And, not least of all, exercise can help relieve stress. No matter what the root cause of your TMD or the events that make it worse, you'll help relieve that pain if you make some moves to relieve stress.

Workouts from the Neck Up

A number of neck exercises for TMD may be especially helpful. "I tell all of my patients to check for correct posture and tongue position and do one of the following neck exercises at least every two hours, and more often if they are really stressed or have a lot of jaw pain," says Bernadette Jaeger, D.D.S., associate professor of diagnostic sciences and orofacial pain at the University of California, Los Angeles, School of Dentistry. "Depending on how long you have been experiencing the pain, you should feel relief with these exercises if you do them consistently over a period of several weeks," she says.

Get it straight. If you jut out your chin and hunch your shoulders forward, as many of us do unconsciously, you're making the muscles around your mouth work harder because it's harder to keep your mouth closed. This extra effort puts extra strain on the temporomandibular (TM) joint. Although it may seem unrelated at first glance, correcting your posture is one of the most important ways to relieve jaw pain, says Dr. Jaeger.

For correct posture, your ears, shoulders, and hips should all be in a straight line, says Dr. Jaeger. You'll achieve that alignment if you move your shoulders back and allow them to relax. Then lift your chest, straighten your hips, and let your knees relax. In this position, your neck and facial muscles as well as your TM joint do only the amount of work needed to hold your head up.

If you're not used to this posture, it may seem uncomfortable at first. And you might want to post some reminders around your home or work area; just a vertical line on a Post-It Note will do the trick. As you work at improving your posture, it will become second nature, says Dr. Jaeger. You may feel an immediate improvement in your jaw, or it might take several weeks.

Open your mouth and say "N." When you work at keeping your jaw closed—actually clenching or grinding your teeth—you'll need to try some tactics to unlock your jaw muscles. Putting your tongue in its proper place can help.

Just say the letter "N," says Dr. Jaeger. When you do, you put your tongue on the roof of your mouth behind your top front teeth. In the "N" position, your upper and lower jaws are slightly apart even if your lips are closed, she notes. When you start checking yourself, you might want to set an alarm or a signal on your computer (if you're near a workstation) for every two hours that will remind you to check whether your tongue is in this position. Of course, when you eat or talk, your tongue is no longer in this position, but get used to returning to "N" when you finish.

Stretch it out. Here's a simple neck exercise recommended by Dr. Jaeger that will really help ease some of the tension that gets into your jaw. To position yourself for this exercise, find a chair with a hard seat, a straight back, and arms. Put your hands on the arms of the chair and hold that position while you straighten up, making sure that your ears are in line with your shoulders. Now you're ready to begin. Do the stretch once slowly.

▶Inhale. As you exhale slowly through your nose, gradually lean your head to the left. Try to bring your ear as close to your left shoulder as you can without raising your right shoulder or rotating your head.
▶When your ear is near your shoulder, hold the position for 30 seconds, breathing normally.
▶Raise your head slowly and reverse direction, leaning your right ear toward your right shoulder.

Look around. Small muscles deep inside the neck can send you sharp reminders of TMD unless you stretch them out. By turning your head

Make Eating a Pleasure Again

"Eating can be a challenge for people with TMD," says Patricia Rudd, director of physical therapy at the Center for Orofacial Pain in San Francisco. But here are some suggestions from Rudd that could help restore your enjoyment of a favorite pastime.

Slow down. Eating more slowly helps prevent overtiring your jaw. It's the difference between running a mile and walking it. If your jaw hurts after gulping down a meal in record time, take it easy the next time.

Eat smaller meals. For a jaw with TMD, eating a full-course meal can be like running a marathon, so try eating several small meals instead. Noting that many people with TMD continue with a meal even when they have jaw pain, Rudd says, "Let pain be your guide, and stop eating when your jaw starts to hurt," she says. You can make your condition worse if you ignore the pain.

Go soft. Eating apples and other hard

foods can intensify TMD pain, says Rudd. Switch to softer foods. Start off with vegetable soup instead of carrots and dip. Try grilled fish instead of grilled steak. Switch to peas and applesauce instead of broccoli and cauliflower. For dessert, go for frozen yogurt instead of gingersnaps.

"The idea is to give your jaw a rest," Rudd says. After you've followed the soft-food routine for a while, you might want to start eating foods that are slightly harder and build up from there. But always stop eating something if it becomes painful, Rudd advises.

Be a real cut-up. If you cut your food into small pieces, your jaw will have to do less work chewing. Cut finger foods into bite-size bits and place them near the back of your mouth so your molars do the work. Your front teeth should do as little biting-off as possible, because that biting motion can irritate your TM joint, Rudd says.

from side to side, you stretch some of those muscles and help relieve the stored-up tension there, says Dr. Jaeger. This exercise should also be done seated upright in a hard-bottom chair, but the chair doesn't have to have arms. Perform this routine once slowly.

►Sitting up straight, with your upper body aligned, inhale. As you slowly exhale through your mouth, turn your head to the right, looking as far over your right shoulder as you can.

►Hold for 30 seconds, breathing normally.

►Face forward again, then pivot your head to look over your left shoulder.

Catch flies. Learning to open and close your mouth without stressing your TM joint can help reduce jaw pain, says Dr. Jaeger. This "fly-catching" exercise will also stretch the muscles that close your jaw and help with jaw muscle coordination, she says. One tip: You may want to practice this maneuver in front of a mirror to make sure that you're opening your mouth in a straight line, not at an angle. Perform this routine one time through, slowly.

►Start in a good posture position, with your tongue in the "N" position to relax your jaw.

►Put your fingers on the TM joint in front of each earlobe.

►Slowly open your mouth straight up and down, lowering your jaw as far as you can while keeping your tongue on the roof of your mouth. Hold this position for 30 seconds.

Rub it out. Massaging your chewing muscles can really help ease your pain, says Patricia Rudd, director of physical therapy at the Center for Orofacial Pain in San Francisco. Whenever your jaw hurts, put your fingers on your TM joints. Then gently rub those areas using a circular motion. Continue for a few minutes, she says.

"I tell some of my patients to massage their chewing muscles in elevators, at traffic lights—wherever and whenever they feel pain from them," says Rudd. "It's a good technique because it relaxes the jaw muscles and improves local circulation." Just be careful not to rub so hard that you aggravate the joint, she cautions.

Spit it out. If you chew gum, you might aggravate your TM joints and muscles tremendously, says Rudd. Gum chewing overtires the jaw muscles, she points out. The instant relief from jaw pain that you'll notice will make it worth giving up the gum.

Tooth Grinding

Calm Down to Unclench

Tooth grinding is a time-honored affliction. It's even mentioned in the Bible. "The children of the kingdom shall be cast out into outer darkness," says the Book of Matthew, "and there shall be weeping and gnashing of teeth." What did the ancients have in common with modern Americans? Stress—the number one contributor to tooth grinding.

Tooth grinding is hereditary, and if you have inherited this tension-relieving tendency, anticipatory anxiety is one of the primary kinds of stress that will start you mashing your molars. "If you have an exam or an important meeting coming up, you're likely to grind the night before," says Bernadette Jaeger, D.D.S., associate professor of diagnostic sciences and orofacial pain at the University of California, Los Angeles, School of Dentistry. Tooth grinding is far less common after a stressful day is over.

Tooth grinding may be hard to detect because it usually happens while we're asleep. Unconscious of our anticipatory anxiety, we may clench and grind as we snooze. All that grinding can contribute to other problems, including wearing down teeth and worsening any existing gum disease. Also, if you have a pre-existing joint condition, it could become aggravated and result in a painful jaw condition known as temporomandibular disorder (TMD).

Avoiding the Nightly Grind

To stop tooth grinding, you have to minimize the stress that sets it off, doctors say. "One of the best ways to relieve jaw pain perpetuated by tooth grinding is to get good, restful sleep," says Patricia Rudd, director of physical therapy at the Center for Orofacial Pain in San Francisco. Here are some tips that will help you doze without doing dental damage.

Walk, run, ride, or swim. Any form of aerobic exercise can relieve stress, says Rudd. She encourages patients to spend 20 to 30 minutes engaged in a low-impact activity at least two to three hours before bedtime. Three or four times a week, you might want to go for a 20- or 30-minute walk or run, ride your bike, or have a swim.

Remember, though, not to work out too close to bedtime, or it will get you keyed up. That could actually increase the chances that you'll grind your teeth.

Picture perfect dental posture. Your mouth actually has a proper resting position. Try visualizing it before you go to sleep. Say the letter "N." When you do, your tongue should rest lightly on the roof of your mouth just behind, but not touching, your top front teeth. That's optimum jaw posture. "The only time your teeth should touch is when you chew or swallow," says Rudd.

Throughout the day, consciously put your jaw in this nonabrasive "N" position until it becomes a habit. Then, just before going to bed, picture yourself sleeping with your jaw in this position.

Steep yourself. Before going to sleep, try soaking in a tub of warm water. Just lie back in the tub and relax, says Rudd. Make your soak as pleasant as possible. Try listening to relaxation tapes, lighting a few candles, or using a little foam bath pillow. To really relieve tension, wrap your jaw and neck with a warm, moist towel. This technique will release tension from the jaw area.

Be a backnapper. Sleeping on your back puts less strain on your neck and teeth than other positions, says Rudd. "Sleeping on your stomach is the worst because it puts your head and neck in extreme positions that cause jaw pain," she says.

Letting Off Steam to Cool the Grinding

Many tooth grinders expect a lot from themselves, says Richard Miller, Ph.D., a clinical psychologist and co-founder of the International Association of Yoga Therapists in Mill Valley, California. When things don't work out—or even when they do—they worry. Combat the daily pressures that grate on you, and you might be able to ease the wear and tear on your molars. Here are the soothing strategies.

Really breathe. Taking deep, even breaths can stave off stress, says Dr. Miller. You can learn to breathe this way if you use everyday events as frequent reminders, he says. Every time the phone rings, the computer notifies you of an incoming e-mail, or a commercial starts on TV, check to make sure that you're taking regular, deep breaths. Expand your diaphragm to take in air through your nose, then relax and expel the air through your nose. You can check to see if your diaphragm is expanding by placing your hands on your sides at about navel height with your thumbs pointing back and your fingers pointing toward the front. Inhale toward your fingers and pull away from them as you exhale.

Clench up. To reduce stress, tightly clench your fists for a count of 10, says Bernadette Jaeger, D.D.S., associate professor of diagnostic sciences and orofacial pain at the University of California, Los Angeles, School of Dentistry. When you release the clench, let your body go limp. Repeat this exercise five or six times during the day or whenever you feel stressed.

Use your imagination. When stress gets you in its tight, sweaty grip, think of yourself in a relaxed setting, says Dr. Jaeger. Close your eyes, feel your arms get warm and heavy, and feel the blood rushing into your fingers. Continue your little mental holiday for three to five minutes.

Hug your headrest. If you can't sleep on your back, position your pillow under your neck and head so that your jaw is supported, says Rudd. "You may need more than one pillow to keep your head from tilting downward," she adds. In that position, give one of your pillows a nice big bear hug while sleeping. "Hugging a pillow keeps your shoulders from rolling forward and your spine from twisting," Rudd says. That also prevents pressure from getting to the neck and jaw.

Be pillow perfect. If you are sleeping on your side, your whole spine should be parallel to the mattress, says Dr. Jaeger. "The goal is to keep the natural curve at the back of your neck," she says. "Don't let your head droop or be propped up. The pillow should support your head and neck so that your spine is straight. If you sleep on your back, the pillow should be under your head and neck, with the corners pulled up over your shoulders to help support and maintain the natural curve of the back of the neck."

Foam pillows can make your head bounce, and that can aggravate tension in your jaw and neck muscles. Instead, choose a pillow made of down or some other malleable filler so you can mold it depending on your sleep position, says Dr. Jaeger.

TYPE-A BEHAVIOR

Ease Anger and Stay Well

Are you forever on a low boil? Angry at that perfectly nice woman in front of you in the grocery line? Enraged because you can't find a parking spot closer to the library? Are you highly competitive? If so, you're probably showing Type-A behavior.

"Type-A behavior is characterized by hostility and irritability," says Peter Halperin, M.D., clinical instructor of psychiatry at Harvard Medical School and staff psychiatrist at Massachusetts General Hospital in Boston. "You may feel a real time urgency to get a lot done and get extremely irritated if someone or something gets in your way."

There are some benefits to this kind of temperament. Type-A behavior often drives people to accomplish a lot. But even if you enjoy being hot-tempered and impatient, your attitude is undermining your health.

Constant rage puts our bodies on perpetual fight-or-flight alert, sending a stress hormone called noradrenaline coursing through our systems. After a while, our bodies become exhausted from the high stimulation, putting us at risk for all kinds of health problems, from infections to high blood pressure and heart attacks.

If you show Type-A behavior, it's important to try to ameliorate your anger and get over im-

patience and impulsiveness, says Jane Sullivan-Durand, M.D., a behavioral medicine physician in Contoocook, New Hampshire. Since stress can either cause or complicate many ailments, you might want to talk to a therapist about some calming techniques. But one helpful strategy—either with or without therapy—is the addition of exercise.

Letting Go of Control

An important component of Type-A behavior is the need to feel in control, explains Dr. Halperin. To wean yourself away from being a control freak, try these workout tips, which may help you give up command.

Be a team player. Joining the local softball or bowling team can be a stress reliever, but only if you resist some Type-A temptations. Don't become the captain. Resist your inclination to fill out the starting lineup or call for the suicide squeeze. There's no need to be a 24-hour in-charge type, says Dr. Halperin. Just try to enjoy the exercise and the simple pleasure of being a member of the team. Remind yourself that you're playing to relax and have fun, not to prove anything about your athletic skill.

Let time stand still. When you're exercising, ignore the stopwatch. Aerobic exercise can release Type-A stress, but not if you're always trying to set a personal best, says Dr. Sullivan-Durand. "Lose the watch. Leave it at home and try to see exercise as something for yourself, not as something at which you need to excel," she says.

Deep-six the monitors, too. If you ride a stationary bike or use a treadmill, turn off the calorie-burning and/or distance readout. Just sweat and smile, advises Dr. Sullivan-Durand.

Play scoreless tennis. To keep your cool, try playing tennis without keeping score. It may be hard for you, as it is for many people who show Type-A behavior. Although you might want to keep close tabs on how you're doing, remember that the point of the game is not to prove you're a better tennis player than your friend. The point is to keep you both in better condition, both physically and mentally.

Be willing to be bad. A good way to give up control is to take up some exercise at which you are flat-out terrible, says Dr. Halperin. After all,

Collect Stamps, Imagine Your Death, Cuddle Your Corgi

Type-A behavior is bad not only for your health but also for your relationships. Here's a trio of exercises that might help reduce the buildup of Type-A stress.

Resume a childhood passion. Take up stamp collecting. It's especially enjoyable to resume this hobby if you did it as a kid. Or try coin collecting or bird-watching. "I recommend that people get back into hobbies that they liked as kids," says Peter Halperin, M.D., clinical instructor of psychiatry at Harvard Medical School and staff psychiatrist at Massachusetts General Hospital in Boston. This is a great way to relax and dissipate your tension.

Write your obituary—twice. Imagine that you've shuffled off this mortal coil and write down two different versions of your obituary, advises Dr. Halperin. In the first obituary, write all of the things that you'd like the newspaper to say about you and your life—

everything that's most important. The second should be what you think the obituary might actually say. If there's a big difference, you know that you need to do some things to move you toward your hoped-for way of life.

"No matter how Type A people are, they usually list themselves first as a loving spouse or parent and place their professional achievement far down on the list," says Dr. Halperin. This exercise helps people recognize what's really important to them.

Team up with a tail-wagger. Spending time with pets helps people calm down, says Karen Allen, Ph.D., a research scientist at the University of Buffalo School of Medicine and Biomedical Sciences in New York. She did a study that showed that people with dogs by their sides had lower blood pressure and heart rates than those who were canine-deprived.

it's tough to be in command if you're trying some Eastern European folk dance that makes you look like a beginner. So take up some form of exercise in which you're starting from square one.

Cooling Compulsiveness and Irritability

If you demonstrate true Type-A behavior, you'll struggle to manage your anger—but it's important that you do. Evidence suggests a link between an angry outburst and a subsequent heart attack, says Dr. Halperin. Remember this: Even if you can't get rid of your rage entirely, you can learn to soften it or express it appropriately. Try these tips.

Defuse the anger bomb with exercise. Your boss just humiliated you in front of your co-workers. Instead of trashing your office in frustration, take a five minute time-out. Head for the secluded back stairs and run up and down a few times. "Exercise helps defuse the intensity of anger," says Dr. Sullivan-Durand.

Swear and sweat. It may not win etiquette points, but yelling, even swearing, during a jog or brisk walk can help your body and mind cope with anger.

"Venting helps you discharge a lot of energy, so it's okay to swear," says Dr. Sullivan-Durand. "By doing so, you may not solve your problems, but you could reduce the intensity of the emotional feelings in your body." Of course, be sure that the object of your frustration isn't within earshot.

Grip if you're miffed. You can release tension by squeezing a rubber hand-gripper or a tennis ball. In the office, they're ideal ways to release some fury and still maintain your cool, says Dr. Sullivan-Durand.

Countdown to avoid blastoff. You can release anger by slowly counting from 1 to 10. As you announce each number aloud, breathe deeply. Then say a positive affirmation such as "I am doing fine" or "I am stronger."

Be reflective. Try the counting technique while looking in a mirror. Speaking in a calm, quiet voice, look at your facial expressions. Loosen the muscles that are forming a scowl and work toward a relaxed visage, says Dr. Sullivan-Durand.

Underarm Sagging

Bye-Bye to Batwings

The nickname for flabby underarms is horrible but all too accurate—batwings. Unfortunately, many of us have this problem. Often, as we age, our triceps muscles, located in the backs of our upper arms, lose their tone. And so we end up flapping away, reluctant to go sleeveless or even raise our arms in farewell.

Well, enough of wearing long sleeves to 4th of July barbecues! Enough waveless goodbyes! With some easy-does-it exercises aimed at the triceps, you can clip those batwings.

Lift Light Weights for Tighter Arms

When it comes to working your triceps and busting batwings, a little resistance work can have a visible effect.

For these exercises, recommended by Michael Yessis, Ph.D., president of Sports Training in Escondido, California, and author of *Body Shaping*, you will need a two- to five-pound weight. If you don't have dumbbells at home, you can use a heavy can of soup or even a weighty book.

French press. This movement works the triceps. Stand with your feet about shoulder-width apart. Hold one dumbbell with both

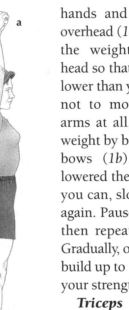

hands and reach straight overhead (*1a*). Slowly lower the weight behind your head so that your hands are lower than your elbows. Try not to move your upper arms at all, but lower the weight by bending at the elbows (*1b*). Once you've lowered the weight as far as you can, slowly bring it up again. Pause for a moment, then repeat 3 to 5 times. Gradually, over a few weeks, build up to 15 repetitions as your strength increases.

Triceps kickback. This exercise targets the uppermost part of the triceps and is a great follow-up to the French press.

Stand beside a bench or a sturdy coffee table (or even a solid kitchen chair). Bend forward at the hips so that your torso is parallel to the floor. Place your left hand on the bench or table for support and hold the dumbbell in your right hand. Begin with your elbow bent and the weight below your armpit (*2a*).

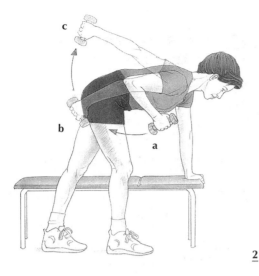

2

3

Slowly straighten your elbow and extend your arm backward toward your hip (*2b*). Keep your elbow close to your body to get the maximum workout for the triceps. Hold for a moment, then pull your arm behind you and above your back (*2c*). Move smoothly—don't swing the weight.

Once you've taken the weight back as far as you can, slowly bring your arm back to the starting position. Repeat 3 to 5 times on each side. Over a few weeks, increase to 15 repetitions per side as your strength increases.

Hate Weights? Tone Your Arms with Tiny Movements

An alternative to using weights is a form of deep muscle movement known as Callanetics. Developed by Callan Pinckney in the early 1970s, Callanetics uses tiny contractions to tone and tighten muscles. This subtle triceps technique works quickly, says Pinckney, author of seven best-selling Callanetics books, including *Callanetics Fit Forever*. The Jell-O jig-

gling of the underarms disappears, and the technique also helps improve posture. In fact, if you do these movements every day, you could see and feel results after just a few sessions, she says.

The underarm tightener. Sit on the edge of a chair with your back straight and your feet planted firmly on the floor, hip-width apart. Extend your arms straight out to the sides, with your hands even with your shoulders and your palms down (*3*).

Roll your hands forward so that your thumbs begin to point toward the floor, then continue to rotate your wrists so that your palms face upward. You'll feel muscles working in several spots in your arms (*4*).

Lean forward. Without rounding your back, gently move your arms back as far as you can, keeping them straight and high. Imagine that you are trying to bring your thumbs together behind you.

Once your arms are as far back as possible, with a very tiny, barely visible movement,

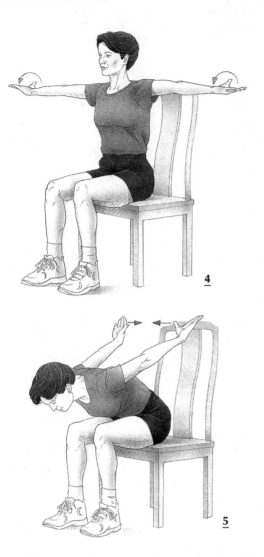

4

5

Prune and Prime Your Triceps

Your triceps get a little work any time you hold your arms up over your head. Here are two chores that can double as batwing prevention, says Dan Hamner, M.D., a physiatrist and sports medicine specialist in New York City.

Practice purposeful pruning. If you have to reach up to prune the trees in your yard, you'll benefit both the landscaping and your arm muscles. So grab the clippers and neaten up that forsythia.

Do some high-altitude painting. If there's a wall to be painted, get the assignment to do the upper half, above the molding. Stretching toward the ceiling will let you tone your triceps as you improve your home.

begin to move them together and apart (5). The slight pressure works the muscles in your arms, especially the underarms. Start with 25 repetitions and gradually work up to 100.

Shoot, Slice, and Swing

Keeping sagging arms at bay is really a matter of keeping your upper arms active, using your triceps whenever you can. Try these tricep-building sports.

Be like Mike. Basketball can help underarm sag, says Dr. Yessis. Tossing a ball into a hoop overhand spotlights the triceps nicely. So make like Michael Jordan and challenge someone to a foul-shooting contest.

Be like Tiger. Any activity that involves straightening the arms will have a beneficial effect on batwings. Golf is a natural arm tightener, says Dr. Yessis.

Be like Billie Jean. Swinging a tennis racquet is yet another great way to tone your triceps. Focus on your serve and your backswing for maximum results.

Varicose Veins

Insure Your Legs with an Exercise Policy

Shortly after she signed on as spokeswoman for L'Eggs panty hose, actress Jamie Lee Curtis insured her eye-catching gams with Lloyd's of London, the insurance company famous for its willingness to write policies that mainstream insurers won't touch.

But Curtis's $1 million policy doesn't cover quite everything. Even Lloyd's won't write leg policies that cover varicose veins. And if Lloyd's won't insure legs against these painful, protruding, blue-green veins, who will?

Well, no one can offer a guarantee that you won't get varicose veins, but a number of doctors are willing to suggest that some regular rounds of mild exercise might help prevent them.

To understand why exercise might help, you need to understand why veins become varicose to begin with. Varicose veins and their cousins, the smaller, red or blue spider veins, appear when tiny one-way valves inside your veins begin to leak. Normally, these valves open and close like the locks in a canal, ensuring that the blood flowing through your veins moves in just one direction—back toward your heart. When the valves start leaking, blood that should flow to your heart gets trapped somewhere in midstream. The blood pools in your veins, stretching them and making them pro-

trude and ache. This pooling usually happens in your legs because your blood has to travel a fair distance through your gams on the way to your heart.

If your forebears had varicose veins, the odds are that you may get them, too. A tendency to develop leaky valves and get varicose veins is inherited. But other factors play a role. You may have noticed that more women than men seem to have varicose veins. That's because female sex hormones seem to play some role in their development. Anything that has to do with the ebb and flow of those hormones—such as pregnancy and the Pill—conspires to help create varicose veins.

Being overweight also raises your risk. Heavier people have more blood circulating through their veins, which puts more stress on the valves, says Luis Navarro, M.D., senior clinical instructor of surgery at Mount Sinai School of Medicine and director of the Vein Treatment Center, both in New York City, and co-author of *No More Varicose Veins*.

But whether you're overweight or not, you're contributing to the risk of getting those veins if you do a lot of uninterrupted sitting and standing. If you're active, the muscles in your calves help pump blood back toward your heart, so it's less likely to pool in your veins and stretch them, says Dr. Navarro. "With every step and contraction, the muscles squeeze blood upward toward the heart," he explains. If you're inactive, however, the blood in your legs doesn't get that extra boost toward the heart, and you have higher odds of developing varicose veins.

Something in a New Vein

To prevent varicose veins, many experts recommend exercise that gets your legs moving, such as cycling, tennis, and walking. First, check with your doctor to get the okay to exercise, then try these hints.

Do it often enough. Get at least 30 minutes three times a week, advises Dr. Navarro. That might be the equivalent of a walk to the store and back. Or you can spend a half-hour on a stationary bike in front of the TV and take in the evening news while you're getting some vein protection.

Build up calf strength. Dr. Navarro also recommends strength training, using weights or exercise bands to provide resistance. Strength-training exercises that beef up your calf muscles are most likely to help prevent varicose veins, he says. (For details about starting a strength-training program, see "Resistance Training" on page 331.)

Take some breaks. If you take frequent, short exercise breaks after you've been standing or sitting in the same place for a while, you get the blood moving again. If you're in the office,

When to Coddle Your Veins

If you already have varicose veins, exercise won't help, and it might make them worse, says Brian McDonagh, M.D., founder and medical director of the Chicago-based Vein Clinics of America. When you start walking or cycling, your calf muscles' pumping action generates even more turbulent blood flow in your varicose veins, distorting and distending them further. This is why young, athletic males have some of the largest varicose veins.

Although Dr. McDonagh still favors exercise, it's because it benefits your overall health, not because it helps your veins.

Comfortably snug support stockings, worn either before or after your workout, may help minimize the negative impact of exercise on varicose veins. But once you have varicose veins, your best bet is to go to a clinic that specializes in treating them.

get out of your chair every hour or so and take a 10-minute stroll, Dr. Navarro says. If you can't get up and walk around, spend a few minutes each hour flexing your calf muscles by pointing your toes up and down repeatedly, he suggests.

WALKING

The No-Excuses Exercise

WHAT IT IS

Walking is the world's most accommodating exercise.

While you walk, you can do all manner of other things. Henry David Thoreau thought lofty thoughts while strolling around Walden Pond. Walking to his day job at the Hartford Accident and Indemnity Company, Wallace Stevens composed Pulitzer Prize–winning poetry.

If you're not in the mood for poetry or philosophizing, take a friend along on your walk. The two of you can catch up on each other's activities while you stroll. Or take your dry cleaning and stop off at the Quick Clean on your ramble.

Walking is simplicity itself. There's no need to count repetitions or keep score while you walk. And you don't have to tote around special equipment or search out special environs. You can walk virtually anywhere. Just pick a destination and a route and set off.

"There's no reason a person can't engage in a walking program on a regular basis," says John Duncan, Ph.D., professor of clinical research at Texas Woman's University in Denton and the author of several studies on walking.

WHAT IT DOES

Walking offers a multitude of health benefits. A regular walking program will help you lose weight and keep it off. It will help control diabetes. One study found that women who spend just three hours a week walking at a moderate three miles per hour are less likely to have heart attacks or strokes. Walking makes your heart work harder, so it grows stronger. And it lowers your blood pressure while also boosting levels of good cholesterol in your bloodstream.

Walking also builds your other muscles and thus lowers your risk of joint trouble. Stronger muscles do a better job of keeping joints in proper alignment, thereby reducing the joint wear and tear that contributes to osteoarthritis. Some studies suggest that walking can keep your bones strong as well and help you outpace osteoporosis.

And if all of those benefits aren't enough, here's the whipped cream on top. Research shows that a regular constitutional boosts mental health, easing depression and anxiety.

"Walking is one of nature's tranquilizers," says Dr. Duncan. "It reduces the production of stress hormones and helps you calm down and put life in perspective."

The faster you walk, the greater the benefits. The more quickly you walk, for instance, the more quickly you'll burn calories and build muscle, Dr. Duncan notes.

"When you're walking five miles an hour, you're getting the same fitness benefits you would if you were jogging—with a significantly lower risk of injury," he says. "The injury rate for walking is far lower than for virtually any other activity."

Unlike jogging, which jostles you up and down like a pogo stick, walking is a low-impact affair. When you walk, one of your feet is always on the ground. No bouncing. No pounding. So walking is kind to your joints and vertebrae.

It's a particularly good exercise choice if you have arthritis or if you're overweight, since extra pounds also put extra stress on your joints. "And walking is an ideal exercise option if you have lower back pain. It'll strengthen your back muscles without giving your disks a drubbing," says Carol Espel, an exercise physiologist, contributor to *The YMCA Walk Reebok Instructor Manual*, and general manager of a health and fitness club, Equinox of New York City in Westchester, New York.

REAPING THE BENEFITS

To start walking, all you need is a good pair of shoes and some pointers on posture and pace. Sore shins, one of the very few hazards of walking, are often the wages of the wrong footwear, says Espel. Keep these thoughts in mind when selecting your shoes, she says.

Shop late. Your feet swell a bit throughout the day, so shoes that fit at 9:00 A.M. may pinch a bit by dusk. Shop in the afternoon or at night, when your feet tend to be largest.

Pace with Your Peers

Peer pressure can work to your advantage. Enlist a walking buddy or sign up with a walking group. When you have a commitment to meet a pal or two for your daily perambulation, you're more likely to show up, says Carol Espel, an exercise physiologist, contributor to *The YMCA Walk Reebok Instructor Manual*, and general manager of a health and fitness club, Equinox of New York City in Westchester, New York. Ask a neighbor or co-worker to join you for a walk at lunchtime or after work. If all else fails, walk the dog! Or contact the American Volkssport Association (AVA) at Phoenix Square, 1001 Pat Booker Road, Suite 101, Universal City, TX 78148-4147. The AVA is a nonprofit association of walking buffs that will give you information about walking events in your neck of the woods.

Pay the freight. Yes, lots of athletic shoes are very expensive. While you don't need the nuclear-powered pair that breaks the $100 mark, you should buy high-quality walking shoes. You may have to spend $50 to $75. Here are some features to look for when you're shoe shopping, Espel suggests.

A heel that slopes up in back. A beveled heel makes it easy to roll your foot from heel to toe with each step. This is the proper stepping technique and will help you avoid straining your shins. Nothing insults a sensitive shin like slapping your whole foot flat down on the ground.

A flexible forefoot. A rolling heel-toe gait also requires a flexible sole. Look for a shoe with a sole that bends easily.

Cushioning under the heel and forefoot. Even though walking is low-impact, your foot still needs some shock absorption.

A spacious toebox. Each time you roll onto your toes, those little digits move forward and far outward. Give them sufficient space in a shoe with a capacious toe.

To get the maximum benefit from walking, you also need proper walking posture, says Espel. Here are two things to keep in mind.

Keep your chin up. When walking, you should hold your head high and look toward the horizon, not down at your feet. "Looking at your feet pulls your upper body down and can cause low back pain," Espel says. Your head should be in line with your spine. And your shoulders should be back so your rib cage is lifted and opened up.

Involve your arms. Your arms should swing as you walk. If they don't, you won't get the upper-body benefit of walking. But beware of flailing your arms. Let them swing freely but not excessively. Keeping your elbows close to your sides as they swing should help you find the right arm rhythm and movement.

TRIAL RUN

According to an ancient Chinese adage, "a journey of a thousand miles must begin with a single step." Here are some tips for your first few steps to lifetime fitness.

Start slowly. Spend the first five minutes of each walk just ambling along at your normal walking pace, says Espel. No strain or huffing and puffing. This will give your muscles an op-

portunity to warm up before you make demands on them. That's important because cold muscles are more vulnerable to exercise-related injuries.

Do some stretches. Here are some simple prewalk exercises.

Hip stretch. While standing with your feet shoulder-width apart, step forward with your right foot about 12 inches. Tuck your buttocks under your hips and pull in your stomach muscles. You should feel a stretch at the front of your upper left hip and your upper left thigh, says Espel. Hold for 15 seconds, then repeat on the other side. You may want to hold onto a table or other sturdy object for support.

Shin saver. While standing, cross your left leg over your right. Bend your left knee, pointing your foot like a ballerina so that only the toe of your shoe touches the ground. Then bend your right knee and press your right shin into your left calf (1). You should feel a mild stretch in your left shin, Espel says. Hold for 15

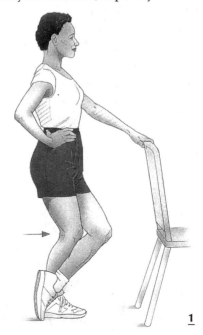

1

seconds, then reverse legs and repeat the move. Again, you may want to hold onto something sturdy for balance.

Pick up the pace. Once you've stretched, you're ready to walk again. Pick up the pace gradually. Don't worry if you're not as fast as the next guy. The important thing when you're starting out is to choose a comfortable pace. You'll pick up speed in due time. "You don't need to huff and puff at first," says Espel. "Walking should be enjoyable, not stressful or painful."

Go around the block. If the last long walk you took was down the aisle of the Hillcrest High auditorium during graduation, start short. Walk around the block or, if you're in the country, past four telephone poles. You want to start with a distance that will take you about 10 minutes to cover, says Espel. Do it three times a week. Remember: Walk at a comfortable pace.

Walk farther. When walking for 10 minutes or so three times a week is no longer a challenge, add distance. Make it 12 to 15 minutes at a pop, Espel says. Gradually work up to 30 minutes three times weekly, then shoot for 30 minutes five times weekly.

If you want to lose weight, shoot for 40 to 60 minutes four or five days a week, says Dr. Duncan.

Walk faster. Over time, you'll find that you can pick up your pace and still feel comfortable. A good rule of thumb: Walk fast enough that you can talk but not sing. To get all of the cardiovascular benefits that walking offers, you have to walk at a pace of at least three miles an hour—about a strolling pace—or faster, says Dr. Duncan. So make that your minimal goal.

Get smaller, get faster. If you continue to pick up the pace, the benefits are compounded, says Dr. Duncan. Paradoxically, the secret to walking faster is taking smaller steps.

"Most people think that to move faster, they should take longer strides, but quicker, shorter steps make you move faster," says Espel. That's because your hips can rotate faster when you shorten your stride. And if you pump your arms faster, your feet have to follow.

Cool your heels. Cool down by spending the last five minutes of your walk strolling, says Espel.

Stretch again. Finally, follow your cooldown with some basic stretches so that you don't get all stiff and rigid. You can repeat the stretches that you did to warm up, plus a couple more.

Foot roll. This gives your hard-working shins and calves a good after-walk stretch. Stand with your feet 6 to 12 inches apart, then slowly roll up onto your toes and hold for a count of two. Then roll over onto the outsides of your feet and hold for two. Next, roll back onto your heels and lift your toes. Hold for a count of two. Finish with your feet flat on the ground. Repeat up to 10 times, says Espel. You may want to hold onto something sturdy for balance. "You need to be careful when rolling the feet and ankles," she says.

Back relaxer. Lie on your back with your knees bent. Place your left hand on the outside of your left thigh (about midthigh) and your right hand on the outside of your right thigh and pull your knees toward your chest. Hold for 15 seconds.

Hamstring stretch. Lie on your back and bend your left knee so the foot is flat on the floor. Extend your right leg straight up toward the ceiling, keeping your back and hips on the

Ski Walking

Imagine cross-country skiing. Without snow. And without skis. Do you have that image in mind? Then you have a good idea what ski walking is all about.

Ski walking offers all the benefits of plain old walking, and then some, says John Porcari, Ph.D., executive director of the LaCrosse Exercise and Health Program at the University of Wisconsin at LaCrosse. Ski walking gives you a more comprehensive upper-body workout than walking, Dr. Porcari says, and it burns more calories per mile.

To do it, you need a pair of lightweight sticks or poles about six inches longer than the distance from your heel to your elbows. Ski poles will do if you attach rubber tips to the ends.

As you move one leg forward while walking, you swing the pole in the opposite hand forward, too. Then you push the tip of the pole into the ground and use it as a lever to propel yourself forward. You continue this way, alternating the leg and arm movements.

Ski walking does look a bit odd at first sight, Dr. Porcari concedes. What if you draw puzzled stares or smart remarks? As a comeback, he suggests, "It's going to snow today"—no matter what the weather.

---

floor. Hold the stretch for 15 to 30 seconds and then switch legs.

Diversify your walking routine. Some variations in terrain and pace will add variety to your workouts.

Hill walking. This is just what it sounds like—walking up and down hills. Repeatedly. Hill walking gives your heart a more strenuous workout. It can also help shape up your tush, says Espel. To do it, find some moderately steep hills (nothing Himalayan—an incline of about 4 to 8 percent will do) and tread up and down them for 20 minutes.

Interval walking. To do this, you walk as fast as you can for 30 seconds, then, over the next 90 seconds, slow down to your usual pace. Then you go as fast as you can for another 30 seconds and drop down again to your regular pace. And then you do it again. You get the idea. Interval walking burns more calories and gives your stamina a bigger boost than the regular variety. Shoot for seven fast-normal cycles in a row, says Espel.

To simulate interval walking on a treadmill, adjust the incline rather than the speed. For a cycle, or interval, workout, you walk on an incline for 30 seconds, then walk at a zero incline, then on an incline again, then on the flat again, increasing the incline up to six times. Begin with the incline at 6 percent for 30 seconds, then drop it to zero for the first cycle. For the second cycle, increase the incline to 8 percent for 30 seconds, then go back to zero. Next, turn it up to 10 percent, then drop it to zero, and so on, up to 16 percent and finally back to flat land. "Your speed should remain at a somewhat challenging pace (3.0 to 4.0), depending on your fitness level and experience," says Espel.

Distance walking. Walking more than four miles at a time qualifies as distance walking,

says Espel. "This is something you can do when you're preparing for an event, like a 5-K walking race," she explains. Walking races add a competitive element to walking—an extra incentive for Type-A types. To find out about races in your area (most running races welcome walkers), give a local sporting goods store or YMCA a call.

Vary your vistas. Since you can walk virtually anywhere, why limit yourself to the same old stretch of sidewalk? Strike out for new frontiers.

▶Take a walking tour of the local zoological garden or, if plants are your thing, the botanical garden.

▶Take the walk of the town. Most cities—even small ones—are home to enough attractions to warrant a self-guided tour or two. Call the chamber of commerce in your town (see the white pages for the number) and ask for a map. Then check out the local sights—the site of Neighborville's first distillery, "gallows hill," and the old Quackenbush homestead.

▶Head into the woods. Odds are that you live within driving (if not spitting) distance of a hiking trail or two. Most state maps show local, state, and national parks, and many of those parks have hiking paths. Stores that sell outdoor sporting goods are sure to have local walking books or detailed trail guides. Or call your state parks department (the number's in the blue pages) to check on free maps of trails near home.

▶Imagine yourself hiking the black sand Kings Highway Coastal Trail in Maui, through lush tropical vegetation to ancient village ruins. Then do it. Vermont-based Country Walkers and New York's Cross Country International—just two of a number of organizations that offer walking tours through the world's most scenic landscapes—will get you there. Walking-tour outfits like Country Walkers offer organized hikes both in the United States and abroad. Cross Country International offers walking and hiking tours of the British Isles. They usually arrange for both accommodations and meals, so all you have to do is put one foot in front of the other. Contact Country Walkers at P. O. Box 180, Waterbury, VT 05676, or ask your travel agent for information about other walking tours. You can write to Cross Country International at P. O. Box 1170, Millbrook, NY 12545.

WATER RETENTION

Sinking the Bloat

About 7 to 10 days before the start of their periods, many women feel as if they're carrying water balloons around their middles. It isn't uncommon for women to retain about three extra pounds of water each month due to abdominal bloating and distention, say medical experts.

You can blame bloating on the hormonal changes that occur during your monthly cycle. About a week before you start menstruating, when your body realizes that an egg has not been fertilized, the level of the hormone progesterone drops, causing your body to store excess salt and fluid.

Bailing Yourself Out

You don't have to take this water torture lying down. Here are three exercise approaches that will fight that floating feeling.

Start sweating. If you want to lose water, one good way is to sweat it out. Perspiring not only gets rid of some water itself, it also gets rid of salt, which tends to hold water. "Anything that helps you sweat and lose sodium will be helpful against water retention," says Dan Hamner, M.D., a physiatrist and sports medicine specialist in New York City.

Aerobic exercise makes the most sense as a sweaty option. When you feel yourself getting waterlogged, be sure to exercise every day for the next three or four days. "Whether it's a brisk walk, jogging, swimming, playing tennis, bicycling, or a session on the stair-climber, aerobic exercise helps that bloated condition," says Dr. Hamner. If you sweat for more than 40 minutes, however, be sure to replace the electrolytes you lose with a drink like Gatorade, he cautions.

Dehydrate with yoga. The art of yoga offers several poses that relieve the discomfort of water retention, says Richard Miller, Ph.D., a clinical psychologist and co-founder of the International Association of Yoga Therapists in Mill Valley, California. You'll get maximum relief if you master the proper breathing techniques that go along with yoga poses.

The modified shoulder stand will get you started. It improves circulation and pulls blood and water out of the legs, says Dr. Miller. "Make sure that your legs and ideally your buttocks are in the air higher than the level of your heart," he says. One note of caution: If you have high blood pressure, glaucoma, or migraines, consult your doctor before doing this exercise.

▶Lie on your back with your legs together and your hands by your sides, palms down (1).

▶Keeping your shoulders relaxed, swing your legs and trunk into the air as you breathe out deeply (2). You can either bend your legs or keep them straight.

▶Move your legs into a vertical position. Gently support the small of your back with both hands (3). Your shoulders, not the back of your neck, should take most of the weight.

▶Stretch your toes and relax your feet.

▶Breathe slowly, rhythmically, and deeply. At first, stay in position for 30 seconds, breathing deeply throughout. Try to gradually build up to one to two minutes.

▶Slowly, on an exhalation, bring your arms down so your hands are palms down on the floor. Let your arms handle some of the weight as you bring your legs back toward your upper body (4), then slowly lower them to the starting position.

Take action with acupressure. Don't underestimate the power of your finger in relieving water retention woes. There are several different acupressure points that can help, says David Nickel, O.M.D., a doctor of Oriental medicine and licensed acupuncturist in Santa Monica.

The beauty of this nearly no-sweat exercise is that you can do it anywhere at any time, with no need to be at a gym or use special equipment. So when you anticipate the onset of water weight, Dr. Nickel suggests this remedy to stimulate your kidneys.

Locate the kidney acupressure point in the center of the hollow just above the entrance to the ear canal (5). Using the tip of your fore-

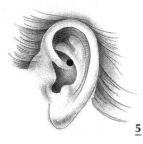

finger, apply steady pressure on that point for five seconds, then gradually release the pressure for five seconds. For better results, exhale through your mouth as you apply pressure and inhale through your nose as you ease pressure; use a deep, rhythmic, relaxed breathing style. Try this pressure-on, pressure-off cycle for one minute.

5

You may experience a hot, stinging, almost burning sensation in your ear, says Dr. Nickel. "You don't need to push so hard that you experience continuous pain or severe discomfort," he cautions. "If you have a long fingernail, you may substitute another finger or use the second knuckle of your index finger," he suggests. You should see results in just a few days.

The kidney point has also been demonstrated to help headaches, forgetfulness, hair loss, ringing in the ears, mineral imbalance, weak bones, and gynecological conditions, Dr. Nickel notes.

WINTER BLUES

Turn Your Brights On

When winter arrives in Alta, Norway, the sun takes a holiday. From the end of November through mid-January, it never rises above the horizon, and the coastal town lies, snowbound and hemmed in by ice, in darkness.

All this gloom and grayness makes Alta an ideal location for studying the low-mood malady called seasonal affective disorder, or SAD. A dreary mixture of depression, fatigue, and sleepiness, SAD seems to be triggered by the winter decline in daylight. While it is most prevalent in northernmost locales like Alta, where the day-shortening is most pronounced, it can be a problem for people in Piscataway and Nashville, too.

Research has long shown that getting lots of bright light can help ease the symptoms of SAD. More recent research from Alta and other winter-dim locations suggests that exercise can help, too, especially if you do it in well-lit surroundings.

"If you have SAD, exercise and bright light are a good therapeutic combination," says Brenda Byrne, Ph.D., director of the seasonal affective disorder program affiliated with the light research program at Jefferson Medical College of Thomas Jefferson University in Philadelphia.

The Effects of Clocks and Chemicals

Why can loss of light leave us so low? The shortening of daylight hours seems to throw a wrench in the works of our biological clocks and alter levels of various mood-governing chemicals in our brains, explains Dan A. Oren, M.D., associate professor at Yale University School of Medicine. Although researchers are still trying to determine which mood-governing chemicals play key roles in the onset of SAD, most of their work has focused on three—dopamine, serotonin, and melatonin.

One of the most significant of these chemicals—the sleep-related substance called melatonin—seems to be partially sustained by exercise. In the Alta study, people who exercised regularly had lower melatonin levels in winter than those who were sedentary. This suggests that regular exercise may make you less susceptible to the winter doldrums.

Your best bet is to exercise in the sun or in the glow of a light box. A study from the University of Tulsa found that people who exercised in the sunshine were less likely to suffer symptoms of SAD than those who got their exercise inside or at night. So, if you're feeling February in your soul, try these sun-soaking exercises.

Do Your Knees Need Some Light?

Showing a little leg while you're out exercising in the sun may help fight seasonal affective disorder (SAD), some preliminary research suggests.

In a groundbreaking study at Cornell University Medical Center in White Plains, New York, researchers trained a bright light on the backs of volunteers' knees and found that the light reset the biological clocks in the volunteers' brains. Since a snafu in your bio-timing contributes to SAD, light on the legs might help banish winter blues. Winter's light shortfall seems to set our internal clocks awry, but doses of bright light—either from the sun or from specially designed light boxes—seem to set it straight again, easing symptoms.

Researchers have long suspected that light entering through the eye helped. The Cornell study suggests that light striking other body parts can also reset the body's clock.

But how does your brain get the behind-the-knee illumination message? Via the hemoglobin in red blood cells, says Dan A.

Oren, M.D., associate professor at Yale University School of Medicine. Just as plants' chlorophyll absorbs light, human hemoglobin appears to absorb and transmit light signals, says Dr. Oren. Light falling on the back of the knee, where the skin is thin and there are lots of blood vessels, could similarly hitch a ride on hemoglobin.

"Light would presumably have a similar effect on other parts of the body where the skin is thin—the palm of the hand, fingernails, and face," Dr. Oren says.

It's too early to say whether you really should go out in the sun with your legs exposed or unveil your knees when working out in front of your light box, Dr. Oren says. "We need to do more studies," he explains, but that might well be standard medical advice some day. In the meantime, there's no harm in wearing shorts when you're exercising in a brightly lit room, and your hemoglobin might get a burst of benefit from the general glow.

Run in the sun. Shoot for at least a half-hour a day, says Dr. Byrne. Although early research suggested that morning light was best for SAD, subsequent studies indicate that midday and even afternoon light can be just as good for some people. So run whenever it's convenient, as long as you do it before dusk.

Skate or ski. Taking up a winter sport can be a real mood changer, according to Dr. Byrne. If you like to skate, try to find an outdoor rink. Or consider cross-country skiing, if you haven't

already tried it. The combination of exercise and bright light can hold SAD at bay.

Follow the cheepers. If you go bird-watching in winter, you'll see all sorts of species that winter in your hometown but are long gone by the time spring rolls around. Grab your binoculars and head out. Your hunt for the elusive arcadian flycatcher may be a wild catcher chase, but don't be surprised if a better mood sneaks up on you while you're in outdoor pursuit, says Dr. Byrne.

Reveal your peepers. Do your outdoor exercising without sunglasses, suggests Dr. Oren. How come? Dark glasses cut down on the amount of light that reaches your eyes, so you don't get the light-related benefits. It's not a good idea to go without sunglasses when the light is blinding or for long periods of time, however, as this can, over time, increase your risk of cataracts and other eye problems. But a half-hour a day during the winter months shouldn't hurt, says Dr. Oren. "Just don't stare at the sun," he cautions.

Sneak in some exercise. Don't have time for a walk or even few solitary figure eights at the lake? Slip out for a minute or two and stroll around the block at lunchtime. A little bit of light is better than none at all. "Get into the light whenever you can," says Dr. Byrne.

Let there be light—in a box. If you work all day and can't get out when the sun is up or you're loath to leave the cozy confines of home for the cold, consider a light box, says Dr. Byrne. The boxes are equipped with sets of fluorescent bulbs mounted behind a screen that diffuses the light and filters out ultraviolet rays. Placed at the proper distance from your eyes, they provide much brighter light than ordinary indoor lamps.

Make sure that you look for a manufacturer who is a member of the Circadian Lighting Association, advises Dr. Byrne. Also work with a medical or mental health professional who is qualified to diagnose and treat conditions related to changes in body rhythm, such as SAD, and make sure that you follow the specific safety recommendations from the manufacturer of your light box, he adds. (One caveat: If you have certain vision problems or are taking medications that make you more light-sensitive,

Summertime and the Living Is Gloomy?

If you feel perfectly fine all winter, but summer gets you down, you may have what's known as summer SAD, the far less common cousin of seasonal affective disorder.

Summer SAD can leave you feeling depressed, agitated, and irritable when the weather is hot and the days are long, says Brenda Byrne, Ph.D., director of the seasonal affective disorder program affiliated with the light research program at Jefferson Medical College of Thomas Jefferson University in Philadelphia.

It's not clear whether summer's light or the heat is the culprit, says Dr. Byrne. But staying out of the sun and keeping cool can help ease the symptoms.

"If you have summer SAD, do your exercise in a cool place," Dr. Byrne suggests. "Swimming and water aerobics are ideal. If you don't want to get in the water, try working out in an air-conditioned gym rather than sweating on an outdoor track."

light boxes may not be for you. Talk to your doctor before buying one.)

Once you have the light box set up, put your treadmill, ski machine, indoor trampoline, exercise mat, or stationary bike in front of the thing and work up a sweat. A half-hour in front of a light box does the mood-boosting trick for most people, says Dr. Byrne.

WRINKLES

Smoothing Out the Decades

"Wrinkles should merely indicate where smiles have been."
—**Mark Twain**

Those skin crinkles twinkling around your eyes and the laugh lines hugging your lips are your face's friendly creases. They symbolize your happy, healthy outlook.

But it's the chisel-deepened frown marks and worry lines plowed across the forehead from years of fretting and pondering that draw the most disdain among people clinging to their fleeting youthful visages. You may have to make peace with your deeper creases, because they're here to stay, but there are ways to help with lighter wrinkles and make your face firmer.

The older we become, the less collagen we produce, which means that ever-renewing layer of skin cells isn't performing up to snuff. As a result, our skin's once-vibrant, bounce-back quality diminishes. When skin loses its elasticity and becomes drier, it's no wonder that it starts to sag a bit.

Wrinkles, like taxes, may be inevitable as we age, but we can take steps to subdue their presence, say doctors. Applying sunscreen daily, drinking at least eight glasses of water a day to keep our skin hydrated, and avoiding smoking may forestall or reduce the presence of wrinkles.

What divides doctors is whether facial exercises truly keep our visages toned, taut, and youthful or whether they actually loosen the skin and make wrinkles more obvious.

Experts suggest the following facial exercises, which may help to slow skin aging without the need for expensive creams or face-lifts.

Ironing Out Wrinkles

Some experts contend that facial exercises fight wrinkles. Facial movements and massages increase blood circulation to the cheeks, chin, and forehead, and that increased blood flow means more nutrients (including antioxidants) are delivered to the skin and more wastes are taken away. This allows the face to better resist the loss of spongy subcutaneous tissue that is associated with aging. Facial exercises produce collagen within the skin, and it is this tissue that enables our facial skin to bend, stretch, and be thicker and firmer. Potentially, more flexible skin means fewer wrinkles.

"In my opinion, stretching and exercising

Massage Your Visage

Many wrinkles are developed and deepened by tension and anxiety, says Dan Hamner, M.D., a physiatrist and sports medicine specialist in New York City.

One smooth option to try is a facial massage, either a self-massage or one performed by your partner or a trained massage therapist, says Dr. Hamner, who regularly relies on Swedish massages for a healthy "face-lift."

"Massage, particularly the pushing and pulling technique of Swedish massage, increases blood circulation and prolongs the longevity of the skin," says Dr. Hamner.

Facial massages also tone and restore muscles to prevent or reduce wrinkle lines, supply nutritional oxygen to the skin and cells, and may help restore some elasticity to facial muscles, he adds.

To rid your forehead of that "just-plowed" look, Dr. Hamner suggests that you try this fingers-to-forehead massage daily. The goal is to reduce tension wrinkles in the forehead by removing waste products from the skin tissue and oxygenating the skin, says Dr. Hamner.

►Clear your forehead of any overhanging hair.

► Place both thumbs on your forehead so that they meet in the middle just below your hairline. Your palms should be facing each other, with your fingers pointing straight up and not touching your face (*above*).

►Slowly and firmly, slide your left thumb horizontally across your forehead toward the left.

►Repeat the same movement to the outside with your right thumb.

your facial skin is good for its overall health and will help reduce wrinkles," says Dan Hamner, M.D., a physiatrist and sports medicine specialist in New York City.

Robert L. Cucin, M.D., a plastic surgeon and assistant professor of plastic surgery at Cornell University Medical College in New York City agrees with Dr. Hamner—to a point. "I'm not a great believer that facial exercises will do a whole lot," he says. "But I think you increase the blood flow to the skin and to your face by exercising generally."

Lighten those laugh lines. Try to do 20 to 30 eyelid lifts each time you brush your teeth to soften those laugh lines around your eyes.

Checking your eye movements in a mirror, use the muscles around your eyes to draw your upper and lower eyelids closer together, but not to the point of squinting. The idea is to try to make your eyes shaped more like almonds than

▶Lower both thumbs slightly and repeat the horizontal movements row by row until you reach your eyebrows.

▶Repeat three times a day.

For light lines around the eyes, try this Swedish massage technique, offered by Dr. Hamner.

▶Place both thumbs in the space between your eyebrows with the pads of your thumbs flat against your face, your palms facing each other, and your fingers pointing straight up (*above*).

▶Smoothly and rhythmically, move both thumbs horizontally on the eyebrow line toward the outside of your eyes.

▶Repeat this movement three times.

▶Place the pads of both thumbs below your eyes, hugging either side of the bridge of your nose. Your fingers should be pointing straight up (*above*).

▶Smoothly and rhythmically, glide your thumbs across the skin just below your eyes and move horizontally to your temples.

▶Repeat three times.

ovals. "You are essentially making the lower eyelid rise," says Dr. Cucin. "Almond-shaped eyes are younger looking than oval-shaped eyes.

"You have ringlike muscles around the eyes and mouth," Dr. Cucin explains. "If you make your round eyes more almond-shaped, you can see extra skin on the upper and lower lid. As you raise the lower lid a bit, almost to a smile, and hold that position for one or two seconds, the skin tends to be flatter and you can make

your eyes look a little better. Try to do this exercise every day."

This mild facial motion won't magically erase those creases around your eyes, but they should make the creases less obvious and your face firmer, he adds.

Whistle away those wrinkles. It seems that every high school class had someone who could whistle like a banshee through his fingers. Well, the same basic motion—minus the whistle, of course—may tone down the wrinkles around your mouth, says Dr. Cucin. Here's how to pucker up.

►Looking into a mirror, make your mouth smaller without pursing your lips.

►For proper position, place the thumb and middle and ring fingers of your right hand in your mouth as if you were going to let loose with a piercing whistle.

►Press your fingers and thumb against the corners of your mouth and check the mirror to see the position of your lips. It should look as if you're about to give a loose, open-mouthed kiss.

►Now form that same expression minus your fingers and thumb.

►Hold for 10 to 15 seconds. Repeat 20 times each time you brush your teeth.

"You'll see as you do that—but don't overdo it—that your laugh lines should flatten a little bit," says Dr. Cucin.

Lead with your chin. Dr. Hamner offers a facial exercise to fight loose, wrinkling skin under the chin.

Look straight ahead, then tilt your head back and point your chin toward the wall. Open your mouth, then stick your chin out. You should feel a slight tug on your chin and neck muscles. Continue feeling that tug by slowly closing your mouth.

"This is a good exercise to keep that turkey wattle from developing under your chin. Loose skin around the face usually gets its start as wrinkles," says Dr. Hamner. "You should wake up each morning and do 50 of these before you get out of bed."

Walk away wrinkles. A 30-minute brisk walk offers many aerobic benefits, including de-livering a healthy, gentle radiance to our cheeks, says Dr. Hamner. "A brisk walk helps your facial muscles because the aerobic activity increases the blood flow to your cheeks," he explains.

Work your cheeks. To combat vertical wrinkles branching down from the nose into the cheeks, pretend you're a grandma who's lost her dentures. Pull your lips over your upper and lower teeth and suck them in so that your cheeks are pulled taut. Close your mouth and then massage the area around the sides of your nose with your fingertips for a minute or so. Do this every time you brush your teeth so you can watch yourself in the mirror.

"This exercise helps flatten out the skin around the nose area and helps reduce wrinkling," explains Dr. Hamner.

Pucker and pout. To tone the muscles around the mouth and diminish the chance of lines around the lips, Dr. Hamner recommends this daily lip workout. Push both lips out as if you were forming a pointed kiss. Then smile like a Cheshire cat, with a full, toothy grin. Repeat the pointed-kiss-to-cat movements 10 times each morning.

As a variation, he also suggests pushing your lower lip straight out and then up so that it covers your upper lip. Hold that pouty pose for 8 to 10 seconds and then relax. Try 10 of these.

Blow off wrinkles. One way to keep facial muscles taut is to do a regular workout that increases blood flow to the skin, says Dr. Hamner. He suggests inhaling deeply from the nose and then puffing out both cheeks as you exhale with force, loudly pushing the air out of your mouth for about five seconds. Keep the opening of your mouth small. You may want to sit in front of a mirror and watch yourself as you do this. Try 10 cheek puffs each day.

"This one may not work for everyone, but it should increase the blood flow into your cheeks, pull out the wastes in those muscles more quickly, and bring in new nutrients," he says.

Hang upside down. Get the okay from your doctor first, but try spending one minute upside down in a safe position, says Dr. Hamner. For instance, lie face-up across the width of your bed and position yourself so that your head and neck are over the edge of the bed but the rest of your torso is safely on the bed. Slowly tilt your head back as far as you can safely and hold that pose for one minute. You should feel the blood rushing to your neck and head. Then slowly raise your head and make sure that you feel stable before getting up.

"Hanging upside down works mainly for the brain, but it helps the facial skin, too," says Dr. Hamner.

WRIST PAIN

Wave Adios to Aches

People who work their wrists hard—guitarists, typists, and carpenters, for example—are at high risk for wrist pain. The wrist is a delicate network of tiny bones, nerves, tendons, and ligaments. Arthritis can attack there, and strains or sprains can make simple, everyday tasks, such as opening a jar or even turning the pages of a book, quite painful.

The key to beating wrist pain is to make sure that your wrists are as limber and as strong as they can be. Experts suggest that you combine stretching exercises (to enhance your range of motion) and strengthening exercises (to build up the tendons, ligaments, and muscles in the wrist area).

"Remember to warm up your hands first. Hands must be warm in order to work correctly," says Robert Markison, M.D., a hand surgeon and associate clinical professor of surgery at the University of California, San Francisco.

Reach for a Warmup

There's nothing like a proper stretch to get the tendons, ligaments, and muscles in the wrist area warmed up, say doctors and physical therapists. Here are some handy stretches that may relieve and possibly prevent further wrist pain.

When to See a Doctor

If you have wrist pain so severe that it keeps you from carrying out daily activities such as eating, dressing, or working or that wakes you from sleep three or four nights in a row, it's time to see a doctor, says Michael Ciccotti, M.D., an orthopedic surgeon and director of sports medicine at the Rothman Institute at Thomas Jefferson University in Philadelphia.

Revisit your childhood. Ask a five-year-old his age and he will likely declare "five," raise his hand, and spread all five fingers as proof. The same gesture can be a curing stretch, says Thomas Meade, M.D., an orthopedic surgeon and medical director of the Allentown Sports Medicine and Human Performance Center in Pennsylvania. Here's how.

Place your hand out in front of you as if you were stopping traffic. With your arm extended, spread your fingers as far apart as possible. Hold that position for at least 20 seconds. Repeat it five times, then relax and do the same stretch with your other hand. "Your muscles and tendons won't remember this stretch unless you hold it for at least 20 seconds," says Dr. Meade. "You should be able to feel your muscles get more rubbery and stretchy with time."

Halt and stretch. You can use that crossing guard "Stop" pose for this stretch, too. First, extend one arm in front of you at shoulder level and parallel to the ground, with your wrist bent and your palm facing out. Then, use your other hand to gently pull back the fingers of your extended hand. Hold for 20 seconds, then relax. Repeat five times, then switch hands and repeat.

The goal is to feel the muscles, ligaments, and tendons stretching in your fingers, hand, wrist, and forearm, says Dan Hamner, M.D., a physiatrist and sports medicine specialist in New York City.

Grab for the door. All you need for this stretch is an imaginary front door, says Michael Ciccotti, M.D., an orthopedic surgeon and di-

rector of sports medicine at the Rothman Institute at Thomas Jefferson University in Philadelphia.

Extend your right arm in front of you with your palm facing the ground, then pretend that you are turning a doorknob slowly to the left, then slowly back to the right. Try 10 of these rotations, then switch hands and do 10 more. "This exercise helps make the joints in the wrist more supple," explains Dr. Ciccotti.

Think sink. Warm water can soothe aches in your wrists, says Jane Katz, Ed.D., professor of health and physical education at John Jay College of Criminal Justice at the City University of New York, world Masters champion swimmer, member of the 1964 U.S. Olympic performance synchronized swimming team, and author of *The New W.E.T. Workout.*

Submerge your hand over the wrist in a sink full of warm water. Pivoting your wrists, do 10 hand circles to the right, then 10 to the left.

This exercise improves the range of motion

Strike Up the Band: Try Wrist Resistance

A thick rubber band can actually be a piece of workout equipment. Try this rubber resistance regimen, suggested by Teri Bielefeld, P.T., a physical therapist and certified hand therapist at the Zablocki Veterans Affairs Medical Center in Milwaukee.

Extend your right arm in front of you with the palm up. Slip a thick rubber band over the crease on the inside of your palm. Placing your left hand under your right hand, grab the other end of the rubber band. As your left hand slowly pulls the rubber band down,

counter by trying to bend your right hand up at the wrist. Feel the resistance. Try 10 times and then repeat, switching hands.

Then turn your right hand palm down and do the same thing. As your left hand slowly tugs downward, try to flex your right hand up, once again using your wrist. Feel the resistance. Try 10 times, then switch hands and repeat.

"You're working on resistance with the wrist motion so that neither the wrist nor the rubber band wins," Bielefeld says.

in your wrist and increases the oxygen-carrying blood flow to your wrists and fingers, says Dr. Katz. "The beauty of exercising in water is that your muscles get resistance from all directions," she adds.

Play shadow games. Relive those childhood days of making shadow animals on your bedroom wall at night. This exercise mixes fun with flexibility, says Teri Bielefeld, P.T., a physical therapist and certified hand therapist at the Zablocki Veterans Affairs Medical Center in Milwaukee. Of course, you can do this anywhere, anytime. (For an illustration of this exercise, see page 109).

► Raise your right hand as if you were being sworn in to testify in court. Keep your fingers relaxed.
► Gently bend your right hand forward at the wrist so that your hand resembles a swan's head.
► Use your other hand to push down gently on the top of the "swan's head."
► Hold for five seconds, then gently bring your hand back into the starting position.
► Hold for five seconds. Repeat four times.

Arm Yourself with Strengthening Exercises

Once you have done stretches to limber up your wrists, step two is to try some exercises that make your wrists and forearms strong. Here are a few.

Tug on a towel. Roll up a hand towel, then grasp it with both hands. Keep your left hand still as you turn your right hand as though you were wringing out excess water, first in one direction, then the other. Repeat, turning with your left hand, says Mary Ann Towne, P.T., a

A Farewell to Wrist Pain

These exercises can help prevent further pain by building flexibility in your wrists, says Teri Beilefeld, P.T., a physical therapist and certified hand therapist at the Zablocki Veterans Affairs Medical Center in Milwaukee.

Wave goodbye to pain. This range-of-motion exercise is designed to stretch the muscles in the wrist area.

Extend your right arm straight in front of you with the palm down. Bend your wrist down, then raise it as if you were doing a slow-motion wave. Do 10 waves with each wrist three times a day, says Bielefeld.

Do the twist. For this flexibility builder, extend your right arm straight in front of you and cup your right elbow with your left hand. Keeping your right elbow still, rotate your right wrist, slowly turning your palm up and then down. Do this for 20 to 30 seconds and then repeat for a total of five repetitions. You should be able to feel the muscles in your arm twisting. Then switch arm positions and repeat with the other wrist. Try this exercise three times a day, suggests Bielefeld.

physical therapist and director of rehabilitation and wellness services of the Cleveland Clinic–Florida in Fort Lauderdale.

"When you are wringing a towel, you are both strengthening and stretching the muscles in your wrist," explains Towne.

Go fly fishing. Fly fishing is a good wrist-strengthening exercise as long as you don't overdo it, says Dr. Ciccotti. "It is a wonderful recreational sport as long as you prepare by doing some wrist stretches. The back-and-forth action flexes and extends the muscles. Using muscles rapidly requires strength and force. You are also stretching the joint capsule around your wrist."

Look Ma, all hands. Try this imaginary hand-pedaling exercise. Extend both arms in front of you and pretend that your fingers are holding onto bicycle pedals. Making a forward circular motion with your arms, move the imaginary pedals forward. Make sure that you flex your wrists as you pedal. You should feel the muscles in your wrists, forearms, and elbows warming up, says Dr. Ciccotti. Gradually work up to three minutes of continuous exercise.

YOGA

A Time-Out to Tune In

WHAT IT IS

This ancient Indian discipline has absolutely nothing to do with former baseball great Yogi Berra or cartoon great Yogi Bear. *Yoga*, which is a Sanskrit term meaning "to join," is a physical practice that blends breathing methods, stretching exercises, and meditation in an attempt to harmonize your mental, physical, and spiritual selves.

A typical yoga class lasts from an hour to an hour and a half. Students wear loose, comfortable clothing—minus socks and shoes. Classes start slowly. And students don't necessarily sit in the infamous lotus position, with their folded legs entwined so the tops of both feet rest high on their opposite thighs. For beginners, the familiar cross-legged pose known as Indian-style is more comfortable—and more realistic.

Some teachers like to begin with a round of breathing exercises called pranayama. These breathing methods help to clear the mind and the lungs. The class then moves to the physical poses, or asanas, the odd positions that most people picture when they think of yoga. While some yoga poses—like the headstand or the full wheel (a complete backbend, with hands and feet on the floor)—are impossible-looking,

a good teacher will start you with the basics.

A variation of the sun salutation is a common warmup for many yoga classes. It's a multistep sequence that begins with simply standing with a straight spine in the mountain pose. On an inhalation, students raise their arms over their heads into a gentle backward bend and then, while exhaling, reach for their toes in a forward bend. The routine moves on through several asanas, including the cobra pose (lying on your stomach with your upper body and head arched up toward the sky). At the end of the sun salutation, the students stand once again in the mountain pose.

After the warmup, the remaining class time is usually spent moving through other asanas, which work different parts of the body. But regardless of the type of asanas being done, breathing and inner concentration always play an important role. In fact, classes typically end with a period of deep relaxation, which calms the mind and, in theory, allows the body to get the most benefit from the work that it has just performed.

Relaxation is usually done in savasana, or the corpse pose. In savasana, the students lie on their backs with their arms resting alongside their bodies and their legs and feet about hip-width apart.

With your eyes closed and your muscles relaxed, your mind can focus on relaxation, and it becomes still. This pose has been called the most difficult because it requires complete relaxation of every part of the body, which is easier said than done.

WHAT IT DOES

Yoga is tranquillity training—one of the most powerful peace producers around. It's so relaxing because it requires sharp concentration and focus, says Lee Lipsenthal, M.D., vice president and medical director of the Preventive Medicine Research Institute in Sausalito, California.

If this seems contradictory, consider the soothing effect of having a hobby. "Any time a person focuses intently on one thing, be it painting model airplanes or playing football or whatever, they tend to become calm," says Dr. Lipsenthal. "Two or three hours can just fly by."

This tension-taming property of yoga is the first in a long line of health-boosting dominoes. A great yoga benefit is that it lowers your levels of stress hormones, called catecholamines. This has happy effects on virtually every system of the body, says Dr. Lipsenthal. Here are just a few examples of what yoga can do.

Enhance immunity. Catecholamines produced by ongoing stress appear to limit the activity of white blood cells, the soldiers of immunity. "Practicing yoga regularly reduces catecholamine levels, freeing up the functioning of the white cells," says Dr. Lipsenthal. "This gives immunity a boost."

Assuage asthma. A study at the All India Institute of Medical Science in New Delhi showed that yoga training can be an advantage for people with asthma. Nine people diagnosed with bronchial asthma spent one week at a yoga camp. After yoga training, their breathing was significantly easier.

Dr. Lipsenthal explains that yoga has an effect on both the physiological and the psychological responses to asthma attacks. Shortness of breath during an attack is stressful and causes an increased consumption of oxygen. The anticipation of an attack makes many people panic, which also uses up oxygen. Practicing yoga on a regular basis and using yoga techniques (deep breathing, muscle relaxation, and meditation) during an attack reduce the stress, the panicky feelings, and the amount of oxygen used up and make breathing easier.

When you feel an attack coming on, take one puff from your inhaler before you begin yoga relaxation techniques, advises Dr. Lipsenthal. This allows the medicine to begin working. Take a second puff a few minutes later. Yoga techniques actually help increase absorption of the medication, he adds.

Supercharge the skin. Anxiety brings an onslaught of catecholamines that cause blood vessels to clamp down. With less blood flow, the skin loses its elasticity and fluid content. The end result for chronically stressed people, says Dr. Lipsenthal, is a gray and wrinkled complexion. Yoga can undo these damaging effects by boosting blood circulation. "Long-time yogis tend to have healthy skin," he says.

Relieve digestive complaints. Stress can cause gastrointestinal trouble—indigestion, diarrhea, or even constipation. But yoga can help.

"I've seen digestive problems diminish rather quickly through yoga," says Dr. Lipsenthal.

Increase flexibility. Yoga asanas are, in essence, slow stretches, and the direct effect of regular stretching is increased flexibility. "You may only manage a tiny stretch the first time you try," says Dr. Lipsenthal, "but keep it up, and you'll notice a profound difference in the long run."

Make sex more fun. Pelvic muscles are frequently tight and have limited movement. Yoga stretches that focus on the hips and groin can limber you up and enhance blood flow to the vagina or penis. Yoga also decreases stress, which is the culprit in a large percentage of cases of male sexual dysfunction, says Dr. Lipsenthal. It also increases the ability to enjoy sex. "If you're peaceful, you can focus more on sex, not the laundry or other chores. Increased focus is good for both you and your partner, " he adds.

REAPING THE BENEFITS

The rewards of yoga are many, but as the saying goes, you've got to play to win. Here are some tips to help get you in the game.

Get started. It's a good idea to start by taking a regular class. At the same time, try to practice daily at home, says Rodney Yee, director of the Piedmont Yoga Studio in Oakland, California, and producer of numerous yoga videotapes. Set aside 10 or 15 minutes, he says, and focus on filling that time slot with yoga.

Stay for six. Unless you give yoga some time, you may not see benefits. Make a six-month commitment, says Yee. "Six months of consistent practice will get you to the place where you really want to make the time," he

says. After a half-year of regular practice—with no excuses—you'll wonder how you ever got along without doing yoga.

TRIAL RUN

So you say you've been out of commission for a while? Not exercising very often, or maybe not at all? Yoga may be a perfect way back into the world of exercise. Just take it easy.

Easy Does It Yoga (EDY) is a user-friendly approach developed by Alice Christensen, founder and executive director of the American Yoga Association in Sarasota, Florida.

Intended for seniors and those with physical limitations, Easy Does It Yoga is based on the three components of traditional yoga—asanas, breathing exercises, and relaxation or meditation, says Christensen.

Beginning EDY poses are done in bed or seated in a chair. Gradually, as the student's strength and balance increases, standing poses and floor work are added to the routine.

Who else might be interested in EDY? It's suitable for anyone who is just starting yoga practice, and Dr. Lipsenthal adds that people with arthritis may find yoga movements especially beneficial. Yoga loosens up the muscles and tendons surrounding the joints, separating the bones and keeping them from rubbing together.

To get a feel for the simple but strengthening moves of Easy Does It Yoga, try the seated leg lift.

▶Start by sitting barefoot in a sturdy, armless chair with your lower back against the backrest and both feet flat on the floor. Grasp the sides of the seat with your hands.

▶Take three slow, deep breaths, inhaling and exhaling fully. On the next inhalation, straighten your right knee and lift your leg as high as you can without straining. Keep your foot active by trying to push your heel away from your body while pulling your toes toward you (*1*).

▶Hold for a count of three or as long as is comfortable, then exhale as you lower your leg. Relax and breathe normally. When you're ready, start again with three slow, deep breaths and repeat with the left leg. Then repeat the whole sequence—right, left, right, left, right, left—for a total of three leg lifts on each side.

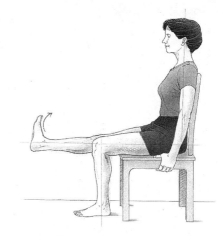

1

CREDITS

The Near Eye-C chart on page 122 was adapted from *The Power behind Your Eyes* by Robert-Michael Kaplan, O.D., M.Ed. Copyright © 1995 by Dr. Robert-Michael Kaplan. Reprinted by permission.

The Rate of Perceived Exertion (RPE) scale on page 185 was adapted from *Medicine and Science in Sports and Exercise*, 14 (1982):377–81.

The Amsler grid on page 245 was adapted with permission from the brochure "Macular Degeneration." Copyright © 1994 by the American Academy of Ophthalmology.

The Snellen eye chart on page 285 was adapted from *The Handbook of Self-Healing* by Meir Schneider, L.M.T., Ph.D. Copyright © 1994 by Penguin Books Ltd. Reprinted by permission.

The following exercises were adapted with permission from the *ACSM Fitness Book*, 2nd edition, by the American College of Sports Medicine: pages 334–36, from "Exercises for Warm-Up and Aerobic Fitness"; pages 336–37, from "Exercises for Muscular Fitness"; and page 337, from "Exercises for Flexibility and Cool-Down." (The information adapted represents only samples of a complete, integrated fitness program.)

The seated underarm tightener exercise on pages 401–2 was adapted from *Callanetics Countdown* by Callan Pinckney. Copyright © 1990 by Callan Productions Corp. Reprinted by permission.

INDEX

Underscored page references indicate boxed text. **Boldface** references indicate illustrations. *Italic* references indicate tables.

A

Abdominal muscles. *See also* Crunches, abdominal
  oblique, 233
  transverse, 310
Acetaminophen, for fever, 84
Achilles tendon, stretching, 65, 343
Acupressure
  as remedy for
    anxiety, 23, **23**
    cold and flu symptoms, 83–84
    dizziness, 203
    fibromyalgia, 129–30
    hay fever, 155
    headaches, 83, 163, 163, **163**
    hot flashes, 254, **254**
    motion sickness, 265–66
    panic attacks, 307, **307**
    stress, 377, **377**
    water retention, 412–13, **413**
  shiatsu techniques, 247–48, **247**, **248**
ADD. *See* Adult attention deficit disorder
Adrenaline
  afternoon slump and, 7
  high blood pressure and, 180
  sinusitis and, 363
  sleep apnea and, 370
  touch and, 348
Adult attention deficit disorder (ADD)
  remedies for
    mental strategies, 2
    physical exercise, 2–3
  symptoms of, 1–2

Aerobic exercise, 4–6
  amounts of, recommended, 5, 171, 181, 298, 369
  as remedy for
    back pain, 30
    caffeine withdrawal, 71
    cellulite, 78
    driver fatigue, 106
    fibromyalgia, 131
    flatulence, 136–37
    forgetfulness, 142–44
    hangover, 150
    high blood pressure, 181
    insomnia, 205
    laryngitis, 223, 224
    sensory loss, 345–46
    water retention, 411
    weight loss, 298
  schedule for, 5
Aerobox, for carpal tunnel syndrome, 73
Afternoon slump
  remedies for
    exercise, 7–10
    naps, 10
    office decor, 9
    schedule adjustments, 9
  sleep deprivation and, 7
Aggression, exercise and, 21
Alcohol use
  hangover from, 149–51, 150
  snoring and, 371
Alconefrin, for sinusitis relief, 362

Alexander Technique, 11–12, **12**
  as remedy for
    fibromyalgia, 131
    laryngitis, 224–25
    stage fright, 373
Allergies
  hay fever, 152–55, <u>153</u>
  home gym location and, <u>58</u>
  laryngitis from, 223
  meditation for, 250
Alveoli, smoking and, 366–67
Amblyopia. *See* Lazy eye, remedies for
Anemia, <u>56</u>, 123
Anger
  avoiding, 171–73
  exercise and, 13–14
  from job loss, 216–17
  pets and, 15
  remedies for
    baths, 16
    cognitive exercises, <u>14</u>, 173–74
    meditation, 16
    music, 15–16
    relaxation exercises, <u>15</u>
    tai chi, 387
  in type-A behavior, 397, 399
Angina
  arterial plaque and, <u>172</u>
  meditation for, 250
Ankles, pain in
  remedies for
    RICE, 17
    strengthening exercises, 19–20, **20**
    stretching exercises, 17–18, **19**
  when to see doctor for, <u>18</u>
Antacids, for heartburn, 168
Antidepressant(s)
  exercise as, 21
  side effects of, 204
Anti-inflammatory drugs, for asthma, <u>56</u>
Antioxidants, levels of, during exercise, 171
Anxiety
  insomnia from, 204
  laughter and 28

during menopause, 253
during menstruation, 259–60
nervous habits and, <u>290</u>
remedies for
  acupressure, <u>23</u>, **23**
  cognitive exercises, 22–23
  physical exercise, 21–22, 405
  reflexology, 329
social, 358–61
Aquatic exercises. *See* Water exercises
Arabesque, to firm buttocks, 327
Arms, strengthening, 275, **275**, 299–300, **299**, **300**, 400–402, **400**, **401**, <u>402</u>, **402**
Aromatherapy
  for erectile dysfunction, 114
  for inhibited sexual desire, 200
Arrhythmia, <u>172</u>
Arthritis, 24. *See also* Osteoarthritis
  effects on, from
    exercise, 24
    laughter, 28
    stretching, 380
  in fingers, 132
  golf guidelines with, <u>28</u>
  in knees, 219
  remedies for
    bodywork, 12, **12**, 27
    cold packs, 28
    flexibility exercises, 24–25
    relaxation exercises, 27
    strengthening exercises, 25–26
    tai chi, 387
    walking, 28
    water exercises, 26–27, **26**, **27**
Asanas, for back pain, 30, **30**
Asthma
  air pollution and, 55–56, <u>56</u>
  exercise-induced, 59
  remedies for
    bodywork, 12, **12**
    breathing exercises, 60–62
    physical exercise, 57–58
    yoga, 427
Atherosclerosis, 112

Attitude
  anxiety and, 22
  effect on healing, 229, <u>242</u>
  in empty nest syndrome, 111
  about exercise, <u>241</u>
  headaches and, 162
  high blood pressure and, 182–83
  in midlife crisis, 261
Autogenics, for Raynaud's disease, 323

B

Back
  anatomy of, 29
  pain in (*see* Back pain)
  strengthening, 274, **274**, 301, **301**
  stretching, 30, **30**, 170, 314, **314**, 343, 381–82, **382**, 408
Back bends, for fatigue, 125
Back extensions, in weight-loss program, 301, **301**
Backpacking, as calorie-burning exercise, 187
Back pain, 29
  activity guidelines for, 32–35, <u>33</u>, <u>34</u>
  during menstruation, 257–59
  preventives for, 31–32, 86, <u>87</u>
  remedies for
    bodywork, 11–12, **12**, <u>33</u>
    cold treatment, <u>33</u>
    exercise, 30–31, **30**, <u>33</u>, <u>35</u>, 86–88
  stretching and, 380
  walking and, 406
  when to see doctor for, <u>31</u>, <u>33</u>
Badminton, exercise-induced asthma and, 59
Balance, techniques to improve
  dance, 95
  standing on one leg, <u>19</u>
  stretching, 380
  tai chi, 295, 387
Barrett's esophagus, heartburn and, 165
Basketball, as remedy for
  high blood pressure, 182
  underarm sag, 402

Baths, as remedy for
  anger, 16
  insomnia, 205
"Batwings." *See* Underarms, sagging
Belching
  eating habits and, 36–37, <u>38</u>
  hypnosis for, <u>37</u>
  stress and, <u>36</u>, 38
Bent-over rows, in weight-loss program, 301, **301**
Beta blockers, low blood pressure and, 237
Bicep curls, in weight-loss program, 299–300, **300**
Bicycles (exercise), in weight-loss program, 303, **303**
Bicycling, 39–42
  contraindications for, 113
  equipment for, 40
  as remedy for
    back pain, 30
    caffeine withdrawal, 71
    clumsiness, 80
    fibromyalgia, 131
    heartburn, 166
    high cholesterol, 187
    knee pain, <u>221</u>
    muscle cramps, 268–69
    rear-end sag, 326
    shinsplints, 353
    sinusitis, 363, 364
  Spinning, <u>41</u>
Bike trainers, for indoor exercise, 69
Biofeedback, 43–45
  as remedy for
    incontinence, 197–98
    migraines, 160–61
    panic attacks, 309
    Raynaud's disease, 322–23
  techniques, 45–46, 160–61
Birdwatching, to improve mood, 415
Birth control pills, varicose veins and, 403
Birthday blues, 47–48
Bladder problems
  controlling, during menopause, 253
  retraining, <u>197</u>

Blood pressure. *See also* High blood pressure; Low blood pressure
    readings, 174, <u>182</u>, <u>237</u>
Blood sugar levels, in diabetes, 101
Blues
    birthday, 47–48
    retirement, 338–39, <u>339</u>
    winter (*see* Winter blues)
Boating, as remedy for
    depression, <u>100</u>
    love handles, 235
Body weight. *See* Obesity; Overweight; Weight control; Weight loss
Bodywork. *See* Alexander Technique; Massage therapy
Bone density
    bicycling and, 40
    dancing and, 95
Bonine, for motion sickness, 265
Boredom, 49–51, <u>50</u>
Boxaerobics, for clumsiness, 81–82
Boxing
    shadow, as remedy for
        afternoon slump, 8
        anger, from job loss, 217
        arthritis, 26–27, **27**
        carpal tunnel syndrome, <u>73</u>
        clumsiness, 81–82
    sparring, for clumsiness, 81
Breasts, sagging, 52–54, **53**
Breathing
    bodywork and, 12, **12**
    driver fatigue and, 105
    during exercise, <u>297</u>
    laryngitis and, 224
    in Pilates Method, 312
    stretching and, 381
Breathing disorders, 55, <u>56</u>
    hypoxemia in, <u>57</u>
    pollution and, 55–56
    remedies for
        breathing exercises, 60
        physical exercise, 57–58

Breathing exercises, 61–62
    in biofeedback, 45
    as remedy for
        anxiety, 22–23
        belching, 38
        breathing disorders, 60
        caffeine withdrawal, 71
        constipation, 92
        driver fatigue, 105–6
        eyestrain, 118
        fatigue, 124
        hangover, 149
        heartburn, 167
        hiccups, 178
        insomnia, 207
        menopause problems, 253, 254
        panic attacks, 307–8
        stage fright, 373
        stress, 379, <u>395</u>
        tooth grinding, <u>395</u>
Breath support, to preserve voice, 224
Bronchitis, chronic, 55, <u>56</u>
    pollution and, 55–56
    remedies for
        breathing exercises, 60
        physical exercise, 57–58
Bronchodilators, for asthma, <u>56</u>
Buddy system, exercise and, 50, 339
Burnout, job, 212–14
Bursitis, 63–65, **64**
Buttocks muscles
    strengthening, 277–78, **277**, **278**, 303–4, **303**, **304**, 325–27, **325**, **326**
    stretching, 376

C

Cabin fever, 68–69
Caffeine withdrawal, 70–71
Calisthenics, for cabin fever, 68
Callanetics, for underarm sag, 401–2, **401**, **402**

Calories, burned by exercise
  in backpacking, 187
  in dance, 96
Calves
  cramps in, 208–9, 267–68
  stretching, 170–71, 305, **305**, 343, 408
Cancer, esophageal, 165
Cardiac hypertrophy, high blood pressure and,
    179
Cardiovascular system
  effects on, from
    bicycling, 39
    running, 341
    walking, 405
  exercise amount for, 5
  hearing and, 347
Carpal tunnel syndrome, 72
  keyboarding guidelines with, 74
  remedies for
    exercises, 73–74, <u>73</u>
    toys and games, 72–73
Catecholamines
  nicotine and, <u>367</u>
  stress and, <u>186</u>
  yoga and, 427
Cellulite, 75–78, <u>77</u>
CFS, <u>126</u>
Charley horses, 269–70
Chest, strengthening, 273–74, **274**, 300–301, **300**
Chest press, in weight-loss program, 300–301,
    301
Chewing, belching and, 37
Chewing gum
  belching and, <u>38</u>
  heartburn and, 167–68
  temporomandibular disorder and, 393
Children, lifting, back pain and, <u>34</u>
Chinning exercise, for headaches, <u>160</u>
Cholesterol levels
  effect on
    erection, 112
    heart disease, 172
  exercise and, 171, 184–86, 341, 405

  healthy, 174
  high, 184–87, <u>186</u>
Chores
  for muscle weakness, <u>272</u>
  for underarm sag, 402
Chronic fatigue syndrome (CFS), <u>126</u>
Chung Moo Doe, for back pain, <u>35</u>
Circadian rhythms, afternoon slump and, 7
Circulation
  effects on, from
    laughter, 227
    running, 340
    sex, 350, 351
  problems, in legs, 208
Cleaning
  for adult attention deficit disorder, 3
  for breast sag, 53
Clothing, tight, heartburn and, 167
Clumsiness, 79
  fine motor skills and, <u>80</u>
  remedies for
    cognitive exercises, <u>81</u>
    physical exercise, 80–82
Cobra pose, in weight-loss program,
    304, **304**
Cognitive exercises, as remedy for
  anger, <u>14</u>, 173–74
  anxiety, 22
  clumsiness, <u>81</u>
  forgetfulness, 142, <u>143</u>
  headaches, 164
  insomnia, <u>206</u>
  panic attacks, 309
  premature ejaculation, 318
  stress, 378–79
Cold, as remedy for
  arthritis, 28
  back pain, <u>33</u>
  bursitis, 64
  shinsplints, <u>353</u>
Colds, common
  exercise and, 83, <u>84</u>, 85
  remedies for, 83–85

Color therapy, for macular degeneration, 244–45

Community involvement, for midlife crisis, 262–63

Commuter's back
  preventives for, 86–88, **87**
  stress and, **88**

Competing response, to nervous habits, 291

Complexion
  sex and, 351
  wrinkles in, 417–21, **418–19**, **418–19**

Compulsiveness, in type-A behavior, 399

Computers, eyestrain and, 116, **117**, **118**

Condoms, premature ejaculation and, 318

Congestion, remedies for, 84–85

Constipation
  vs. normal bowel habits, 89
  remedies for
    biofeedback, 44
    breathing exercises, 92
    physical exercise, 89–92
    reflexology, 328

Conversation skills, shyness and, **360**

Cooldowns
  for bicycling, 42
  for strength training, 337
  for walking, 408

Corticosteroids, anxiety and, 21

Cortisol
  aerobic fitness and, 13
  laughter and, 227, 376
  massage and, 247
  music and, 16

Cramps, menstrual, 256–57, 350

Cramps, muscle
  cautions about, **268**
  from intermittent claudication, 208–9
  preventives for, 268–69
  remedies for
    bicycling, 268–69
    massage, 268
    stretching, 267–70, **268**, **270**

Creativity block, 93–94, **94**

Cross-training, adult attention deficit disorder and, 3

Crunches, abdominal
  compared to Pilates Method, 311
  instructions for, 302, **302**, 337, **337**
  as remedy for
    breathing disorders, 58
    constipation, 90

Curl-ups
  to prevent commuter's back, **87**
  in weight-loss program, 302, **302**

D

Dance, 95–96
  aerobic, 181, 205
  ballet movements, 327
  instructions for, 96–97, 143
  as osteoporosis preventive, 294
  as remedy for
    cabin fever, 68–69
    clumsiness, 80–81
    creativity block, 94
    forgetfulness, 142–44
    high cholesterol, 187
    intermittent claudication, 209
    love handles, **235**
    menopause problems, 253
    stress, 375

Daydreaming, for boredom, **50**

Decongestants, for sinusitis relief, 362

Dehydration. *See also* Hydration
  from alcohol use, 149
  constipation and, 90
  as cause of
    fatigue, 123, 125
    low blood pressure, 237, 238

Depression
  effects on, from
    exercise, 21, 98–99
    laughter, 28
  fatigue from, 123
  remedies for
    boating, **100**
    massage, 247

nature, 98
pets, 99–100
tai chi, 387
walking, 99, 405
types of
birthday blues, 47–48
from chronic pain, 128
menopausal, 253
retirement blues, 338–39, <u>339</u>
serious, 98
winter blues, 414–16, <u>415</u>, <u>416</u>
Desensitization, mental, for hay fever, 154
Dexedrine, for adult attention deficit disorder, 2
Dextroamphetamine, for adult attention deficit
disorder, 2
Diabetes
erection and, 112, <u>114</u>
exercise for, 40, 102–3, <u>102</u>
foot care in, 103–4
types of, 101
when to see doctor for, <u>103</u>
Diet. *See also* Eating habits
constipation and, 89
exercise motivation and, <u>297</u>
heart disease and, 169–70, 174
macular degeneration and, 242
Digestion
belching and, 37
relaxation and, 330
Digestive problems, remedies for
reflexology, 329
yoga, 427–28
Dimenhydrinate, for motion sickness, 265
Dizziness, 202–3, 236–38
Dopamine
appetite and, 296, <u>367</u>
seasonal affective disorder and, 414
Dramamine, for motion sickness, 265
Driving
back pain and, <u>33</u>, 86–88
fatigue from, 105–6
headaches and, 157–58
Drugs. *See* Medications, side effects of; *specific
medications*

Drumming, for anger, 15–16
Dumbbells, buying, <u>77</u>. *See also* Strength
training
Dyna-Bands, for strength training, 76, <u>77</u>, 272.
See also Strength training

E

Ears, acupressure points on, 23, **23**, <u>307</u>, 307,
412–13, **413**
Easy Does It Yoga (EDY), for beginners,
428–29, **429**
Eating habits
effect on
belching, 36–37, <u>38</u>
heartburn, <u>167</u>
in temporomandibular disorder,
<u>392</u>
Eccentric viewing, for macular
degeneration, <u>243</u>
Education
during midlife crisis, 264
about sports, 49
EDY, for beginners, 428–29, **429**
Ejaculation, premature, 317–18
Elbow
anatomy of, 107
golfer's, 108–9, **109**
tennis, 107–8, **108**
Emotions, heart disease and, 169, 171–74.
See also Anger
Emphysema, 55, <u>56</u>
air pollution and, 55–56
remedies for
breathing exercises, 60–62
physical exercise, 57–58
Empty nest syndrome, 110–11
Endorphins
effect on
adult attention deficit disorder, 3
anxiety, 21
arthritis, 28
depression, 98

Endorphins (*continued*)
  effect on (*continued*)
    headaches, 157, 159, 162, 164
    social anxiety, 359
    stress, 374
  released by
    laughter, 227
    sex, 351
Endurance, strength training and, 332
Enzyme deficiency, muscle cramps and, 268
Epinephrine
  breathing exercises and, 62
  sinusitis and, 363
Equipment
  bicycling, 40
  indoor exercise, 69
  strength training, 77, 217, 332
Erectile dysfunction, 112–13
  remedies for
    aerobic exercise, 113
    aromatherapy, 114
    meditation, 250
    sex, 113–14
  when to see doctor for, 114
  worry and, 114–15
Exercise. *See also specific types*
  appropriate, choosing, 239–40
  attitude and, 241
  as family event, 111
  goals for, 50, 240–41, 339
  monotony in, 127, 213–14, 263
  motivation for, 49–51, 297, 339, 406
  records of, keeping, 51, 241
Eye charts
  to improve farsightedness, 122
  to improve nearsightedness, 285
Eye-hand coordination, improving, 347–48
Eyestrain
  computers and, 116, 117, 118
  remedies for, 116–18
Eye tracking ability, improving, 348

F

Facial pain, biofeedback for, 44
Fainting, from inner ear problems, 202, 203
Farsightedness
  exercise and, 119–20
  remedies for, 120–22, 121, 122
Fat, body, heart disease and, 173
Fat, dietary
  heart disease and, 169–70, 172
  macular degeneration and, 242
Fatigue
  acceptable, 124
  chronic, 126
  driver, 105–6
  remedies for, 123–27
  sleep apnea and, 370
Feet
  acupressure points on, 84, 155, **329**
  care of, with diabetes, 103–4
  pain in (*see* Foot pain)
Fencing, for clumsiness, 82
Fever, remedies for, 83–84
Fiber, dietary, heart disease and, 170
Fibromyalgia
  diagnosing, 130, **130**
  remedies for
    acupressure, 129–30
    massage, 129
    physical exercise, 128–29, 129, 131
    relaxation exercises, 131
Fine motor skills, clumsiness and, 80
Fingers, stiffness in, 132–34, 132, 133, 134, **134**
Fishing, to strengthen wrists, 424–25
Flatulence
  remedies for
    aerobic exercise, 136–37
    massage, 135–36
  statistics, 135
Flexibility, techniques to improve. *See also* Stretching exercises
  dance, 95
  Pilates Method, 310–12
  yoga, 428

Flies (exercise), in weight-loss program, 301, **301**
Flu, 83–85, <u>84</u>
Fluoxetine hydrochloride, insomnia and, 204
Football, exercise-induced asthma and, 59
Foot pain, 138, 269. *See also* Feet
   remedies for
      massage, 140, 141
      strengthening exercises, 140, **141**
      stretching exercises, 138–40, **139**
   shoes and, <u>139</u>
   when to see doctor for, <u>138</u>
Foot pedal (exercise), for fatigue, 125
Footwear. *See* Shoes
Forgetfulness, <u>2</u>, 142–44, <u>143</u>
French press
   for underarm sag, 400, **400**
   in weight-loss program, 300, **300**

G

Gardening
   as anti-burnout activity, 212–13
   as exercise, 53–54, 370
   guidelines, with back pain, 32
   warmup for, 381–82, **382**
Glaucoma
   remedies for
      cognitive exercises, 147, <u>147</u>
      conventional, 145
      physical exercise, <u>146</u>, 147–48
   risk factors for, 145
Glucocorticoids, stress and, <u>186</u>
Glucose, diabetes and, 101
Gluteal muscles. *See* Buttocks muscles
Goals, exercise, 50, 240–41, 339
Golf
   compassionate, 182
   exercise-induced asthma and, 59
   guidelines with
      arthritis, 28
      back pain, 33–34

   as remedy for
      love handles, 235
      midlife crisis, 263–64
      underarm sag, 402
   walking during, 370
   warmup for, 382, **382**
Gym
   air-conditioning in, sinusitis and, 364
   home, allergies and, <u>58</u>

H

Habits, nervous
   awareness of, 289–91
   competing response to, 291
   remedies for
      meditation, 249–50
      relaxation exercises, <u>290</u>
Hammer curls, in weight-loss program, 300, **300**
Hamstring curls, in weight-loss program, 304, **304**
Hamstrings
   strengthening, 326, **326**
   stretching, 269–70, **270**, 305, **305**, 342, 381, **381**, 408–9
Hands
   acupressure for
      allergy attacks, 155
      anxiety, 23, **23**
      dizziness, 203
      headaches, 83, <u>163</u>, **163**
   switching use of, burnout and, 213
Hangover, remedies for
   breathing exercises, 149
   hydration, 149
   massage, 150–51
   physical exercise, <u>150</u>, 151
   sauna, 150
Hay fever
   exercise and, <u>153</u>
   remedies for
      acupressure, 155
      imagery, 152–54

HDL cholesterol. *See* High-density lipoprotein
cholesterol
Headaches, 70–71, 157
preventives for, <u>159</u>, 161–64
remedies for
acupressure, 83, 161, <u>163</u>, **163**
biofeedback, 44, 160–61
bodywork, 12, **12**, 70–71, 83, 247–48, **247**,
248
chinning exercise, <u>160</u>
laughter, 159–60
relaxation exercises, 157–58, 319–21
sex, 161, 351
types of, 156, 157, 160
Hearing, improving, 346–47
Heart attack
anger and, 399
symptoms of, <u>172</u>
walking and, 405
Heartburn
exercise and, 166
laryngitis and, 223, <u>224</u>
preventives for, 167–68, <u>167</u>
weight loss and, 165–67, **166**
Heart disease
anatomy of, <u>172</u>
preventives for
diet, 169–70, 174
doctor visit, 174
meditation, 175–76
physical exercise, 170–71, 173, 176
sex and intimacy, 176
social support, 170, 174
sleep apnea and, <u>370</u>
symptoms of, <u>56</u>
Heart rate, target, 171, <u>185</u>
Heat, as remedy for
arthritis, 27
back pain, 31, 257
Heels, pain in, 138–40, **139**
Hemorrhoids, 90, <u>91</u>
Hiccups
long-lasting, <u>177</u>
remedies for, 178, <u>178</u>, **178**

High blood pressure
cardiac hypertrophy from, 179
diabetes and, 103
exercise and, 113, 171, 179, 405
remedies for
aerobic exercise, 39, 181, 405
attitude adjustments, 22, 182–83
biofeedback, 44
competitive sports, 181–82
meditation, 249, 250
reflexology, 328
relaxation exercises, 22
risk factors for, 179–80
sleep apnea and, <u>370</u>
weight lifting with, <u>180</u>
High-density lipoprotein (HDL) cholesterol
exercise and, 171
cholesterol levels and, 184
Hiking
as calorie-burning exercise, 187
as remedy for
lung damage from smoking, 368
osteoporosis, 293–94
Hips
pain in, 188, <u>191</u>
strengthening, 26, <u>189</u>, 190–91, **191**
stretching, 25, 65, **65**, 188–90, **189**,
407
HIV, massage and, 246
Hobbies, type-A behavior and, <u>398</u>
Home projects, job loss and, 218
Hormone imbalances, shortness of breath
and, <u>56</u>
Hormone replacement therapy (HRT), for
menopause problems, 252
Hormones. *See also specific hormones*
afternoon slump and, 7
anger and, 13
stress, 62, 170, 180
Hot flashes, remedies for, 254
Hot tubs, sex in, for back pain, 35
Housework
for adult attention deficit disorder, 3
for breast sag, 53

HRT, for menopause problems, 252
Humidity
  asthma and, 59
  congestion and, 85
  laryngitis and, 225
  sinusitis and, 363, 364
  snoring and, 371
Humor. *See* Laughter
Hydration
  laryngitis and, 225
  while exercising, 40, 125, 238
Hyperactivity
  in adult attention deficit disorder, 2
  in children, 328–29
Hypertension. *See* High blood pressure
Hypertrophy, cardiac, high blood pressure and, 179
Hyperventilation, panic attacks and, 307
Hypnagogic state, creativity and, 94
Hypnosis
  as remedy for
    belching, 37
    erectile dysfunction, 115
    headaches, 158
    smoking, 366
Hypoglycemia, 102
Hypoglycemics, oral, effect on glucose, 101, 103
Hypotension, orthostatic
  remedies for, 237–38
  symptoms of, 236
Hypoxemia, breathing disorders and, 57

Ice. *See* Cold, as remedy for
Imagery, 192–93
  as remedy for
    anxiety, 22–23
    belching, 38
    breathing disorders, 60
    clumsiness, 81
    fibromyalgia, 131

glaucoma, 147, 147
hay fever, 152–54
incontinence, 198
insomnia, 206
menstrual discomfort, 258, 259–60
panic attacks, 308
stage fright, 372–73
tooth grinding, 395
techniques, 45–46, 193–95
Immunity
  bicycling and, 40
  laughter and, 227
  yoga and, 427
Immunoglobulin A, laughter and, 376
Impotence. *See* Erectile dysfunction
Inactivity, results of, 123, 219
Incontinence
  remedies for
    biofeedback, 44, 197–98
    bladder retraining, 197
    Kegel exercises, 196–97, 253
  urge, 197
Inderal, insomnia and, 204
Indigestion, meditation for, 250. *See also* Digestive problems
Inertia, boredom and, 49
Injuries, athletic, 329
Inner ear problems, 202–3
Insomnia, 204
  exercise-induced, 371
  remedies for
    baths, 205
    biofeedback, 44
    breathing exercises, 207
    cognitive exercises, 206
    meditation, 250
    physical exercise, 21, 205–7
    progressive muscle relaxation, 319–21
Insulin
  diabetes and, 101, 103
  massage and, 247
Intermittent claudication, 208–9
Internal clock, jet lag and, 210
Intimacy, heart disease and, 176

Irritability
  during menstruation, 259–60
  as type-A behavior, 399
Irritable bowel syndrome, 250, 329
Isometric exercise, for knee pain, 222

J

J'ARM (jog with arms), for headaches, 162
Jet lag, 210–11, 211
Job loss, 215–18, 216
Jogging. *See* Running
Joints
  bicycling and, 39–40
  walking and, 405
Journaling, laughter and, 228
Jump ropes
  for caffeine withdrawal, 71
  cost of, 187
  using indoors, for cabin fever, 69

K

Karate
  for back pain, 35
  to build confidence, 218
Karkicks exercise program, for drivers, 86–87
Kegel exercises, as remedy for
  constipation, 91
  incontinence, 196–97, 253
Keyboarding, carpal tunnel syndrome and, 74
Kite flying, for birthday blues, 48
Knees
  anatomy of, 219
  exposing to sun, for seasonal affective disorder, 415
  pain in, 219–20, 222
  strengthening, 25–26, 221–22
  stretching, 25, 220–21, 220, 221, 221, 342–43
Knee squeeze, in weight-loss program, 304–5, 305

L

Lactic acid
  alcohol use and, 149, 151
  menstrual cramps and, 257
Laryngitis
  preventives for
    breath support, 224–26
    tension reduction, in neck, 226
  reflux and, 223, 224
  remedies for, 225
  types of, 223
Lateral raise, in weight-loss program, 299, 299
Latissimus dorsi muscles, strengthening, 234
Laughter, 227–28
  as remedy for
    adult attention deficit disorder, 2
    arthritis, 28
    headaches, 159–60
    panic attacks, 308
    stress, 374, 376–78
Lazy eye, remedies for
  exercises, 230–32, 230, 231, 232
  traditional, 229–30
LDL cholesterol. *See* Low-density lipoprotein cholesterol
Leg curls, in weight-loss program, 304, 304
Leg extensions, in weight-loss program, 304, 304
Leg lifts
  to prevent commuter's back, 87
  yoga exercise, 428–29, 429
Legs, strengthening, 276–77, 276
Libido, inhibited
  during menopause, 255
  remedies for, 199–201
Lifting, back pain and, 31, 34
Listening skills, improving, 346–47
Love handles, remedies for
  exercises, 233–34, 233, 234, 235
  sports, 235
Low blood pressure
  remedies for, 237–38
  symptoms of, 236

Low-density lipoprotein (LDL) cholesterol
  cholesterol levels and, 184
  heart disease and, <u>172</u>, 173
Lunges, in weight-loss program, 303–4, **304**
Lungs
  healing damage to, from smoking, 366–68
  reflexology for, 328

M

Maalox, for heartburn, 168
Macular degeneration
  detecting, <u>245</u>, **245**
  remedies for
    attitude, <u>242</u>
    color therapy, 244–45
    exercises, <u>243</u>, 244, <u>244</u>
Martial arts. *See also* Tai chi
  for back pain, <u>35</u>
  to build confidence, 218
Massage therapy, 246–47
  as osteoporosis preventive, 294
  as remedy for
    arthritis, 27
    back pain, 31, <u>33</u>
    dizziness, 203
    farsightedness, 120–22
    fibromyalgia, 129
    flatulence, 135–36
    foot pain, 140, 141
    glaucoma, <u>146</u>
    hangover, 150–51
    hay fever, 155
    headaches, 70–71
    laryngitis, 226
    muscle cramps, 268
    temporomandibular disorder, 393
    tennis elbow, 108
    wrinkles, <u>418–19</u>, **418**, **419**
  shiatsu techniques, 247–48, **247**, **248**
Masturbation, erectile dysfunction and, 114
Meclizine hydrochloride, for motion sickness, 265

Medications, side effects of. *See also specific medications*
  erectile dysfunction, 112, <u>114</u>
  headaches, 157
  low blood pressure, 237
  snoring, <u>371</u>
Meditation, 249–50
  as remedy for
    anger, 16
    anxiety, 23
    boredom, <u>50</u>
    driver fatigue, 106
    heart disease, 175–76
    menstrual discomfort, 259–60
    nearsightedness, <u>284</u>
  techniques, 250–51
Melatonin, seasonal affective disorder and, 414
Memory, improving, 44, 142–44, <u>143</u>
Menopause, 252–53
  problems, remedies for
    acupressure, 254, **254**
    breathing exercises, 253
    physical exercise, 253–54
    yoga, 255
  sex during, 255, 350–51
Menstrual discomfort
  back pain, 257–59
  cramps, 256–57, 350
  mood swings, 259–60
Mental focus
  anger and, 13–14
  improving, in adult attention deficit disorder, 3
  running and, 341
Methylphenidate, for adult attention deficit disorder, 2
Midlife crisis
  activities for
    community involvement, 262–63
    education, 264
    exercise, 261–64, <u>262</u>
  attitude and, 261
Migraines, 44, 160–61

Mood
  laughter and, 227
  menstruation and, 259
Motion sickness, 265–66
Motivation, for exercise, 49–51, <u>297</u>, 339, <u>406</u>
Mountain climbing, for birthday blues, 47–48
Mucus, clearing, 85
Muscles. *See also specific muscles*
  achy, 85
  cramps in, 267–70, <u>268</u>, **268**, **270**
  tension in, 44, 88
  weakness in, 217 (*see also* Strength training)
Muscle-toning exercises
  bicycling, 39
  as remedy for
    arthritis, 24
    breast sag, 52
    hip pain, <u>189</u>, 190–91, **191**
    knee pain, 221–22
  tai chi, 387
Music, 279–80
  effects on
    anger, 15–16
    birthday blues, 48
  exercising to, 127
Myopia. *See* Nearsightedness

N

Nail biting. *See* Habits, nervous
Naps, for afternoon slump, 10
Nasal spray, breathing disorders and, 59
Nature, as remedy for
  depression, 98–99
  midlife crisis, 262
  stress, 379
Nearsightedness, remedies for
  exercises, <u>282</u>, 283–85, <u>283</u>
  meditation, <u>284</u>
  surgery, 281

Neck
  acupressure points on, <u>377</u>, **377**
  pain in, 286, <u>287</u>
  strengthening, 288
  stretching, 286–88, 375–76, **376**
  tension in, laryngitis and, 226
Neo-Synephrine, for sinusitis relief, 362
Nervous system disorders, 249
Neurogate acupressure point, for anxiety, 23, **23**
Neurotrophins, anxiety and, 21
Nicotine, appetite and, <u>367</u>
Noradrenaline, anger and, 397
Norepinephrine
  breathing exercises and, 62
  massage and, 247
Nose, stuffy, 84, 362–63

O

Obesity. *See also* Overweight
  diabetes and, 102
  exercise and, 171
  reflexology for, 329
Optometrists, behavioral, approach to vision problems, 119–20
Orbit Meditation, breathing exercise, 62
Organization skills, for adult attention deficit disorder, <u>2</u>
Orgasm, pain-relieving qualities of, 257, 351
Osteoarthritis
  exercises for, 24
  hip pain from, 188
  preventives for, 40, 332, 405
Osteoporosis
  balance and, 295, 387
  exercising with, <u>293</u>
  preventives for, 40, 292–95
  sex and, <u>294</u>
Out-of-doors, as remedy for
  depression, 98–99
  seasonal affective disorder, 415
  stress, 379

Overhead press, in weight-loss program, 299, **299**
Overweight. *See also* Obesity; Weight loss
    exercise prescription for, 296, 298–305, **299,**
        300, 301, 302, 303, 304, 305
    as risk factor for
        heart disease, 170, 171, 173
        high blood pressure, 180
        knee pain, 219
        varicose veins, 403
    snoring and, 369

P

Pain
    chronic, 128, 204, 258, 320, 351
    during exercise, 273, 387
Pain-management techniques
    progressive muscle relaxation, 319–21
    sex, 351
Painting, for underarm sag, 402
"Palming" exercise, for
    eyestrain, 118
    farsightedness, 121
    glaucoma, 146
    lazy eye, 231
    macular degeneration, 244
Panic attacks
    remedies for
        acupressure, 307, **307**
        breathing exercises, 307–8
        cognitive exercises, 309
        relaxation exercise, 308
    when to see doctor for, 306, 307
Pectoral muscles, strengthening, 52–54, **53**
Pelvic pain, when to see doctor for, 258
Perimenopause, 252
Pets
    effect on
        anger, 15
        depression, 99–100
        pain, 131
        stress, 131, 398
    walking with, 368

Phenylephrine, for sinusitis relief, 362
Philtrum, as acupressure point, 161
Photorefractive keratectomy (PRK), for
    nearsightedness, 281
Piano playing, for finger stiffness, 134
Pilates Method, for strength and flexibility,
    310–12
Plantar fasciitis, remedies for, 138–40, **139**
Plaque, arterial, 172, 175
Plié, to firm buttocks, 327
PMR, for muscle tension, 319–21
Pogo sticks, caffeine withdrawal and, 71
Pollution, breathing disorders and,
    55–56
Posture
    dental, 394–95
    effects on, from
        strength training, 332
        stretching, 380
    fatigue and, 124
    as preventive for
        back pain, 31, 86
        neck pain, 287
    proper, 313–14, **313, 314,** 391
    problems, remedies for
        bodywork, 11–12, **12**
        exercises, 314–16, **314, 315**
    temporomandibular disorder and, 391
    while walking, 407
Prayer, anxiety and, 23
Pregnancy, varicose veins and, 403
Premenstrual tension, 250
Presbyopia. *See* Farsightedness
PRK, for nearsightedness, 281
Problem-solving, laughter and, 227
Progesterone, water retention and, 411
Progressive muscle relaxation (PMR), for
    muscle tension, 319–21
Propranolol, insomnia and, 204
Prostaglandins, menstrual cramps and, 256
Prozac, insomnia and, 204
Pruning, for underarm sag, 402
Public speaking, fear of, 372–73
Pulse, positive thinking and, 22

Pushups, as remedy for
   breast sag, 52–53, **53**
   cellulite, 76–78
   driver fatigue, 106

Q

Qigong, breathing exercises in, 61, 62
Quadriceps
   strengthening, 326, **326**
   stretching, 270, 305, 336, **336**,
     342–43

R

Racquetball, for high cholesterol, 187
Radial keratotomy (RK), for nearsightedness,
   281
Rage. *See* Anger
Raking, for breast sag, 53–54
Range-of-motion exercises, as remedy
   for
   bursitis, 64–65, **64**
   wrist pain, 424
Rate of Perceived Exertion (RPE) scale,
   185
Raynaud's disease, 44, 322–24
Rear end, sagging, remedies for
   dance, 327
   exercise, 325–27, **325**, **326**
Records, keeping, for exercise, 51, 241
Reflexology, 328–30, **329**
Reflux, laryngitis and, 223, 224
Reflux belching, 37
Reinforcement, for exercise, 51
Relaxation exercises. *See also* Biofeedback
   progressive muscle relaxation, 319–21
   as remedy for
     anger, 15
     anxiety, 22–23
     arthritis, 27

   back pain, 88
   belching, 36
   fibromyalgia, 131
   laryngitis, 226
   motion sickness, 266
   nearsightedness, 283
   nervous habits, 290
   panic attacks, 308
Repetitive strain injury, 12, **12**. *See also*
   Carpal tunnel syndrome
Resistance bands, for strength training,
   19–20, **20**, 76, 77, 423
Resistance training. *See* Strength training
Retinopathy, diabetic, 103
Retirement blues, 338–39, 339
RICE treatment, for ankle pain, 17
Ritalin, for adult attention deficit
   disorder, 2
RK, for nearsightedness, 281
Rope jumping, as remedy for
   cabin fever, 69
   caffeine withdrawal, 71
   high cholesterol, 187
Rowing, as remedy for
   flatulence, 136–37
   high cholesterol, 187
RPE scale, 185
Running, 340–42
   after age 50, 344
   aqua, 26
   as remedy for
     afternoon slump, 8
     anxiety, 21–22
     arthritis, 26
     creativity block, 94
     driver fatigue, 106
     erectile dysfunction, 113
     flatulence, 136
     hangover, 150
     high cholesterol, 187
   shinsplints from, 352–54, 352, 353,
     354
   shoes for, 341–42, 352
   warmup for, 342–43

S

SAD. *See* Winter blues
Sauna, as hangover remedy, 150
Scar tissue, sprains and, 17
Schedules
    afternoon slump and, 9
    exercise, 5
Scoliosis, Alexander Technique for, 12
Scuba diving, for birthday blues, 48
Sea-Band bracelets, for motion sickness, 266
Seasonal affective disorder (SAD). *See* Winter blues
Self-esteem, low, 239, 240, 358
    exercise plan for
        attitude and, 241
        choosing exercise in, 239–40
        goal-setting in, 240–41
Self-hypnosis, for headaches, 158
Self-image
    in empty nest syndrome, 111
    when job hunting, 217–18
Self-talk, stress and, 378
Sensory loss
    hearing, 346–47
    smell and taste, 345–46, 348
    touch and, 348–49
    vision, 347–48
Serotonin
    appetite and, 296, 367
    massage and, 246
    seasonal affective disorder and, 414
    sex and, 161
Sex, 350–51
    guidelines with
        back pain, 34–35
        osteoporosis, 294
    improving, with yoga, 428
    during menopause, 255
    problems with
        erectile dysfunction, 112–15, 114
        premature ejaculation, 317–18
    as remedy for
        anxiety, 21–22
        heart disease, 176

insomnia, 207
migraines, 161
stress, 377
Sexual desire, inhibited
    during menopause, 255
    remedies for, 199–201
Shins, stretching, 353–54, 354, 407, 407
Shinsplints
    remedies for, 352–54, 353, 354
    shoes and, 352
Shoes
    fit of, 104, 139
    running, 341–42, 352
    walking, 406–7
Shortness of breath. *See* Breathing disorders
Shoulder hugs, in weight-loss program, 302, 302
Shoulders
    pain in, 355, 355
    strengthening, 274–75, 275, 299, 299, 357
    stretching, 25, 125, 170, 355–56, 356
Shoulder shrugs, as remedy for
    driver fatigue, 105
    headaches, 162–63
    jet lag, 210
    shoulder pain, 357
Shoulder stand, as remedy for
    constipation, 91–92
    water retention, 411–12, 412
Showers, as remedy for
    anger, 16
    arthritis, 27
    back pain, 31
    sinusitis, 363
    tendinitis, 66
Shyness
    exercise and, 358–59
    sociability and, 359, 360–61, 360
Singing, panic attacks and, 308
Sinusitis
    mucus with, 362–63
    remedies for
        exercise, 363–64
        humidity, 363
        reflexology, 328

Situps, modified, for back pain, 30. *See also* Crunches, abdominal
Skating, for winter blues, 415
Skiing, for winter blues, 415
Skydiving, for birthday blues, 47
"Skying" exercise, for farsightedness, 120
Sleep
    contraindications for, 363
    exercise and, 21
    positions for
        back pain, 31
        sinusitis, 362–63
        snoring, 371
        tooth grinding, 395–96
    as preventive for
        driver fatigue, 105
        headaches, 164
        tooth grinding, 394
    as cold and flu remedy, 85
Sleep apnea, 106, 370
Sleep deprivation, results of, 7, 123. *See also* Insomnia
Slinky toy, for wrist fitness, 72
Smell, sense of, 345–46, 348
Smiling
    headaches and, 162
    stage fright and, 373
Smoking
    diseases caused by, 56
    effect on
        belching, 38
        erection, 112
        menstrual cramps, 257
        snoring, 371
    quitting, 44, 365–68, 365, 366, 367
    as risk factor for
        heart disease, 169, 172
        high blood pressure, 180
        laryngitis, 223, 225
        macular degeneration, 242
        strokes, 175
Snoring, 371
    remedies for
        sleep positions, 371
        weight loss, 369–71

sleep apnea and, 370
    smoking and, 371
Sociability
    heart disease and, 170, 174
    at parties, 359
    practicing, 360–61, 360
Softball
    exercise-induced asthma and, 59
    as remedy for
        anger, 15
        midlife crisis, 261–62
Sore throat, 84
Speech pathologists, finding, 385
"Speech reading," to improve listening skills, 347
Spelunking, for birthday blues, 48
Spine, lengthening, 11–12, **12**
Spinning (bicycling workout), 41
Sports, competitive. *See also* specific sports
    for anger, 15
    modified for
        high blood pressure, 181–82
        type-A personalities, 397–98
Sports education, as exercise motivation, 49
Sprains, ankle, 17–20, 18, 19, **20**
Squats
    as remedy for
        cellulite, 75–76
        hip pain, 191
    in weight-loss program, 303, **303**
Stage fright, 372–73
Stair-climbing
    as osteoporosis preventive, 294
    as remedy for
        afternoon slump, 8
        inhibited sexual desire, 200
        job loss, 217
        rear-end sag, 326–27
        shinsplints, 353
        smoking, 368
        stage fright, 373
Stair stretch, for Achilles tendon, 65
Stenosis, spinal, 314
Steps, for step aerobics, 69

Strength training, 271–73, 331–32. *See also*
  Weight lifting
  cooldown for, 337, **337**
  daily chores as, 272
  equipment for, 76, 77, 272, 297, 332
  guidelines for, 297, 332–33
  pain during, 273
  as preventive for
    osteoporosis, 292–93
    varicose veins, 404
  as remedy for
    ankle pain, 19–20, **20**
    cellulite, 75–78
    constipation, 90
    elbow pain, 109
    fatigue, 125
    finger stiffness, 134, **134**
    foot pain, 140, **141**
    heartburn, 166–67, **166**
    heart disease, 176
    low blood pressure, 238
    neck pain, 288
    shoulder pain, 357
    underarm sag, 400–401, **400**, **401**, **402**
    wrist pain, 423, 424–25
  warmup for, 334–36, **334**, **335**, **336**
  workouts
    general, 336–37, **336**, **337**
    for muscle strengthening, 273–78, **274**, **275**,
      276, **277**, **278**
    for weight loss, 299–305, **299**, **300**, **301**,
      302, **303**, **304**, **305**
Stress
  in adult attention deficit disorder, 2
  belching and, 38
  as cause of
    commuter back, 88
    fatigue, 123
    flatulence, 137
    headaches, 157–60, 164
    insomnia, 204
    stuttering, 383–85, 383, 384, 385
    temporomandibular disorder, 12, 320,
      390–93, 392, 394
    tooth grinding, 394–96, 395
  effects on, from
    exercise, 21, 171, 341
    laughter, 227
  remedies for
    acupressure, 377, **377**
    biofeedback, 44
    breathing exercises, 61–62, 253, 379, 395
    cognitive exercises, 378–79
    conflict management, 174–75
    dance, 95–96, 375
    imagery, 395
    laughter, 227, 374, 376–78
    meditation, 249
    progressive muscle relaxation, 319–20
    reflexology, 328
    running, 341
    sex, 350
    stretching exercises, 375–76, **376**
    tai chi, 387
    walking, 374
  as risk factor for
    heart disease, 170, 174
    high blood pressure, 180
    high cholesterol, 186
    jet lag, 211
  smoking and, 365–66
  from telephone interruptions, 375
Stretching exercises, 380–81
  post-workout, 42, 297, 337, **337**, 408
  as remedy for
    afternoon slump, 8–10
    ankle pain, 17–18, **19**
    arthritis, 24–25
    back pain, 30, **30**, 33, 87–88
    breathing disorders, 60
    bursitis, 65, **65**
    cabin fever, 68
    driver fatigue, 105, 106
    fatigue, 125
    fibromyalgia, 128–29, 129
    finger stiffness, 132–33, **132**, 133, **133**
    foot pain, 138–40, **139**
    golfer's elbow, 108–9, **109**
    hangover, 151
    hip pain, 188–90, **189**

Stretching exercises (*continued*)
  as remedy for (*continued*)
    insomnia, 205–6
    knee pain, 220–21, **220**, **221**
    menstrual discomfort, 257–59
    muscle cramps, 267–70, **268**, **270**
    posture, 314, **314**
    shinsplints, 353–54, **354**
    shoulder pain, 355–56, **356**
    stage fright, 373
    stress, 375–76, **376**
    tendinitis, 65–67, <u>66</u>, 66
    tennis elbow, 107–8, **108**
    wrist pain, 422–23
  as warmup for
    gardening, 381–82, **382**
    golf, 382
    running, 342–43
    strength training, 334–36, **334**, **335**, **336**
    walking, 170–71, 381, **381**, 407–8, **407**
  in weight-loss prescription, 302, **302**, 304, **304**, 305, **305**
Stroke
  preventives for, 405
  risk factors for, <u>175</u>
  sleep apnea and, <u>370</u>
Stuttering, 383–85, <u>383</u>, <u>384</u>, <u>385</u>
Sulfonylureas, effect on glucose, 103
"Sunning" exercise, for
  farsightedness, 120
  macular degeneration, 244
Support groups, for stutterers, <u>385</u>
Swimming. *See also* Water exercises
  exercise-induced asthma and, 59
  as remedy for
    back pain, 30, 31
    breast sag, 52
    erectile dysfunction, 113
    high blood pressure, 181
    high cholesterol, 187
  sense of smell and taste and, 346
  sinusitis and, <u>363</u>

T

Tai chi, 386–88
  balance and, 295, 387
  cheer move, 388–89, **388**, **389**
  pain during, 387
  as remedy for
    back pain, 30–31
    clumsiness, 82
    fibromyalgia, 131
Target heart rate, 171, <u>185</u>
Taste, sense of, 345–46
Temporomandibular disorder (TMD)
  chewing gum and, 393
  eating habits and, <u>392</u>
  remedies for
    bodywork, 12
    correct posture, 391–92
    massage, 393
    neck exercises, 392–93
    progressive muscle relaxation, 320
  tooth grinding and, 394
Tendinitis, 63, 65–67, <u>66</u>, 66
Tennis
  exercise-induced asthma and, 59
  modified for
    high blood pressure, 181–82
    type-A behavior, 398
  as remedy for
    love handles, 235
    underarm sag, 402
Tension. *See* Stress
Testosterone, effect of exercise on, 113
Threshold IMT, for breathing disorders, 58–59
Thyroid gland
  fatigue and, 123
  insomnia and, 204
Time management, adult attention deficit disorder and, 2
TMD. *See* Temporomandibular disorder
Toe raises, in weight-loss program, 305, **305**

Toes
  acupressure point on, 84, 155, **329**
  cramps in, 140, **141**
Tooth grinding
  complications from, 394
  remedies for, 394–96, 395
Touch, to improve sensitivity, 348–49
Tourette's syndrome, meditation for, 249
Triceps, toning, 400–402, **400**, **401**, **402**
Triceps kickback
  for underarm sag, 400–401, **401**
  in weight-loss program, 300, **300**
Trichotillomania. *See* Habits, nervous
Twists
  oblique, 302–3, **302**
  spinal, 90–91
  trunk, 233–34, **233**, **234**
Type-A behavior, 397–99, 398

U

Underarms, sagging, 400–402, **400**, **401**, 402, **402**

V

Vacuuming, for breast sag, 53
Varicose veins, 403–4, 404
Vascular disease, 114, 208
Very low density lipoprotein (VLDL),
  cholesterol levels and, 185, 186
Videos, exercise, 69
Vision problems, 347–48. *See also*
  specific problems
Visualization. *See* Imagery
VLDL, cholesterol levels and, 185, 186
Vocal exercises, for stuttering, 384–85,
  384
Volunteer work
  burnout and, 214
  retirement and, 339

W

Walking
  cooldown for, 408
  belching and, 37
  as headache preventive, 163–64
  as remedy for
    adult attention deficit disorder, 3
    anxiety, 21–22
    arthritis, 28
    back pain, 30, 33
    creativity block, 93
    depression, 99–100
    fatigue, 124–25
    fibromyalgia, 131
    heart disease, 170–71
    high blood pressure, 181
    inhibited sexual desire, 200
    intermittent claudication, 208–9
    menopause problems, 254
    overweight, 370–71
    sinusitis, 363, 364
    smoking, 367–68
    stress, 374
    wrinkles, 420
  shoes for, 406–7
  types of, 409–10, 410
  warmups for, 170–71, 381, **381**, 407–8, **407**
Warmups
  for bicycling, 40–41
  for running, 342–43
  for strength training, 334–36, **334**, **335**, **336**
  for walking, 170–71, 381, **381**, 407–8, **407**
Water exercises. *See also* Swimming
  asthma and, 59
  osteoporosis and, 295
  as remedy for
    arthritis, 26–27, **26**, **27**
    hip pain, 190, **190**, 191
    intermittent claudication, 209
    low blood pressure, 237–38
    midlife crisis, 264
    shinsplints, 353
Water retention, 411–13, **412**, **413**

Weight control, 224, 253, 340–41
Weight gain, quitting smoking and, 367
Weight lifting. *See also* Strength training
  contraindications for, 90
  equipment, 77
  hemorrhoids and, 90
  with high blood pressure, 180
  as remedy for breast sag, 53, **53**
Weight loss
  activities for, 39, 332, 369–71
  exercise prescription for, 5, 298–305, **299, 300, 301, 302, 303, 304, 305**, 369
  as preventive for
    back pain, 31
    heartburn, 165–67
  sense of smell and, 348
  sexual desire and, 199
Weight training. *See* Strength training
Window washing, for breast sag, 54
Winter blues, 414–16, 415, 416
Workouts, strength-training
  general, 336–37, **336, 337**
  for muscle strengthening, 273–78, **274, 275, 276, 277, 278**
  for weight loss, 299–305, **299, 300, 301, 302, 303, 304, 305**

Wrinkles, 417–21, 418–19, **418–19**
Wrists
  acupressure points on, 265–66
  pain in, 422–25, **422, 423, 424** (*see also* Carpal tunnel syndrome)
Writer's cramp, 44

X

Xertube, for strength training, 272

Y

Yardwork, as exercise. *See* Gardening
Yoga exercises, 426–28
  breathing, 38, 71
  for eyes, 230
  during menopause, 255
  during midlife crisis, 263
  seated leg lift, 428–29, **429**
  shoulder stand, 91–92, 411–12, **412**
  spinal twist, 90–91
  stretches, 30, **30**, 205–6
Yo-yos, for carpal tunnel syndrome, 73